AF251798

GEOLOGY AND SETTLEMENT

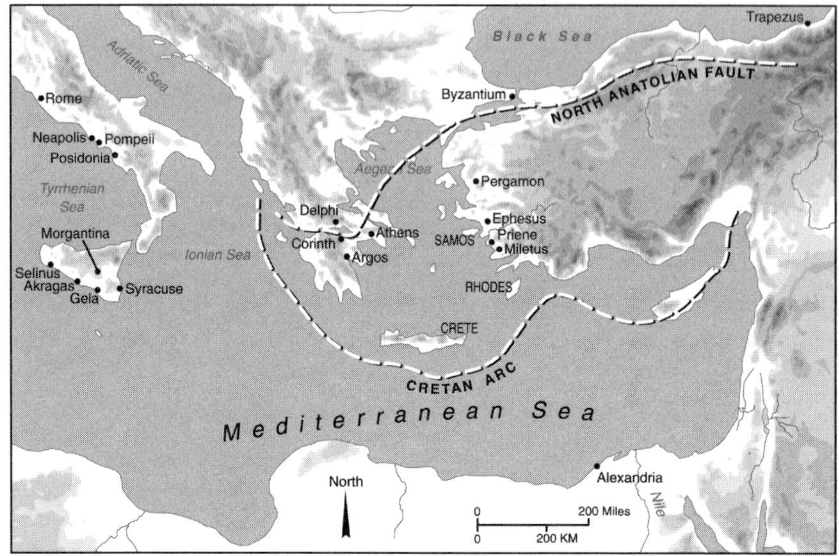

The Greco-Roman world and two major fault systems: The North Anatolian and the Cretan

GEOLOGY AND SETTLEMENT

Greco-Roman Patterns

DORA P. CROUCH

With scientific contributions from
Aurelio Aureli, Giovanni Bruno, Laura Ercoli, Talip Güngör,
Marina de Maio, Paul G. Marinos, Charles Ortloff, Ünal Özis,
and Wolfgang Vetters

and the assistance of
Ahmet Alkan, Ayhan Atalay, Yüksel Birsoy, Pietro Cipolla,
Poppy Gaki-Papanastasiou, Eleni Kolati, Roberto Maugeri,
Paolo Mazzoleni, Antonio Pezzino, Sprios Plessas,
Rossario Ruggieri, and Antoinella Sciortino

OXFORD
UNIVERSITY PRESS
2004

OXFORD
UNIVERSITY PRESS

Oxford   New York
Auckland   Bangkok   Buenos Aires   Cape Town   Chennai
Dar es Salaam   Delhi   Hong Kong   Istanbul   Karachi   Kolkata
Kuala Lumpur   Madrid   Melbourne   Mexico City   Mumbai   Nairobi
São Paulo   Shanghai   Taipei   Tokyo   Toronto

Published by Oxford University Press, Inc.
198 Madison Avenue, New York, New York 10016

www.oup.com

Oxford is a registered trademark of Oxford University Press

Library of Congress Cataloging-in-Publication Data

Crouch, Dora P.
    Geology and settlement : Greco-Roman patterns / by Dora P. Crouch.
       p. cm.
    Includes bibliographical references.
    ISBN 0-19-508324-5
       1. Engineering geology—Rome.   2. Engineering geology—Greece.   3. City
    planning—Rome.   4. City planning—Greece.   I. Title.

TA705.4.R66 C76 2002
711'.42'0937—dc21   2001021613

9 8 7 6 5 4 3 2 1

Printed in the United States of America
on acid-free paper

This long labor began in curiosity and ends with thanks to those
who revealed to me the patterns of the world's unfolding
in both geological and human history.

Preface

The water supply and engineering questions I asked earlier (Crouch 1993), with the geology questions of this book, shed new light on Greco-Roman cities. By reflecting upon data and insights from additional disciplines, we have a larger matrix for ancient cities than when archaeology alone deals with explication of a site. Old methods can offer archaeological evidence of Greek walls to contain the river at Argos, or historical documents such as lists of all earthquakes in western Turkey since Roman times, with dates and description of perceived severity (*Earthquake Catalogue*), to make dating more manageable. Checking each site for geological evidence of datable events has improved the comparisons we make. By comparing ten cities we introduce generalization, revealing more than would any one individual case history (Finley 1977: 314). At Argos, for instance, there is dramatic evidence of flooding in the hinterland as well as in the agora at the center of the city, whereas at Miletus the very process of modern excavation has had to be timed with the annual flooding pattern in mind.

When the right kinds of questions are asked, there is an abundance of material for answers, even allowing for our tendencies to apply our classifications onto the ancient world (Gordon 1979). Inferred parallels from insufficient data are likely to be closer to the truth than wild guesses based on what we think *ought* to have been the case. The specificity of the geological settings and elements of water systems in these ancient cities is gratifying to me as one who bases history on tangible objects, and dear to me because I am "allergic or totally deaf to ideal types" (Finley 1977: 316). That tangibility helps to avoid some of the "elusiveness of truth" confronting those—deconstructionists and others—who deal with text and context at the verbal level (Galloway ca. 1992).

Rigid boundaries between disciplines—seismologists not knowing the ancient literature and the results of modern archaeology, or archaeologists barely acquainted with geomophology, seismology, and other scientific disciplines—interfere with team work. This work cannot be expected to be smooth and easy, although we can achieve illumination through discussion of "facts," methods, and discoveries. Different disciplinary standards of "truth" and "results," and characteristic national behaviors among the investigators affect the evaluation—but also the residual ambiguity—of interdisciplinary studies. Multidisciplinary teams also have to work out communication problems arising from disciplinary jargons in urban demography, karst geology, sociology, classical archaeology, or climatological history. Not every scholar who works on ancient cities, for instance, is fluent in Greek and Latin. One simple precaution would be to provide translations for all foreign languages quoted, even for mathematics and engineering formulas, which are foreign languages to humanists (Glossary, Appendix B).

Working at two intellectual levels simultaneously has been extremely challenging. On the one hand, to master—even with team work—the scattered data from ten ancient cities has been nigh impossible. On the other hand, to assimilate the methods of sets of scientific fields that have changed faster than we could learn has been daunting. Modern traditions and methods may constitute a filter between us and the remnants of the past, facilitating or inhibiting our understanding of that past (Keller 1985). Our team has done its best to acknowledge our filters consciously and to look beyond them. As pioneers, we hope to have the remaining blind spots forgiven us.

Acknowledgments

Although this book is history and geology, not archaeology, it has been necessary to have official permission to work at archaeological sites. In Sicily, we thank excavator D. Mertens who allowed our work at Selinus. The superintendent at Syracuse gave permission, while G. Bongiovanni, R. Maugeri, and R. Ruggieri helped us to see the geology connected to the form of the city. The superintendent at Enna and several chief archaeologists at Morgantina gave permission. Dott.essa G. Fiorentini, Superintendent at Agrigento, made the library of the Superintenzia available to us, and local experts there, Emma and Giovanni Trasatti and C. Micceché, shared their insights.

Access to the Morgantina excavation room at Princeton University and early encouragement were kindly given by Dr. William Childs. We further acknowledge the help of the Department of Civil Engineering and Structural Geology at the University of Palermo, Prof. Dr. R. Schiliro and Dr. P. Atzori of the Institute for Mineralogy at the University of Catania, and the Institute for Applied Geology and Geotechnics at the Technological University at Bari, where Egidio Messina and Mrs. M. Rosaria Paiano were most helpful.

In Greece, our team cleared the work with Mme. Anne Pariente of the French School at Athens for the sites of Argos and Delphi. At Corinth, former director Charles Williams and associate director Dr. Nancy Bookidis were helpful. The Department of Civil Engineering and Engineering Geology at the Technical University at Athens gave notable assistance for all three Greek sites, through its chair Prof. Dr. Paul G. Marinos, his staff, and his graduate students, especially Dr. Spyros Plessas.

In Turkey, our team worked under the aegis of the Departments of Geology and Civil Engineering at Dokuz Eylül University, informed by the long-term expertise of Prof. Dr. Ü. Özis and field visits with him and Prof. Dr. Y. Birsoy and their graduate students Dr. A. Alkan and Dr. T. Güngör. Dr. Güngör's priceless knowledge of the geology and Alkan's of the terrain were supplemented importantly by geologist Dr. W. Vetters' life-long attention to Ephesus. The excavator

Prof. Dr. W. Koenig enrolled us in his team for Priene, and his counterpart at Ephesus, Prof. Dr. S. Karwiese, helped us obtain permission for work on Ephesus under the Austrian Archaeological Institute, with permission continued under Prof. Dr. F. Krinzinger. The staff of the Ephesus excavation were generous in sharing their information. Prof. Dr. W. Müller-Wiener encouraged and assisted our examinations at Miletus in the 1980s, and the present excavator, Prof. Dr. V. von Graeve, and his staff, especially Dr. G. Tuttahs, were helpful.

The assistance of the German Archaeological Institute and its magnificent libraries at Istanbul under Prof. D. W. Radt and at Athens under Prof. Dr. H. Kienast can never be repaid with enough thanks. The French and American school libraries in Athens have been extremely helpful, especially Librarian Nancy Winters of American School of Classical Studies. The libraries and reference librarians of the University of California at Los Angeles made important contributions, as did the Geology Library at Stanford University and the library at the Archaeological Institute of Austria in Vienna.

The long-term stimulus of the Fontinus Geshellschaft, which studies ancient water systems, has greatly influenced me. I am especially grateful for the initial acceptance of Prof. Dr.-Ing. G. Garbrecht and the enduring friendship of Prof. Dr.-Ing. H. Fahlbusch.

The enlightened and energetic guidance of Saskia de Melker and Daniël Koster, archaeologist and historian, respectively, started this long work in Greece and Turkey in 1984–85; the effort would have been impossible without them, and I thank them with all my heart. At a later point in the research, a donation from Janann Strand provided for the field expenses of graduate students in Turkey and Greece, without which the results would have been much poorer.

Many scientists and engineers are accustomed to doing funded research, but the colleagues who worked with me gave freely and abundantly to this quest for knowledge, motivated solely by their fascination with the topic and an opportunity to bring together their many interests into one humanistic topic. These geologists and engineers knew better than I what we were seeing, and they enlightened my ignorance, corrected my mistakes, and increased my understanding. Even though unified conclusions are unlikely at this time, all of us tried for accuracy. I thank these colleagues humbly, and ask their pardon for any remaining errors.

Contents

PART IV. APPENDICES

BACKGROUND

Introduction

Rationale

Cities are a constant interplay between tangible and intangible, visible and invisible factors. Long-lived cities can provide data to compensate for the brevity of our modern urban experience (Croce 1985). To overcome these gaps in research, just beginning to close, the city is a most useful unit of study. Ancient cities can serve as four-dimensional models (length, width, height, and time) of how humans survived in their ecological niches. Yet comparative studies of groups of cities—such as Rorig's (1967) of German medieval trading cities of the Hanseatic League, Andrews's (1975) of the urban design history of Maya cities, and Hohenberg and Lee's (1985) of the economic history of European cities—ignore the geological setting.

The setting of our study is the Mediterranean periphery where cities are united by their Greco-Roman historical and cultural relationships. From the twenty-five Greco-Roman sites studied in *Water Management in Ancient Greek Cities*, we[1] have selected for further study 10 sites with sufficient geological information to form a basis of comparison. Our comparisons are based on the physical aspects—both form and function—of the local area, not the particular object. There are exciting possibilities, both intellectual and practical, in such an approach.

Until recently, ancient Mediterranean cities have been investigated mainly by ancient historians and classical archaeologists. Cities, however, are so complex as to require every possible sort of investigation. Because each model and methodology leaves out too much, the use of a single model from one discipline, whether archaeological, mathematical, engineering, or historic, has limited usefulness. The documents of the classicists and the physical remains located by archaeologists seem to an urban historian like myself to be useful but incomplete sources that take for granted the geographical base, assume a past social organization, and may ignore the technological and scientific aspects of ancient urban life. As classicist M. H. Jameson (1990) has written, "The surviving literature from

Classical Greece sheds light only incidentally on practical matters such as patterns of settlement and domestic architecture . . . [yet] conceptions drawn from literature, sometimes with dubious justification, continue to prevail." The Mediterranean area continues to fascinate not only classicists and archaeologists, but also persons interested in urban history, in the ecology of human settlements, and in the scientific understanding of human life. This book attempts to reach a mixed audience from these fields. To facilitate that effort, a glossary of technical terms in geology, archaeology, hydraulic engineering, and architecture can be found in Appendix B.

The ancient cities themselves vary in the amount of written history and archaeological information available from standard archaeological and philological research. Adding insights from geology, engineering, and urban history, we can contribute to a better understanding of each of these cities. Miletus, Syracuse, Ephesus, and Delphi have rich documentary histories and extensive excavations. Agrigento and Selinus were grand cities, but they have not received the same depth of attention until recently. Corinth has been studied by archaeologists and biblical scholars, who published with great thoroughness, though with proportionally little attention to either chronology or engineering.

Precise observational data of all the events and processes we should like to know about has been accumulated for less than 200 years and is far from comprehensive. It is schematically possible, however, to date geological events such as earthquakes, landslides, floods, and sea intrusion, particularly at sites where dated evidence of human occupation correlates with geological events (see tables 1.1 and 1.2).

Cities are embedded in geological matrices. The search for information about the influence of geology on human settlements occupied our research teams during 1992–99. We looked for evidence of geological processes (erosion, subsidance), geological events (earthquakes, volcanic eruptions), and physical constraints on engineering solutions for human construction (topography, materials). Construction and destruction revealed in the archaeological and documentary record at these sites parallel the natural events revealed in the geological record. Preliminary correlation of the interrelationship of human construction and history with geological events and processes over relatively long time periods by human standards is one outcome of this study (see Appendix A).

The chosen period, 800 B.C.E.–600 C.E. (B.C.E., before the common era; C.E., common era), is long enough for ancient historians to have noticed that the earth has changed, yet most history is still written as if the earth were static. Today's historians of Greek and Roman times, like their classically educated ancient and modern predecessors, tend to see the earth as inert, passive, and changing little. Urban historians who deal with other periods or cultures have not paid much attention to the geological base of the sites they study. Nor have modern geologists customarily dealt with the geological base of human settlements, instead dismissing human history as a mere blink of the eye in comparison with the vast extensions of geological time. To begin to understand why geology and archaeology have developed largely in isolation from each other, I examined the historiography of each.

TABLE 1.1 Earthquake Data: Systems for Evaluating the Severity of Earthquakes

Richter Scale	Intensity Equivalence	Mercalli Scale
2	I and II	I. Seldom felt.
		II. Felt by a few on upper floors.
3		III. Felt noticeably indoors; vibrates.
4	IV–V	IV. Felt outdoors by few. Vehicles rock.
		V. Felt by most. Some breakage. Pendulums stop.
5	VI–VII	VI. Felt by all. Slight damage.
		VII. People run outside. Worst structures much damaged. Noticed in moving cars.
6	VII–VIII	VIII. Considerable damage in ordinary buildings; partial collapse of chimneys. Heavy furniture over-turned. Sand and mud ejected. Changes in well water.
7	IX–X	IX. Considerable damage in best structures; build-ings shift off foundations. Ground cracks. Pipes break.
		X. Masonry and frame structures destroyed with foundations; ground badly cracked. Rails bent. Landslides. Water splashed over banks.
8+	XI–XII	XI. Few masonry structures standing. Bridges de-stroyed. Broad fissures in the ground. Under-ground pipelines out of service. Earth slumps. Rails bent greatly.
		XII. Damage total. Waves seen on ground surfaces. Lines of sight, levels distorted. Objects thrown upward.

Scientific studies based on physical remains (not words) have not been as energetically pursued at Mediterranean sites as in Mesoamerican sites. At New World sites, where the surviving documents were nonexistent or until very recently unreadable, investigators were forced to devise nonlinguistic research tools (Ambraseys and White 1996; Bousquet et al. 1989; Sanders et al. 1989; two Mediterranean studies and one Mesoamerican). These tools are applicable to Mediterranean sites and are now being adopted by classical archaeologists.

The works of two important scholars are illustrative: R. Martin (1956) has for 50 years or so illumined for us the ways that Greek cities were organized. Although excellent in pioneering the new field of ancient urban history, his work gives scant attention to the physical base of cities. Even W. L. MacDonald's exemplary work (1986) on Roman urbanism considers neither underlying geology nor ancient technology as determining urban form.

Asking geological questions about urban development is our enterprise: How and to what extent did the physical bases of Greco-Roman cities determine their urban development? Our new synthesis of the human history and geology is intended to complement the data from older studies with new information from our

TABLE 1.2 Earthquake Data: Earthquakes at the 10 Mediterranean Sites Studied

Date	Place	Source[a]	Comment
Historic			
before 700 B.C.E.	Ionia, Lydia	(23, 21)	Changed swamps to lakes; tsunamis.
		(12)	Delphi destroyed.
560–550	Selinus		Possible major earthquake.
480	Delphi	(13)	
426	Corinth	(6)	f[b]
	Selinus		Damages.
419–20	Corinth	(6)	f; little damage.[b]
		(25)	Disrupted war conference.
393	Corinth		
388	Argos	(6, 14, 28)	f; aborted Spartan invasion.[b]
373	Delphi	(14)	Comet, rockfalls, damage;
		(12, 20, 17)	tsunami in Gulf of Corinth.
354–46	Delphi	(11, 23, 2)	Phocian soldiers repelled; no damage.
304–3	Ionia	(9)	
279–278	Delphi	(22, 20)	Quake, storm during battle; snow, frost broke off crags; falling rocks killed many Gauls.
227	Hellenic arc	(23, 22, 11, 20, 14)	
227 or 225	Sikyon	(12)	Destruction.
225	Doric Greece	(7)	
201–197	Samos	(6)	
198 August	Rhodes		Shattered.
	Asia Minor	(6)	f; many cities ruined.[b]
	Cyclades Is.	(23, 21, 14).	New island near Thera.
27–24	E. Mediterranean Cos, Tralles	(23, 3)	
23 B.C.E.	Qt. Cybiritica in Asia Minor;	(24, 7)	
	Aegean esp. Egio	(12)	
17 C.E.	Calabria, E. Sicily, Selinus		
17 C.E. or 28	Asia Minor	(24, 7, 6)	12 cities on Gediz R ruined.
68	Gt. Meander valley	(3)	
77 June 10	Corinth	(12)	Destruction.
3rd quarter 1st c.	Syracuse		
140	Asia Minor		
238	Gt. Meander valley	(3)	
244	Gt. Meander valley	(3)	
262	Gt. Meander valley	(3, 7)	
306–10	Tindari & North Africa		
310–65 series		(3)	
350–550			One of most seismically active periods in last 2000 yrs.
ca. 350	Ephesus		Earthquake and fire.
362	Ephesus	(26)	Possible earthquake.
363 or 364?	Sicily		
365 21 July	Crete, Greece, Sicily		Major earthquake. Tsunami felt from SW Peloponnese to Alexandria, Egypt.

TABLE 1.2 (*Continued*)

Date	Place	Source[a]	Comment
		(16, 7, 4)	Destructive in Sicily, islands; epicenter in Cretan trench.
364–455	E. Mediterranean	(6)	Frequent major earthquakes.
ca 400	Tindari	(27)	Contaminated water supply in Syracuse.
551	Corinth	(12, 7)	Great earthquake like 303 B.C.E. destroyed towns N of Gulf of Corinth.
6th c.	Selinus	(10)	Damage.
8–9th c.	Selinus & Castelvetrano	(19, 5)	Worst quake in 852; overthrew Selinus walls.
12th c.	Selinus		Temples of E. Hill destroyed.
Modern			
1653	Gt. Meander valley		IX; ruptured N edge 70 km.
1693	SE Sicily		IX; cities near Morgantina destroyed.
1891	Gt. Meander valley	(3)	
1899	Gt. Meander valley	(3, 29)	IX; ruptured N edge 50 km.
1905	Delphi		Destroyed Temple of Athena Pronaia.
1955	Gt. Meander valley		Destroyed Balat village in Miletus ruins.
1968	Belice Valley E of Selinus	(5, 8)	Major quake
	Partanna NNE of Selinus	(18)	Ruined

a. Sources for earthquake data:
    2. Aelian. Var. IX.421
    3. Altunel, citing *Earthquake Catalog*
    4. Ammianus Marcellinus
    5. Amadori et al.
    6. Ambraseys:
    7. Bousquet, Dufare, and Péchoux
    8. Bosi et al.
    9. *Chronicle of Paros:* O1.CXXXV or CXXV/IG XII.v.444.
    10. de la Genière
    11. Diodorus Sicilicus. XVI. lvi. 7–8; XXVI.8.
    12. Galanopoulos
    13. Herodotus. vIII. or vii. 37–39 VI. 98.
    14. Inscriptions. F911/F91, CIL x, 1624.
    15. Jacques and Bousquet
    16. St. Jerome
    17. Justinian the historian. XXIV.viii.9.
    18. Kininmouth
    19. Naselli
    20. Pausanias. X.xxiii; X.xxiii, 1–4 (dating by 125th Olympiad = 278) and X.xxiii.9; XXVI.8; xi. 7.
    21. Pliny. NH ii.93; lxxxix
    22. Polybius. I.v.5; V.lxxxviii–lxxxix.
    23. Strabo. I.iii. 16 and 161; IX.421; XIV.ii.5; XII. viii. 18; II.viii.4.
    24. Tacitus
    25. Thucydides. V.1.4, II.viii.4.
    26. W. Vetters
    27. Wilson
    28. Xenophon. IV.vii.44–5
    29. Yeats et al.
b. Severity, according to Ambraseys: f = felt; F = strongly felt.

multidisciplinary teams. The broadest concepts of how human society relates to its physical environment mesh with the most exacting attention to underlying geological structure and processes and with computerized analysis of engineering solutions for ample water resources, essential for the survival of any settlement larger than a hamlet.

The intellectual setting of this study transcends the limits of any one discipline. Our team draws on ancient and urban history, archaeology, hydraulic and structural engineering, and many subspecialities of geology: petrology, topography, geomorphology, seismicity, engineering geology, structural geology, hydrogeology, sedimentology, archaeological geology, eustatic sea-level changes at coastal cities, and the study of weathering. Specific geological information makes the urban history and archaeology of each site more exact and plausible, giving us new insights into the process and constraints of urbanization.

Making correlations from archaeological to historical to scientific evidence for the past can be a questionable undertaking, according to geologist Ambraseys (1971). He thinks that seismologists who have used literary sources for information on historical earthquakes (before 1900) have often accepted these accounts uncritically. His own study of tectonics in the eastern Mediterranean basin drew on the *Teubner Series, Patrologia Graeco-Latina, Corpus Scriptorum Historiae Byzantinae,* Greco-Roman writers (e.g., Pausanias, Jerome), Syriac sources, Arabic sources, Armenian and Georgian sources, medieval and later manuscripts (especially their marginal notes and colophons), monastery chronicles, coins and inscriptions, and archaeological evidence from recent excavations. From all of this, he recorded over 300 earthquakes from 10 C.E. to 1699 — a total of 20 times the number listed in modern catalogs for the same period. Such a marked difference required reappraisal of the data, after which Ambraseys postulated that single earthquakes, particularly those of large magnitude affecting a wide area and hence generating multiple accounts (such as the major quake of 1870, felt in both Greece and Turkey), had been listed as multiple earthquakes. He concludes that Corinth, Delphi, Ephesus, Miletus, and Priene but not Argos are in areas of strong seismicity.

Not only seismologists but also archaeologists may have accepted the ancient written evidence uncritically. Further, the latter have sometimes accepted modern scientific evidence with greater confidence than the accuracy of the data warranted, hampered by the fact that it is hard to maintain necessary scholarly skepticism in a field where one is relatively inexperienced.

Only recently has the investigation of ancient cities and their settings expanded both chronologically and scientifically. While realizing that communal life in the Mediterranean world has been a continuum from the Stone Age until now, we concentrate here on the Greco-Roman world, acknowledging that the cultures were set in geography of much longer duration. The 1400-year period from 800 B.C.E. through the sixth century C.E. was long enough for earthquakes, volcanic eruptions, erosion, sea transgression, floods, alluviation, and other geological events and processes to affect both hinterlands and settlements, and to be reflected in literary and archaeological records. In our set of cities, similar petrologies and similar geological processes and events have contributed to a family resemblance manifested in similar cultures.

In this study, we attempt to correlate what is known of the seismic and sedimentary history of each site with what is known of the building chronology. The resulting insights (not "facts") emphasize interactions between geological events and human history.

Hypotheses

We postulate three simple hypotheses:

1. Similar geology fosters similar urban development, even though not every urban difference has a geological basis.
2. Geological differences are likely to result in developmental differences.
3. Geology forms an active backdrop to human actions.

The gradual pace of much geological change, such as the laying down of a strata of limestone or the uplift of a mountain range, is commonly but erroneously thought of by nongeologists as unrelated to historical change. Yet our building materials are the results of these processes. Our landscapes are temporary pauses or slowing in geological activity. Aggradation and other processes gradually change the world we try to master. Earthquakes and landslides change it abruptly and are rare enough to be nearly invisible to the historical record. Geology thus furnishes not merely a slow-motion, long-term backdrop to human action, but also an energetically changing milieu with which we interact.

The underlying geology is of great importance for urban development because of its effect on water resources, topography, geomorphology, the nature of the soil, and the availability of building materials. The chemical composition of stone and its history of deformation or metamorphosis determine its fitness for different structures. Tectonic structure, seismology, and more subtle processes such as the silting up of estuaries affect settlements and must be taken into account for a fuller understanding of each place.

This study is a story woven from hints, not a report of replicable experiments or isolated scientific analyses. On principle, I cannot ignore the human factor—although Gerschenkron (1968) called for a history that has been "purged of emotion and preconception"—because both historians and their audience are human beings complete with emotions. Rather, I strive to acknowledge and make explicit those nonquantifiable factors when they occur. The difficulties of narrowing a subject, selecting evidence, and weaving the data and insights together to create a balance among (1) the original actors with their beliefs, values, and intentions, (2) the social structure that restricted them, (3) the sources that recorded their actions but introduce biases of their own, and (4) the audience of readers for whom the historian writes are in themselves a cautionary tale (Hopkins 1978b) for anyone attempting broad comparisons.

For modern investigators, the Greco-Roman sites have the advantage of long duration and depth of study. Fourteen hundred years of urbanization gave the Greeks and Romans time and incentive to observe results and plan improvements—and that time span gives us the opportunity to observe the planned and unplanned events of urbanization. Thanks to the fascination the Greco-Roman world has exerted on modern peoples for the past 600 years, a great deal of information has accumulated about that period.

The city was the basic unit of the Greco-Roman world during the entire period. In the fifth and fourth centuries B.C.E., the Greeks believed that 5000 men were not enough to be considered a city, whereas 100,000 were "a city no longer" (Aristotle 1327a). By "city" (*polis*, Greek; *urbs*, Latin) was meant the densely settled core, the dependent villages, and the farmlands of an individual polity.

The process of urbanization was both physical and cultural. Urban design is the conscious arrangement of elements of a city so that maximum efficiency and utility are combined with maximum beauty and agreeable provisions for day-to-day living. Greek urban design was based on isolation of major monuments and angular (not axial) views of them from a distance. (Doxiadis 1972; Greek street patterns, see Crouch 1993, Chapters 5 and 6). The Romans used two major urban patterns: the regular grid mostly associated with veterans' colonies and the towns that developed from them, and the jostle of monumental buildings set close together and at angles to one another without a regular pattern of streets to set them off, as in the capital at Rome (an "armature," according to MacDonald 1986). In cities of the eastern part of the empire, Roman density was mitigated by the older Greek tradition of isolating major buildings and using topographical features to set them apart. (For a list of the names of periods and their approximate dates, see the introduction to Appendix A, Chronologies.)

A Greco-Roman urban place was expected to have a standard ensemble of buildings and spaces. The Roman travel writer Pausanius (10.4.1) of the second century C.E. defined a city as having certain facilities: streets, an aqueduct and fountains, sewers, government buildings, temples, theater and often an amphitheater and a stadium, shops, houses, and a rampart. Temples rose on hills or in plains, depending on the deity worshiped, and were surrounded with multipurpose precincts having stoas, altars, and buildings for general use. Public plazas (*agora*, Greek; *forum*, Roman) were furnished with porticoes, shops, and recreational buildings. The plazas provided open space for markets, military reviews, political meetings, and other public activities. Favorite sites for plazas were the intersections of major roads at or near the center of the town or in the vicinity of important gates. The major streets, particularly in cities of the western empire, had stores and workshops tucked into the street facades of houses. Fountains enlivened streets, temple precincts, and public plazas. By early Roman times, public plazas were also equipped with latrines. The economic burden of building all of these structures and maintaining them weighed on the local government, though it was often eased by imperial contributions.

In 1992, realizing that I had questions but neither answers nor methodology for acquiring these answers I enlisted practical scholars from Turkey, Sicily, Greece, and the United States to undertake joint field work and the scientific analyses for this study. The geologists and engineers assisted me in seeing the physical world at the selected sites, in particular how the sites' physical features were based on karst geology and amenable to hydraulic engineering, how the topography and building materials determined the urban design, and how geological events and processes altered human plans. For each site, we have photographs and geological maps correlated with maps of Greco-Roman buildings, streets, ramparts, and so on—the visual equivalent of our intellectual understandings. Chronologies correlate geological and human historical events at each site.

I had initially expected to concentrate on urban-karst relations but broadened the scope of work in response to the research interests of my colleagues. Although geologists, most of the scientific contributors were not karst experts. To utilize their services appropriately, it was necessary to deal with a wider range of geological questions. Would research into nonkarst geological questions dilute the focus of the book? This seeming disadvantage became a source of enrichment for the project. Interesting and useful geological questions such as the nature of materials, tectonic and structural questions, geomorphology, petrology, sedimentology, and speleology suggested a number of potentially fruitful investigations. This breadth is not out of place in an introductory work.

Setting

To this day, Greek terrain seems more wild than Italy's, although people have lived in both countries for about the same number of years. Turkey seems richer than Greece and more like Italy, divided into a coastal area settled by Greeks and a higher plateau that occupies the center. The Romans incorporated the entire peninsula of Anatolia into their empire.

Most of the rocks in the Mediterranean area are carbonate (limestone, marble, dolomite, calcite, calcarenite), but there are interlayered clays, marls, conglomerates, and sands, as well as some sandstone and some volcanic stone such as basalt. Carbonate stones commonly contain karst shafts and channels, and have proved to be ideal bases for settlements with the technology to tap them for water, mine them for precious metals, and quarry them for building stone and clays for making pottery, roof tiles, and pipes. Even the volcanoes of the area proved to have two beneficial features during the time of our study. First, as lavas weather, they become excellent soil for farming. Second, volcanic stones can be useful for construction and for hand mills to grind grain, and in the form of *pozzolana* (volcanic tuff or ash), they contributed to the excellent hydraulic cement the Romans were famous for. Details of the geology will become evident as we discuss our ten cities.

Karst

Karst, a key constraint and advantage of this area, shows local variations that expand the definition of this geological type (Crouch 1993, with bibliography). Karst aquifers are distinguished by large void spaces, high hydraulic conductivity, flat water tables, and extensive networks of solution channels, often exhibiting turbulent flow. Carbon dioxide given off by the roots of plants dissolves limestone and thus enlarges cracks to shafts and caverns; calcium bicarbonate dissolved in water enters a cave where it gives off carbon dioxide, which deposits calcium carbonate on the surfaces of the cave, forming stalactites and stalagmites (A. T. Wilson 1981b; 218, his fig. 1). Karst phenomena dominate Greece (Burdon 1964; Morfis and Zojer 1986). In southern Italy and Sicily, the Greeks chose karst terrain for their settlements in the eighth and seventh centuries B.C.E. Sicilian karst is locally developed in the Madonie Mountains of the interior, at Palermo in the northwest, in the hills above Syracuse in the southeast, and in the south-central area; some of these karsts occur in gypsum outcroppings as at Akragas (Dall'Aglio and Tedesco 1968; Belloni et al. 1972). Karst still produces flowing water in southern and western Turkey (Özis 1985), is found in Ionia as far north as Troy, and was even more important in antiquity before deforestation.

The cities of this study all depended on karst or karstlike geology for their water (Crouch 1993, Chapter 7). Karst systems can be used directly or tapped for long-distance waterlines. At Selinus, for instance, off-site sources of water and stone related the city strongly to its hinterland. At Argos, the hydrogeology made possible long distance water lines that supplied water to the city. At Miletus and Priene, two different forms of karst were tapped. Priene's water and drainage system relied on the water stored in the adjacent karstified marble and limestone mountain, readily available from springs, but the older karst with fewer on-site springs at Miletus required development of long-distance water supply lines as early as the sixth century B.C.E.

Controlling water in karst terrane beginning in the seventh century B.C.E., was a significant human accomplishment. The fountainhouse at Megara and the famous tunnel at Samos, of the seventh and sixth centuries, respectively (bibliography in Crouch 1993), are evidence that engineers and politicians mastered the complex interaction of karstic geological processes with human behavior, for urban purposes. Urbanization itself, however, produces problems in karst areas, such as greatly increased runoff.

Other Geological Matters

Nonkarstic geological processes such as sedimentation also affected cities located next to rivers, and sea intrusion influenced coastal cities. Mediterranean rivers are limited in length, but they transport huge loads of suspended materials, leading to landlocked cities, especially in Turkey, where the process has affected all three of the Ionian cities we study. Ancient ports such as Ephesus became separated from their coasts beginning in the eighth century B.C.E. (Altunel 1998; Furon 1952–53; cf. Meiggs 1960 for a comparable problem at Ostia, Italy). Along these coasts, the sea has risen or the land has sunk slightly but continually in the last 2000 to 5000 years.

We also examine tectonic problems related to city building in this terrane. Expert information from engineering geologists has helped us especially understand the site histories of our Greek cases, notably Delphi.

Materials are another aspect of the geological base affecting urban development. Noticing the different proportions of buildings and the visibly different qualities of stone, I began to wonder whether all differences between a Doric building at Selinus and one at Athens were matters of style and the diffusion of taste or whether physical constraints such as the quality of the stone determined the pro-

FIGURE 1.1. Columns partially covered with stucco, at Selinus. (Photo by Crouch)

portions. Recent studies have begun to clarify this issue (Wycherly 1978; Carapezza et al. 1983; Amadori et al. 1992). Particular materials, local stones and clays especially, not only controlled the mechanics of construction, but also determined the visual qualities of each place. The strength and beauty of limestones and marbles, and their wide availability, made them the materials of choice, although sometimes the builders had to utilize the weaker calcarenite of a site such as Agrigento. A finish surface of stucco made of limestone dust over easily crodable stone still preserves the underlying stone 2500 years later at Selinus (Fig. 1.1).

Urban Form

In addition to geology of settlement, we focus on the sociophysical development of the cities studied, as exhibited in their tangible form. A modern reevaluation of the significance of urban form must consider both the amount of wealth tied up in streets, fountains, plazas, public buildings, houses, and ramparts, and the durability of these features (E. Abrams 1994). Ancient building traditions both derived from and informed the physical manifestation of the community in the structures of the city, embodying the social values of the people, their understanding of the physical properties of the stone, and their appreciation of water. Because of the high cost of transporting building materials, stone, wood, and brick were usually locally produced, so that each city had a specifically local visual appearance. Aqueducts, baths, and fountain houses displayed the water supply in the public areas of Greco-Roman cities, as did the domestic cistern in the privacy of the home, with the appropriate articulation in architectural form (Crouch 1996). Each city's pattern of streets, plazas, and buildings embodied the urban concepts that were current when it was first being laid out, in a unique combination of stone, soil, water, climate, and orientation, as these interacted with the ethos of the people. Even if grid-platted Miletus (fifth century B.C.E.) had come later to prefer the scenographic urbanism of Pergamon (third century B.C.E.) with its wedges of terraces fanning out from the central focus of the theater, the geological and economic constraints on remodeling their grid plan into a system of radiating terraces precluded such alteration.

Superimposing the archaeological map of each site on the geological and hydrogeological maps made possible new understandings of each site (e.g., Fig. 3.18). The "fit" between the three kinds of data is evaluated in the individual case studies. Was the bath placed in proximity to a spring, as at Priene? Did a basic stratum of firm rock govern the placement of large temples, as on the east hill at Selinus? Such an analysis enables us to have a more realistic understanding of how and why a city developd in one way or direction rather than another.

The studies of Van Andel et al. (e.g., 1986) were among the first to show the real limits placed on settlements by geology. The amount of rainfall and extent of karstification together with the amount and quality of arable land at a particular site were significant for ancient Greek urban development. Defense can no longer be regarded as the singular factor determining placement of ancient cities. Rather, an ancient Greek town was located amid the area with the best soil where the most work was done, while farthest away was the worst soil, to be allotted the least work. The ancient settlers chose fertile soil if they could not have water and fertile soil at the same spot. Several construction developments made this choice logical: the cistern for storing rainwater at the point of use; wells in the torrent beds, in alluvial fans, and in perched water tables; and long-distance waterlines that made moving water feasible in a way that moving soil could never be. Ancient settlers had more leeway in choice of settlement location than one might at first think, because their water management technology freed them from total reliance on water sources at the same location as their farmland or settlement. Building materials were important economic resources for developing cities. Within these constraints, they chose sites for their beauty whenever possible.

The cities we have chosen to study exhibit a full range of population and site sizes. Akragas, Corinth, Miletus, Selinus, and Syracuse were large in area and had large populations also. Priene and Morgantina were the smallest (Doxiadis 1972). Ancient populations are usually estimated from gross area or scattered literary data such as numbers in military service or census data of the city of Rome in later imperial times. But Crouch (1972) calculates population by density, by area, by the usual proportion between rural and urban numbers in antiquity, and by size of army; the point at which these numbers converge is likely to have been the actual population.

The ten chosen cities represent the geographical spread of the Greco-Roman world (frontispiece). Three of these settlements—Corinth, Argos, and Delphi—are from the center; three are from the east: Ephesus, Priene, and Miletus; and four are from Sicily in the west: Agrigento, Morgantina, Selinus, and Syracuse. Three were capitals, five were ports, one was an inland hill town. The cities have different foundation dates and periods of habitation. Argos was one of the oldest and the longest in duration, being settled during the Bronze Age and currently occupied, whereas Morgantina had perhaps the shortest run on its final site, from 450 B.C.E. to the late first century B.C.E. Ephesus was inhabited by Greeks between the eighth century B.C.E. and the eighth C.E. at different places on the site, but before and afterward by other peoples. Argos and Syracuse are cities with modern populations living above the ancient ruins, whereas at Akragas/Agrigento, Corinth, Delphi, Ephesus, and Priene, the modern population lives near the ancient site.

The cities varied in their functions. The fertile plains of Sicily, with their crops of grain, exploitable minerals, and international trade in luxury items, provided wealth for the building of the three large Sicilian cities. Ephesus and Miletus benefited from similar settings—rich plains and mountainous upstream areas with forest and mining products—to which Ephesus added the religious attractions of the sanctuary of Diana/Artemis and Miletus the religious satellite town of Didyma where Artemis's brother Apollo was worshipped. The adaptable site of ancient Corinth enabled that city to flourish during three different periods: it was first an important pottery center in the seventh and sixth centuries B.C.E., then a trade and religious center in the fifth to third century, then a Roman colony, provincial capital, and trading center from the first century B.C.E. on. Trade was of great importance at Syracuse, Corinth, Miletus, and Ephesus but less so at Agrigento. Religion was a primary economic force at the pilgrimage sites of Ephesus, Delphi, and Corinth, but somewhat less at Agrigento and Selinus, although their splendid temples drew pilgrims. Agrigento, Morgantina, and Selinus emphasized export farming. Fishing, grazing, and forest products were the main concerns of Priene, with shipping handled through its nearby port. Argos combined fishing, grazing, and farming on the Argolid plain, and the city probably shipped of forest products as well.

One feature of Greco-Roman life that has received relatively little attention is the diversity of the population, both rural and urban. Now that population diversity is becoming a modern issue, we may see renewed study of this factor. In California, for instance, there is no longer a majority ethnicity; Sicily today has many black Africans, Germany has Turks, and so on. As long ago as the seventh century B.C.E., the colonizing Greeks accommodated themselves and their culture to the non-Greeks they settled among (Descœudres 1990). Seven hundred years later, the Romans engaged in a deliberate policy of importing slave populations and exporting veterans to form colonies. These policies may help to explain the occasional lacunae in local resource-management knowledge as observed, for instance, at Morgantina (Hopkins 1978b).

The next chapter expands the discussion of our hypotheses and methods, with accounts of the historical development of several disciplines that have contributed to our research. Then follow the ten case studies, grouped geographically. Because not all the same questions are addressed in each case study, it has not been possible to maintain parallel organization of the chapters. We conclude with reflections on our scientific and historical findings and their meanings for the histories of these particular settlements, for the history of Greco-Roman urbanism as a whole, and for the development of geology as a humanistic and historical discipline.

History, Geology, Engineering, and Archaeology

It remains a deplorable fact how little historians have understood the myriad meanings for the human race of its own planet. . . . In tossing out the 19th century idea that God made this place for human use, the geologists also tossed out the examination of this place as setting for human life.

M. J. S. Rudwick, *The Great Devonian Controversy*

Geology concentrates on individual events and this is more like history than like physics.

P. Gay, *Style in History*

For convenience, human knowledge, especially in the German and American educational systems, has been separated into disciplinary packages. Thus chemistry, for instance, is defined by certain analytical actions taken toward certain materials, to answer a particular group of questions. Unfortunately, many topics are not amenable to isolating methodology. Cities, for instance, are so complex that understanding them requires coordinated research by historians of many specialties, by architects and planners, by sociologists and psychologists, and by statisticians and geographers, all of whom also benefit from the insights of scientific disciplines. Planet Earth is even more complicated and calls for every field of expertise to examine it and to synthesize results. Four disciplines that contribute to this study are history, geology, engineering, and archaeology.

History

History may be the most recalcitrant of the humanist disciplines, notorious for partial or complete gaps in understanding and fated to reinvestigate earlier situations and earlier research. History reconstructs contexts in which past reality can be comprehended. Although human knowledge is so variable that no scientific law can sum it up (Wright 1975), we can tell stories that reveal our insights. For every attempt at general history and geography such as that of Herodotus, there are dozens or hundreds of more or less philosophical and dramatic memoirs such as those of Julius Caesar. Sometimes the only surviving data is from dynastic chronologies, for example, I and II Samuel in the Bible. Little of this ancient history involved "research" as we now understand the process (Gabba 1981). Ancient history is still monopolized by philologists, whose first passion is language and who are often out of phase with important methodological developments in history (Ramsey 1890, Ch. 3, History in Classics). Like their ancient predecessors who concentrated on the elite, modern historians of Greco-Roman times may fail to investigate ordinary people. Modern cities are places where people dance, use computers, sing, bathe, and vote. Realization of our diversity of behavior impels us to expand the study of ancient cities to incorporate their similar diversity.

After the medieval era of hagiography (the Venerable Bede's seventh-century *History of England*), the idea of "universal history"—what happened everywhere, told as objectively and factually as possible (Huppert 1970)—developed from the seventeenth century on. The seventeenth century was the first to use coins, inscriptions, and other physical objects as historical evidence (Collingwood, 1946–1967). At the same time, the distinction was first made between primary and secondary historical sources (Tholfsen 1967); coins and archaeological findings are primary sources, but the explication of them is secondary. In the last two decades of the twentieth century, new methodologies proliferated, including statistical and sociological investigations, specialist studies (history of the labor movement, history of bridges), and finally new efforts at synthesis. Unlike the history of religion or of capitalism, which use generalizing methodologies, the histories of architecture, of cities, and of geology are specifically local (Crouch and Johnson 2001, intro. and chap. 15, Class, Gender, and Ethnicity).

The best history does not deal with the facts, events, or processes separately but with their interconnectedness. Personal involvement can actually assist in understanding (MacMullen 1990: 28). "The values and experience of the knowing subject [the historian] are not 'subjective' obstacles to be overcome," writes Tholfsen (1967: 206, 224–5) but are indispensable tools for the study of the past. Yet the historian's activity is a good deal more vulnerable to extraneous considerations than is the geologist's: the historian's attitude toward political behavior will necessarily lack the detachment of the geologist's attitude toward erosion.

"The enlargement of historical knowledge comes about mainly through finding how to use as evidence this or that kind of perceived fact which historians have hitherto thought useless to them" (Collingwood 1946/1967). The new data from scientific studies of the sites supplement the old data from ancient history and archaeology, and along with the insistence on the importance of everyday life, makes a more complete history possible.

It is true that the timescales in history and geology are drastically different. It is our contention, however, that coordination of the two timescales is not impossible, merely difficult. (Geological timescales per se have been discussed in Cullingford et al. 1980, Gage 1978, and Thornes and Brunsden 1977.) The impulse to sort evidence into chronological order and then reflect on that order is equally strong in geology and in art, architectural and urban history, qualifying all four as types of history.

Geology and Engineering

Whereas history has become more and more differentiated and comprehensive, the discipline of geology evolved from descriptive to historical (the relative order of the great eons of the geological record) to analytical to the cultural or humanistic geology of today (Marsh 1874; Stent 1972; Vetters 1994, 1996). Although it is difficult to date geological processes that take millions of years to happen, we can still examine their current results and the way these affect the human environment. In antiquity, "large-scale natural phenomena were regarded as factors in human history" (Gabba 1981). Though the ancients were more interested in the psychological and social effects of disasters, we are more interested in the physical effects.

Geology began as a descriptive discipline (G. Owen of Henley's, 1595). In the seventeenth century, Nicholas Steno wrote the first "history of the earth" based on observations and deductions about rocks. By 1700, terrestrial forces were seen as systems, at least by some scholars, whereas earlier the earth had been thought of as inert and passive, changing only a little (Owen and Steno cited in Porter 1977: 45, 72–74), except for the ancient Deluge.

From the eighteenth century on, the new science of geology, useful to engineers and businessmen, became the passionate hobby of amateurs (Fig. 2.1). This broadened social base increased the flow of data about irreducibly diverse local phenomena, because rocks do not have worldwide extent or thickness but must be matched chronologically by their fossils (Hallam 1989). By 1800, the knowledge had trickled upward to Oxford, Cambridge, Edinburgh, and Dublin universities (North 1928; Porter 1977). By the 1820s, geology was the science most pursued by the general public, most debated in the literature for the next 50 years, and most feared in both religion and social orthodoxy.

Among the learned gentlemen who traveled in Europe and analyzed the terrain they were seeing, a major figure was Sir Charles Lyell (1797–1875), the first geomorphologist. From his study of history, Lyell brought into his new work in geology the ideas of the importance of sequence and causal connections (Schneer 1969). He perceived the earth's history as a continuous causal chain in dynamic flux; his early description of visiting Mt. Etna was the first modern understanding of volcanoes. From linguistics he brought the concept that the visual evidence had meaning when the fossils and strata were read by the geologist. He also wrote of how people's physical characteristics and social organization relate to their geographical districts about 100 years before the origins of environmental medicine and the birth of ecology (Rudwick 1976, 1985).

At first, geology was thought to validate the biblical account of the flood (*diluvialism,* L. E. Page 1969), but additional fossil evidence fostered critical examination of that position. During the following controversy-filled period, scientists worked out an orderly geological chronology, vastly extending Earth's history. The

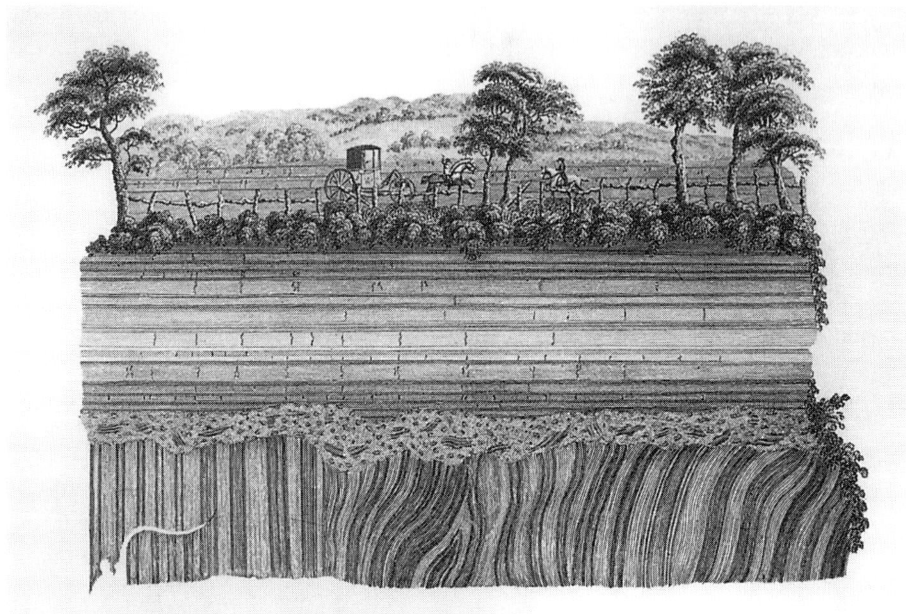

FIGURE 2.1. Unconformity at Jedburgh, Scotland. Drawing by James Hutton, for his *Theory of the Earth*, 1788. Near-vertical beds of Silurian sandstone are topped by angular dedritus, over which lie flat beds of Devonian sandstone, topped by soil with vegetation, humans, and horses. Hutton was one of the first to notice the specific features of the geology around him and to speculate on their meaning.

most important intellectual tool in this new way of looking at the earth was stratigraphy, which postulates that the oldest stone in a sequence of layers is at the bottom, unless the stack has been deformed through folding or faulting. Stratigraphy gives relative ages, based on layers of stone and not on years, and is closely related to sedimentology. For nearly a century, emotional discussion pitted a literal acceptance of the Genesis account of creation against the new physical knowledge — and this discussion has still not completely abated (Collingwood 1946; Oldroyd 1990; Yeats et al. 1997: 425–26). Earth is now thought to be at least 700,000 times as old as people thought 300 years ago (Albritton 1980).

This new information forced both history and geology to stop being merely descriptive and assume the modern task of explaining "the pattern of events over time." [But] "the science of the Earth divested itself of the appeal to written testimony [authority of the text] long before the science of human history" (Porter 1977: 109, 215, 219, 221).

Analytical geology, according to North (1928), owed its impetus to the shift in demand for supplies of petroleum. A recent analytical development, plate tectonics theory (W. J. Morgan 1968), describes spreading and extending surfaces on land and at the bottom of the oceans (Fig. 2.2), accounting for continental drift. Where

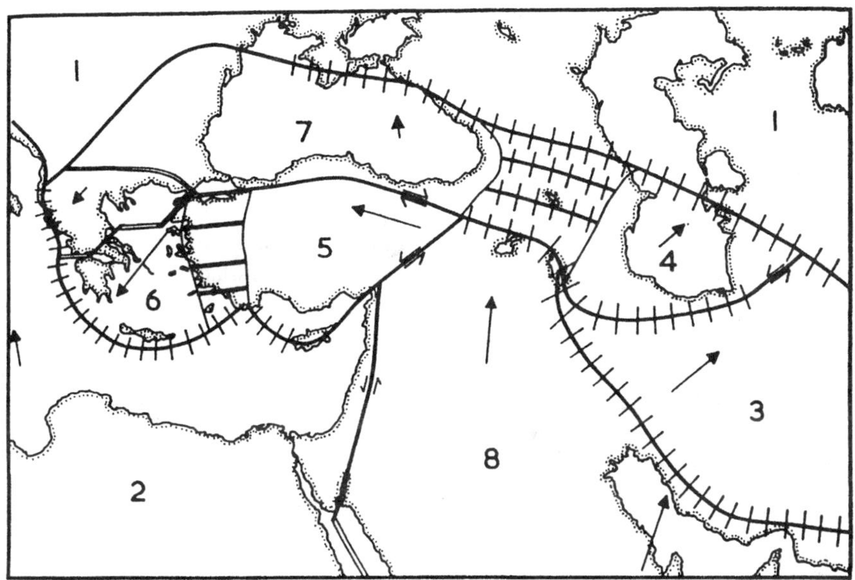

FIGURE 2.2. Plate boundaries and motions in the eastern Mediterranean area (Mac-Kenzie 1972). Plate edges in western Turkey and the Caucasus are shown only gener ally. Arrows show direction of motion, their lengths proportional to the relative veloc-ity. Double lines indicate extension across plate boundaries. A single heavy line indicates a transform fault. Crosshatching represents boundaries across which shorten-ing is occurring. Plates are numbered: 1. Eurasian, 2. African, 3. Iranian, 4. South Caspian, 5. Turkish, 6. Aegean, 7. Black Sea, 8. Arabian. This simplified map suggests the complications of plate movements in the area.

two plates come together, zones of compression result in earthquakes and provide opportunities for magma from deep layers to reach the surface in volcanoes.

The next step in developing as full as possible a study of the operative systems affecting environment and topography (Nash and Petraglia 1987) is to separate the natural from the human objects in depositional pattern, dealing in turn with ar-tifact, site, and region.

When I was trying to learn about how the ancient Greeks managed their water systems (Crouch 1993), I found a previously unnoticed set of clues to the development of those cities. The geological base that determined the water re-sources also defined the topography and geomorphology as well as the available building materials. We have learned that Greco-Roman peoples exhibited an en-vironmental sensitivity in practice that had not been realized previously from ex-amination of ancient literature, because those who composed dramas and political treatises were not the practical ones who carried out city building.

The geology of these sites controlled the resurgence and abundance of water, of soils suitable for agriculture and forestry, of building materials, and the location and type of transportation facilities such as ports and roads. The wide availability of water in a seemingly arid environment is due to the karst that underlies the sites (Crouch 1993: Karst Basis of Urbanization). From a general interest in how water was supplied to and drained from ancient cities, I gradually came to understand the importance of water in determining urban form.

It seemed likely that other aspects of the physical environment were equally important in development. The carbonate rocks of the karst, whether calcarenite at Agrigento, sandy limestone at Morgantina, or marble at Priene, provided not only water but also distinctive building materials for local architects. Beyond water, soil, and building materials, the geology determined surface topography. Hills here, valleys over there, gentle slopes, abrupt cliffs, or fertile plains provided different environments and different urban settings. Through the process of weathering, geology also determined the soil. Limestone became the common red soils. Alluvium could be mainly sand and clay, which during millennia of pressure and cementation became sandstone, shale, or marl. Deforestation and erosion contributed pebbles, stones, and boulders as well as silt to the centuries-long process of filling valleys and building deltas — the impoverishment of the hills becoming the enrichment of the plains. The contest between the silt-laden rivers and the sea into which they dumped the sediment determined the conditions that provided seafood to balance the largely cereal and vegetable products of the land. Mountain pastures, as at Priene and Argos, provided meat and milk products, another variation of diet.

These physical features, in turn, depended on how the land had been formed over long time periods by both gradual processes such as uplift, metamorphosis, and the intrusion of the sea over the land, and abrupt changes like earthquakes and landslides. Localized study expanded our knowledge of the origin and development of landforms, the relation of landforms to underlying geologic structure, the history of geologic change as recorded in landforms, and the relation of local tectonics, both events and processes, to local geomorphology.

Political problems of the late Roman Empire converged "accidentally" with the silting up of the bays on which Miletus and Ephesus in Turkey were built. This meant that the harbors could no longer function. In addition to economic depression, malaria developed in the new marshes. The difficulty of maintaining a water and drainage system under the new physical circumstances and with less money contributed to the decrease in population and importance.

Thus the geological base was key to understanding both the basic problems — food, water, and shelter — that each human settlement had faced and the solutions that the ancient people found. Yet the few existing studies of the geology of Greek regions (Schröder 1990; Schröder and Yalçin 1991, 1992a & b, Hayward 1995; Higgins and Higgins 1996) are not well integrated with historical and archaeological information. Our ten case studies examine this interaction between settlement and geology.

Technology of the ancient world was fairly sophisticated by 2000 years ago, especially that related to water management. Some pioneering modern studies of ancient water technology are now seen as incomplete, because their authors based their conclusions on an inadequate number of case studies in technological history (compare Ashby 1935 with Çeçen 1996). We have investigated the hydraulic history of all ten cities, finding not only site-specific information, but also new information about the practice of engineering in the Greco-Roman world. See the case studies of Ephesus, Miletus, and Syracuse and related publications (Ortloff and Crouch 1998, 2001; Tuttahs 1998).

Archaeology

Archaeology is the study of the material remains of the human past by specially trained persons who excavate and who use the data they find to make historical judgments. Indeed, for the period earlier than about 5000 B.P. (before the present), there are no other sources of information about human life than material and environmental ones. Archaeology was for a time hampered by uncritical reliance on the surviving literature of antiquity and by bias in favor of beautiful objects that could be shown in museums. The excavators of Mediterranean sites, blessed and cursed with ample literary evidence of the Greco-Roman past, slowly adopted modern scientific methods. Such sites as Ephesus, Athens, and Corinth have, for instance, over 100 years of archaeological and philological data, but relatively little geological data. In contrast, Mesoamerican sites had until recently no readable premodern documents. Their excavators were forced to develop scientific technologies (Sanders et al. 1989) since the evidence was not "preverbalized" for them. The discipline is notably adapted to the discovery of art and technology, but by inference archaeology can provide information on social arrangements, religion, and economics.

As an intellectual discipline, modern archaeology is based on the following:

- Stratigraphy. Its application in archaeology made relative and then absolute dating feasible.
- The antiquarian revolution. From about 1815 on, the successive technological stages in human development such as the Bronze and Iron Ages were noted (Porter 1977).
- Evolution. Darwin's (1856) method of "collecting and classifying information without any particular aim until at last a grand pattern emerged" (McNally 1985) has been adopted by archaeologists and urban historians who are willing to suspend pattern definition as long as possible.
- Scientific anthropology. This field is especially rich in results when assisted by physics, chemistry, geology, and engineering.

Following the example of the early nineteenth-century geologists who studied stone tools (McNally 1985), modern archaeologists study objects. Uncritical acceptance of correspondences between past and present in the forms, types, and classes of human life was an early feature of their expanding understanding; as an unexamined assumption, this idea was one of archaeology's greatest problems (Taylor 1948). Archaeology now utilizes sciences such as geology, physics, chemistry, and other analytic fields. Outreach to scientific disciplines may have been furthered by the fact that Americanist scholars (those who study Native American remains) have most often been pragmatic American "technologists"—for example, engineer

E. Squier (1877), who made seminal contributions to the study of Native American architecture and engineering—rather than classically trained European gentlemen.

Urban History

Less common than the study of objects is the examination of entire ancient cities, subsuming the individual objects under urban systems, for example, water system, defense system, and educational system. Little of such system work had been done on ancient Greek cities when I began my study in 1970. Urban historians rejoice in the enhanced meaning of an object when its function in a system is exposed. For instance, one can validly study a collection of coins as art objects, but additional insight comes from realizing that they functioned as a medium of exchange. System concepts can provide connective tissue for archaeological study of singular items, whether house walls or signet rings. Systems can be valuable even when the parts are of no monetary or aesthetic worth: the water system of a city and its constituent clay pipes and cisterns come to mind, usefully studied by modern hydraulic and fluids engineers. Because water is still doing what it has always done, engineers have few cultural presuppositions to interfere with their analysis of the ancient water systems, whereas modern religious beliefs or economic assumptions may interfere with the study of ancient shrines or markets.

Yet the urban historian can founder when too much attention to detail swamps the overall understanding. Finley (1977: 308 n. 3), for instance, cites a study of 100 variables defining a city and another that found 333 variables, numbers better suited to statistical than historic manipulation. To find the human meaning of what is said, done, or made, you must know the question to which it is the answer (Collingwood 1939: 31), then notice and analyze new data, which can give new meaning to physical objects and to behavior.

II

CASE STUDIES

The underlying geology and visible geography constrained ancient city builders as they do modern engineers. To examine our hypotheses in ten case studies has been a challenge; we compared western, central, and eastern Greco-Roman cities in terms of their geology. More extensive geological and ecological studies of each city would give fuller results, but we hope that this introductory work will be useful in raising additional questions as well as in summarizing what is known to date. By the accumulation of details in these ten case studies, we will gradually see which elements of the eventual urban setting were natural and which cultural.

Even if we had many times as much information as is presently available, however, we still could not assume that Akragas or any other of these cities presented the same geological details in the eighth century C.E. as they do today, subject to the identical forces. Similarities and differences of detail will emerge from our study of the ten cities.

The geological discussion of each site depends on published and on-site research, which is both uneven and sometimes dated. Only Selinus, Argos, Delphi, Miletus, and Ephesus have recent geological studies, but none so far have studies that incorporate all aspects of geology. Through the efforts of Italian hydrogeologists and applied geologists, the study of the natural aspects of several ancient Sicilian sites is yielding significant findings.

Since karst is the usual geology at these sites, a description is warranted. Karst was first named in the Trieste area of Slovenia and Italy. It is defined as a region underlain by limestone or other carbonate stone, where landforms result from chemical weathering involving reactions between dilute carbonic acid and a mineral such as calcium carbonate (e.g., limestone), making large and small void spaces and extensive networks of solution channels in the rock, with rapid hydraulic conductivity. The waters dissolve the stone upstream and redeposit the dissolved carbonates downstream. A common pattern in karst is vertical shafts and horizontal channels with caverns of different sizes. (One notable result of such a process is Mammoth Cave in the United States.) These shafts and channels connecting to caves can be used as natural reservoirs and delivery systems, as Fekete (1977) has pointed out for modern Pécs in Hungary. The karst geology offers three advantages: first, no leakage, which means small or no loss from beginning to end, resulting in a system that is both cheap and clean; second, equalized flow year round; third, the relative absence of pollution. Fekete cautions that setting up such a water system utilizing karst needs patience, since the elements of the system are largely invisible, and even today trial and error cannot be avoided. It is extremely important to keep the catchment area (from which a surface stream or groundwater system derives its water) free of pollution, lest the invisible waters be contaminated. Throughout our study, we will return to water management as indicative of geological constraint and as amenable to human manipulation.

In Sicily we study four sites: Agrigento on the south-central coast, Morgantina in the center, Selinus on the southwest coast, and Syracuse on a great harbor in the southeast. Morgantina was a small hill-town and the other three were port cities with adjacent plains for growing grain. Today, Morgantina and Selinus are deserted archaeological sites, Agrigento a lively modern town surrounding its archaeological site, and Syracuse an active port drawing many tourists to its archaeological sites.

We then examine three cities on the Greek mainland, Argos, Corinth, and Delphi. Delphi, the only inland-mountain city of the three was also unusual in being mainly a sanctuary. Argos is very old, Corinth perhaps a thousand years younger; both are the foci of similar areas with mountain backdrops, plains near the city, and the sea more distant.

Finally, we study three Ionian cities: Ephesus, Miletus, and Priene. Miletus was the archaic leader of Ionian cities, Ephesus more important in Roman times. Priene was always much smaller, but it had an important role in the Hellenistic period. All three were constrained by the relatively rapid changes in their physical environment caused by delta-building of the two Meander Rivers.

3

Western Greco-Roman Cities

Agrigento

Introduction

The polity of Rhodes, with Cretan assistance, founded Gela on the south coast of Sicily in 688 B.C.E.[1] (Herodotus, VII, 153) and assisted in the foundation of Akragas/Agrigento farther northwest on the same coast in 580 B.C.E. Akragas's foundation was part of the second wave of Greek city building in Sicily, about 150 years after the founding of Syracuse and other east coast settlements. Much of the Rhodian situation was replicated in the new cities. Settlers found familiar terrain like Gela, on a steep ridge facing the sea, surrounded by generous plains. At Gela, the acropolis at the east end is near the River Gelas, which waters the plains. Agrigento is bracketed by two rivers with plains to the south, and its lower ridge is visually equivalent to the site of Gela. An irrigation system of the Greek period like that known a little to the east at Camarina could have facilitated growing food in the alluvial soil between the two rivers, to the south of the temple ridge (Di Vita 1996: 294).

If we notice geological similarities and extrapolate too freely from them to architectural similarities, we may introduce chronological fuzziness to our study. The island-wide Rhodian tradition of dealing with water resources was carried to Sicily by the colonists along with other aspects of the culture. Exchange of ideas continued during the centuries between the founding of Akragas and the synoecism of Rhodes City centuries later. For instance, the grottoes of the acropolis of the city of Rhodes are "cut into the bioclastic limestones of the Rhodes formation, with, in some cases, the floor cut down into clayey and marly units that correspond to a line of seepage" (E. Rice, personal communication). At Akragas as at Rhodes, the builders cut down through the stone to the impermeable clay and marl units, to tap the line of seepage. With similar geology, it is not surprising that many elements of the water system of the two places were similar, developed independently from the old tradition. New concepts of water management were carried from place to place by expert builders, from the seventh through the fifth century B.C.E.

With caution, we may use ancient literature (Cicero, *II Verres* II. 4,43 and II.75) and travelers' accounts from the sixteenth century on, especially those illustrated with drawings (now gathered in *La Alta Valle dei Templi* ca. 1996), which present now-vanished details, both natural and cultural, of this site. These travelers were not twentieth century scientists; their data is pictorial and impressionistic, but not necessarily incorrect. Some of their illustrations are cited subsequently.

Geology

In the urban nucleus of ancient Akragas,[2] a series of sedimentary deposits are recognized (Fig. 3.1) (Motta 1957: Ruggieri and Greco 1967). The stratigraphy of Agrigento is as follows, beginning with the topmost modern surface layers:

> Crossed beds of silty clay with two types of calcarenite.
> **a.** visually distinct carbonate rocks: white to yellow biocalcarenite (quarries on Rupe Atenea and Girgenti, to extract yellow calcite)
> **b.** buff-colored sandy calcarenite (called limestone), shading into clayey sediments below,
>
> Base of impermeable blue silty clay of the upper middle Pliocene and Calabrian (Early Pleistocene) epochs; hardened to marl.
>
> Surrounding plains of disintegrating lava, tuff, and volcanic ash.
>
> Local area yielded oil, bitumen, sulfur, salt, and possibly iron.

Piled up in crossed beds, the strata formed an asymmetrical syncline dipping from northwest to southeast, visible at the north and southeast of the site (Fig. 3.2). Impermeable clay alternates with visually distinct sandy-carbonate rocks, which are not microscopically distinct, having the same large and small grains, the same porosity and cementization, and shading into clayey sediments. The calcarenite's grainy quality works against the hypothesis of karst activity here, since karst more commonly occurs in limestone with fractures. Yet the many passages (*ipogei*) of the hill seem karstlike, requiring further study.

Features of this complicated geology include many sets of discontinuities (breaks in sedimentation) of tectonic origin (Fig. 3.3), related to the uplift of the syncline. Ercoli (1997) divides the site into four unequal quadrants in terms of the lithography, with the northwest and northeast relating to the upper calcarenite layer, and the southwest and southeast to lower calcarenite layers.

Relations between discontinuities and passages—each strongest in the quarter where the other is weakest—are shown in Figure 3.4. Three kinds of discontinuities are present: (1) vertical or subhorizontal discontinuities are basically uniform, sloping between 0 and 15–20°; (2) subvertical discontinuities between 70° and 90° are persistent and reach the contact with clay, where water dissolves the calcarenite; and (3) medium discontinuities range from 15–20° to 65°. Joints that run

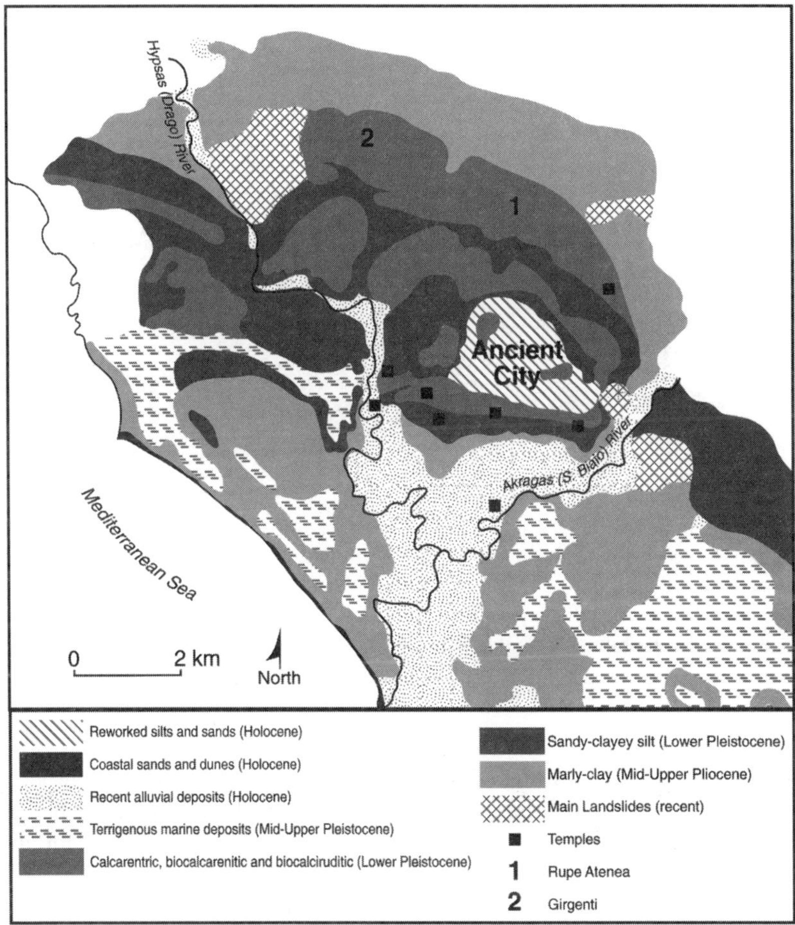

FIGURE 3.1. Geology of Agrigento. The Greco-Roman city may have occupied Rupe Atenea and Girgenti, or more likely used them as acropoli. "Ancient City" was the main residential quarter. Important temples occupied the lower ridge. (Redrawn by C. Langford from Cotecchia 1996)

across the stratification indicate crossed sedimentation. The most variable set of discontinuities is Family F in the northeast, a "blind spot" due to the orientation of the cliff face on which the measurements were taken. The northwest sector has the only group of passages lying east-west at 8° not associated with discontinuities. This group may relate to an east-west fault below the south edge of the western summit.

The discontinuity patterns relate to the hydrogeological potential and geomorphological evolution of the site as shown in Ercoli and Crouch's study (1998) of the mesostructural framework and in landslide analysis (see also Schubring 1865;

FIGURE 3.2. View of lower ridge from southeast (W. H. Bartlett 1853). At center is the final temple of the row along the lower ridge, and at left a better preserved temple. Between the central temple and the dimly seen buildings at the top of the site stretches the syncline site of the Greco-Roman city. In the middle distance is the valley of the ancient Akragas River. The travelers in the foreground are using a road that runs eastward along the coast.

G. Ruggieri 1996; Arnone 1948). A nearly perfect correspondence between the pattern of the passages and the mesostructure of the rock mass has been detected, since the passages follow the natural drainage network, approximating the directions of the vertical discontinuities of the rock mass (Fig. 3.4).

The rock under Agrigento stores notable quantities of groundwater, in spite of high porosity from 24% to 37% (Rossi-Manaresi and Ghezzo 1978). Because of the many discontinuous impermeable limey-clay layers sandwiched between permeable rock bodies, no large aquifer can be formed, but the discontinuous layers create many restricted and suspended water tables, which supply springs and wells, the latter common in the medieval part of the town, northwest of the site (Girgenti).

Distinguishing natural resurgences from the points of outflow of *ipogei* requires further study.[3] At least 23 points of water (springs and the outfalls of underground galleries) survive within the site of Agrigento (Crouch 1993; Ercoli 1997). (Fig. 3.3). There were five springs at the eastern and western edges of the site, one on the northeast summit (Rupe Atenea, cliff or rock of Athena), at least four on the northwest hill called Girgenti, one outside the town near the top of the north slope, thirteen within the "Hellenistic quarter," and one at the Temple of Asklepios in the plain to the south. Most springs of Agrigento functioned in the Greek or Roman periods; however, some may very well be Arab or medieval (Crouch 1993, Chap. 15). Topography and human manipulation together determine where springs appear. It is natural for outlets to shift location from time to

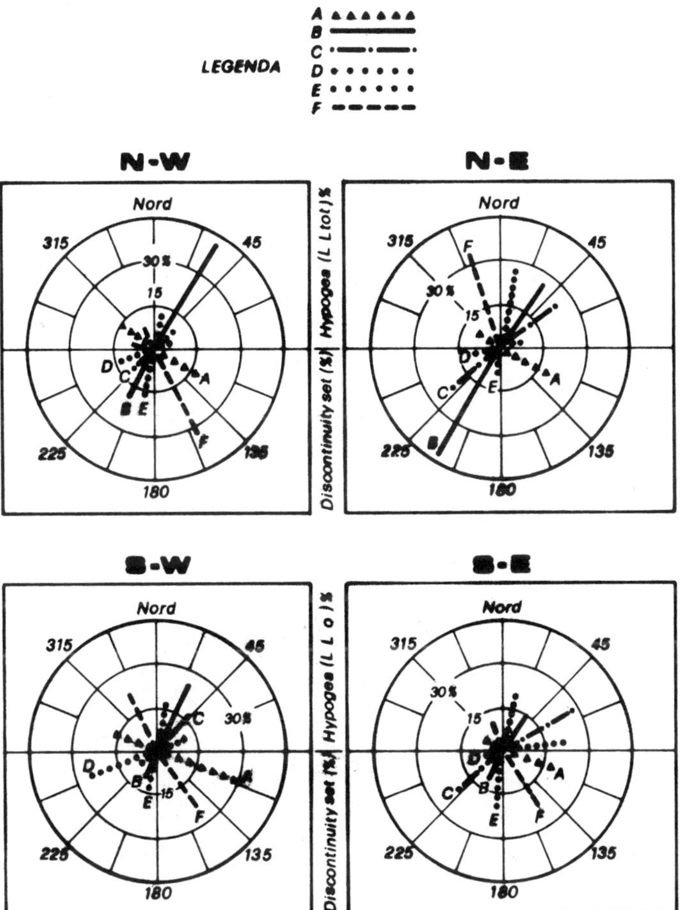

FIGURE 3.3. Comparison of directions of underground passages and discontinuities at Agrigento (Ercoli 1994) in the four quarters of the city. In each diagram, discontinuities of that quarter are shown in the bottom half of the circle, and directions of the hypogea in the top half; the two sets of directions are reciprocal. (Legenda=Legend; Nord=North.)

time, reacting to landslides, earthquakes, and simple dissolution of the rock. Together the original subsurface and the ongoing processes of the site have determined the changing layout of the town.

The daily water requirements of the large ancient population of 80,000 in 489 B.C.E. (De Waele 1985; Diodorus [Sicilus] XIII.84.1) can be estimated at 4000 m³. The tyrant Theron decided to supplement the on-site springs with a long-distance waterline. He called in the engineer Phaiax who built a new water system for the city in 480–75 B.C.E. Phaiax linked his underground waterlines to the mesostructural setting of the site, not to the street pattern. Being technologically sophisticated

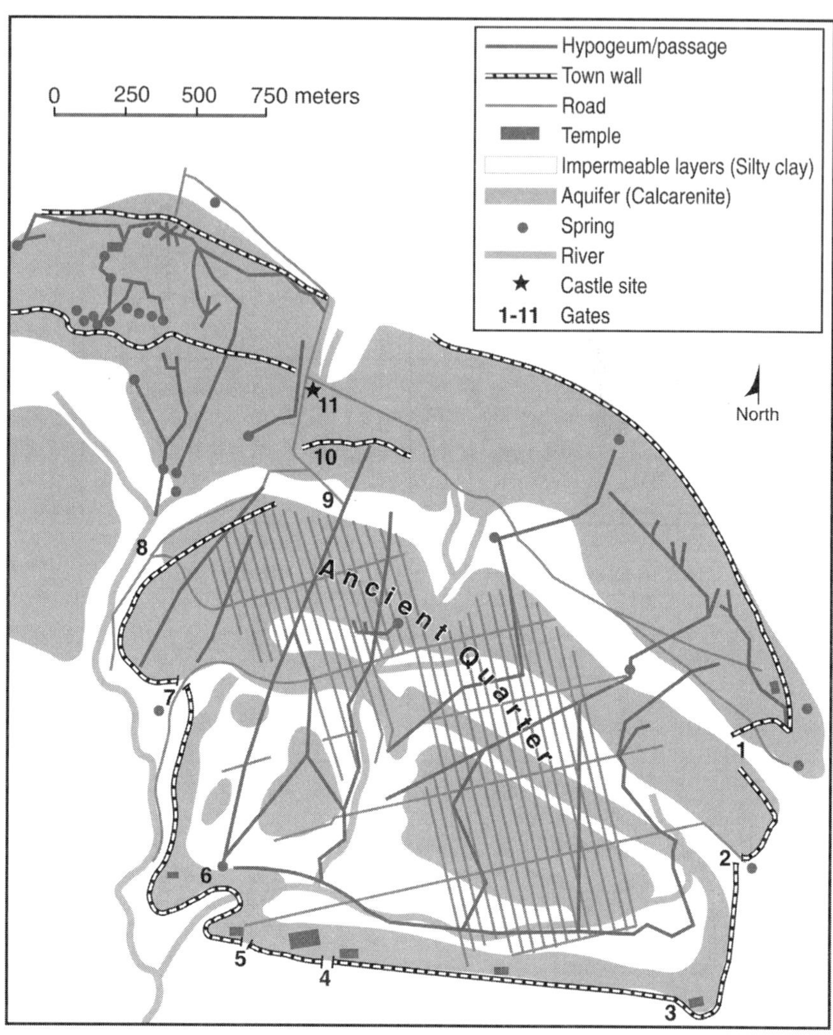

FIGURE 3.4. Site map of Agrigento, placing the underground passages and springs in lithological context (Ercoli 1998), along the seam between rock and clay. Compare with Figure 3.1. Most passages relate to the springs in the southwest area. Some passages seem to begin in stone, some in clay. Underground passages along the sloping bedding planes conduct water to the lower parts of the city. Landmarks, top to bottom, left to right: Girgenti hill with temple of Athena; Rupe Atenea with the Temple of Demeter (later church of Santo Biagio); rampart/town wall with gates numbered 1–11—the wall near gate 8 may be for flood control; ancient quarter with Greco-Roman street pattern; ridge of temples at bottom; ★ location of destroyed medieval castle over manmade cave, next to possible gate 11.

FIGURE 3.5. Purgatorio Passages next to Purgatorio church. View of interior. A natural passage, probably formed by karst activity, was enlarged possibly by the Greeks or Romans, certainly by the Berbers from the seventh century C.E. on. By the late 1980s, the passage was almost completely blocked by drainage deposits and fallen rock (Arnone 1948).

in spite of his lack of "scientific" knowledge, he utilized accumulated data about the behaviors of the stone and the water to guarantee ample supply for Agrigento. Completion of the water line was celebrated as one of Theron's accomplishments in an ode by Pindar in 475 (Olympian ode III).

The aqueduct seems to have consisted of a long-distance line and a delivery system within the site. Long-distance waterlines from the hills north of Akragas were well within the technical competence of the ancient Greek builders. Compare this with the pressure pipe known from the sixth century B.C.E. aqueduct (not its fourth-century enlargement) at the archaic town of Olynthus in northern Greece (Crouch 1993; Robinson 1935; Robinson and Graham 1938) and the fifth-century aqueduct at Syracuse. For the new channels, Phaiax utilized preexisting fractures in the rock mass, formed through natural enlargement of discontinuities, and amplified them for human purposes. The fractures had been slowly enlarged by dissolution of the rock, although the relation of discontinuities to faults suggests an additional tectonic element in development of the passages.

The passages have a median height of 1.9 m and width of about 90 cm (Fig. 3.5). Mostly dug in rock, they sometimes are made of rock slabs set in earth fill. Some still contain terracotta pipes, visible, for example, in a cliff-opening behind the Rupestrian Sanctuary (at the dot NE of 1 on Fig. 3.4). Probably not all passages were built for the same function—some are true ooze-galleries or *qanats*, some are now and probably always were sewers for wastewater. The intracity galleries not only delivered the aqueduct water, but also collected and delivered the water from urban springs, for human and animal use, as was also done at Syracuse and Corinth. Comparison of different passages as they relate to the position of springs in the southwest area shows that most of them are situated along or a little below the contact between calcarenite and clay. Some passages seem to begin in stone, some in clay.

The rural long-distance waterline reported by Pancrazi as it existed in the second quarter of the eighteenth century may have been the ancient line. The aqueduct came from the hills north of the acropolis, splitting so as to serve both sides of the hill (Pancrazi, plate III No. 5). A map of 1937 in the Aureli collection (found in Foglio della Carta d'Italia dello I.G.M., 267, Aragona), based on one of 1896, shows the streams and aqueducts of the area, with only one aqueduct coming to the Agrigento hill from the north, and others serving the city from the flanking hills to the east and west. Pancrazi also reports an underground line in the Lodovico garden and another at the hill of Vulcan.

Redundancy of supply—long-distance aqueducts, wells, cisterns, and local springs—was a consistent feature of Greco-Roman water systems. Local water from springs and wells and imported water from the aqueduct could supply the large population of the mature city.

Since the ancients deliberately preserved, protected, and funneled water as needed to increase their supply, the visible "springs" of today are likely to be ancient transformations of the natural supply. The Tamburello Ipogeum, for instance, exhibits repairs and corrections over time, as Ercoli verified, so that even if this passage began as a natural feature, it is now a human construct. The Bonamorone waterline begins as a spring in the cliff behind the hospital, that is, on the south flank of Rupe Athena (Fig. 3.3). Channelized, it appears again near the hospital, and then runs underground to its lower manifestation above the orchards in a street fountain bearing the name, rebuilt in the 1980s, and then disappears underground again in the direction of the lower ridge. Unfortunately, this water, though suitable for irrigation, is now too contaminated to use for anything more domestic than washing cars.

During the Hellenistic period, the drainage from the western groups of *passages* was led to an ancient basin or artificial lake called Colimbetra (Colymbethra). The lake was a major urban design and engineering feature of the city. It was about 1300 m in circumference and 15 m deep, famous for its black swans and many fish, as reported by Diodorus (XIII.84.1). Today there is no lake because the ancient dam is gone—probably washed out—but the site was most likely at the confluence of the Valley of Pipitusario and the Hypsas River, according to engineer Trasatti and geologist Ercoli. The alternate site between the promontory with the Vulcan temple and the spur west of the great Zeus Temple, where the Castor and Pollux Temple stood, as suggested by some scholars, is edged by porous rocks with many visible drainage holes, so it could not have held water. Some of these holes seem to be manmade passages leading to Colimbetra on the other side of the ridge; they may be flood-control devices. There is still an immense underground aquifer under the town, draining to Colimbetra. The water treatment plant in this area processes 250 liters of wastewater per second, although the supply is officially only 80 liters per second. When the treatment plant was being expanded, a bulldozer that broke through into the top of the aquifer was only saved with difficulty (treatment plant engineer G. Trasatti, personal communication, 1985).

The modern problem of insufficient water supply seems not to be the result of dewatering per se. Rather, there has been an unfortunate combination of solubility and friability of the calcarenite of the hill, solubility of the clay, and pollution of the available waters. Failure to repair damage to sewer pipes and lack of control of construction of both houses and commercial buildings have exacerbated natural dissolution.

I first went to Agrigento to study water management issues that have been my key to understanding the geological base for urban development. But the deeper my associates and I dug into the geological questions that related to water, the more it became apparent that this was only one of the useful geological questions. The site has already repaid study by sedimentologists, seismologists, karst geologists, and structural geologists.

Settlement and Land Forms

The geological advantages of the site must be set against both natural and human deficiencies. Springs and other natural resources made survival possible on the summits and lower ridge during the city's long history. The pre Creek inhabitants of Sicily consisted of the Elymi or Elamites in the northwest, the Sikans (Sicans) in the southwest who seem to have been the earliest inhabitants, and the Sikels at Akragas and the rest of the island (Antonaccio 1997). A prehistoric (Stone Age) settlement lasting until Neolithic times has come to light toward the western end of the lower ridge. From the Bronze Age, Mycenaean and Cypriot sherds of the fourteenth and thirteenth centuries B.C.E. (Vagnetti 1996) indicate international trade here. The Neolithic village was resettled by indigenous people in the seventh century B.C.E., near shrines of the mother goddesses. The Rupestrian Sanctuary at the east edge of the site was another pre-Hellenic sanctuary, tucked into a fold of the cliff just beyond what became Gate I, devoted to a goddess later assimilated to Demeter. Although de Waele (1971) prefers an early fifth-century B.C.E. date for this sanctuary, A. Siracusano (1983) dates it from the second half of the sixth century B.C.E. on the basis of wall technique. Other remnants from this pre-Greek people were the foundations of another sixth century temple lying under the Temple of Vulcan, and possibly a precursor to the odd collection of ruins today called the Temple of Castor and Pollux.

From the eighth or seventh century B.C.E., Agrigento was in frequent trading contact with North Africa and then the eastern Mediterranean (Antonaccio 1997). The first Greek traders came here from Gela and farther east. Their grave goods near the shore at present-day Montelusa are the among the oldest ceramic materials found at Agrigento, dating from 580 B.C.E. and later, which agrees with traditional dates for founding of the settlement. An *emporium* (trading area) was a standard feature of Greek contact with indigenous peoples and remained as the locus of trade; several Mediterranean cities still carry the name Emporium (Johnston 1993–94). A good location was at the mouth of a river, especially a long one like Akragas's Hypsas that extended inland 144 kilometers. Although dry in summer, it provided a route between Agrigento and the hinterland. The mid-sixth century port where early Greek traders pulled their boats out of the sea was probably on both sides of the river mouth, below the confluence of the Akragas and the Hypsas rivers (Di Vita 1996: 295). The ancient harbor is known from aerial photos and from excavations of the 1920s and 1950s (R. J. A. Wilson 1990). Sixth-century graves with pottery at Montelusa on the coast to the north and at San Leone to the south delimit the area. This harbor began to silt up in the fifth or sixth century C.E.

Where the plain meets the first ridge, the early agora area (under the modern vehicle parking for the temples on the lower ridge) was the locus of trade in both mineral and agricultural products. The first monument of the Greek settlement was a temple of Herakles of the late sixth century, set on the ridge above the agora (just above 5 on Fig. 3.4), where the god could supervise the commercial area.

Perhaps associated with the geological activity and minerals of the area was a local shrine to Hephaestus (Roman: Vulcan). There are no modern volcanoes, geysers, hot springs, or other evidence of infernal activity in the Agrigento area. Yet the sixteenth century C.E. appearance and disappearance of volcanic islands in the sea a few miles south of Agrigento indicate the possibility of a similar event during the Greco-Roman period.[4] In any case, something in the environment was dramatic and threatening enough to warrant a temple to placate this god of fire.

Oil slicks have been noted in the area of the Roman Temple of Asklepios. "Bitumen, also used in lighting, came from a spring near Agrigento; the substance is called 'Sicilian oil,' a name which implies that it was known outside the island and therefore had an export market" (R. J. A. Wilson, 1990, citing Pliny *Natural History* XXXV 179, 184, and Dioscorides I.73.1).

Colonists followed soon after the traders. One major motivation for the Greek settlement of Sicily and later for the Roman conquest of the area was to obtain the wheat and other grains grown on the broad plains. In most Mediterranean areas, it was easier to grow barley than wheat in the dry conditions from sea level to 4900 feet; areas where wheat could be grown were highly prized. The cities of Asia Minor and the Aegean Sea settled colonies in the wheat lands of southern Ukraine and also founded colonies in Sicily. Because they were made of disintegrating lava, tuff, and volcanic ash, the Sicilian plains were fertile and moist (Theophrastus VI. 6).

Aware of their vulnerability from North African rivals and displaced native peoples, the residents of Akragas built ramparts most of the way around the site, usually dated to the Greek period. The steepness of the slopes along the northwest-north-northeast hillsides meant that a defensive wall was not necessary there. Indeed, it seems likely that the upper slopes were steepened by the human builders, providing additional stone for construction. Along the southern edge of the temple ridge, the extant ramparts lean against the foot of the ridge, making the temples on the ridge more visible from below; this visual effect, however, may be the result of demolition of upper layers of the wall. Like other Greek cities, Akragas has river walls, now visible between Gates 5 and 7 on the western side of the city; these closely resemble the river walls of Argos and Selinus, both dated to the fifth century B.C.E. (Gaki-Papanastasiou 1994; Mertens's excavations in the 1990s).

During the classical, Hellenistic, and Roman eras, residents lived between the two ridges, where the strata are subhorizontal and the slope gentle. It was easier to build a traditional Greco-Roman house on such gentle slopes than on the summits, and easier to go from there out to the fields or down to the port. Excavations confirm that the regular street layout of the ancient core dates from the sixth century B.C.E. (Calderone et al. 1990). This plan and house allotments persisted through the Roman Imperial era—nearly 1000 years. Archaeological discoveries include walls, floors, coins, statues, and objects of everyday life, especially pottery, mostly in the form of sherds. Urban design of the area is discussed subsequently.

The city achieved such prosperity that it became the object of conquest first from the Carthaginians and then the Romans. The Etruscans and the Greeks, including men from Agrigento, joined together to fight Carthage in the great battle of Himera in 480 B.C.E., winning dominance of the trade in the western Mediterranean. Akragas then prospered till the end of the fifth century B.C.E., devoting its resources to a building boom, of which the Phaiax water system was an important component. Numerous temples were built. The stone used for temples is a brown-yellow biocalcarenite, like the underlying rock (Rossi Manaresi and Ghezzo 1978).

In 406 B.C.E., however, after a long siege, Akragas was conquered by Carthage and razed to the ground. After 338, Akragas attracted new Greek settlers and was rebuilt during a general renaissance of Sicilian cities. The fourth-century drought in Greece may have stimulated emigration, as we will see in our discussion of Priene. The population rapidly increased. New houses and other buildings were erected to accommodate this increase. Between the summits above and the temple ridge below, the Hellenistic residential quarters extended onto the sloping area, in the same area as the classical Greek city.

After 75 years, the new arrangements were imperiled by Rome's struggle with Carthage, which resulted in a Roman sack of the city in 262 B.C.E., followed by a Carthaginian sack in 255. This conflict expanded into the Punic Wars, leading to the Roman conquest of Akragas in 210, symbolized by its new Romanized name, Agrigento, again the residential area of the city was occupied by Romans and Romanized Greeks. Sicily became a land of vast slave-run estates (ancient agribusiness).

The Second Punic War was possibly triggered by the importance of wheat. Rome had begun purchasing Sicilian grain not only for the city of Rome but also for the army. As Semple points out (1931: 369), Egypt and North Africa were so important for their wheat that the conquest of one was the equivalent of conquering Italy. After vanquishing the Carthaginians of Tunisia in the late third century, the Romans went on to absorb Egypt in the late first century.

Roman conquest of Sicily began a long period (200 B.C.E. to after 500 C.E.) of prosperity but little political importance for Agrigento. The city was chartered as a Roman municipality by 14 C.E., a status lasting until the late Roman Empire. It was an important road station, the focus of routes to Palermo and Catania, and indirectly to Syracuse via Acrae. Unfortunately, "very little archaeological fieldwork to trace the actual course of these roads on the ground has ever been undertaken" (R. J. A. Wilson 1990).

During the Roman period, Agrigento was noted for its minerals. The sulfur industry flourished from the second century to the first quarter of the fourth century C.E. Sulfur from the hinterland, a major staple of Agrigento's trade, was in great demand for lighting, medicine, and metallurgy. Indeed, Agrigento—and not the area near Mt. Etna, more obviously volcanic in historical times—was a principal source of sulfur for the Roman world; there were additional "solfatras" at Puteoli and Cumae on the Bay of Naples (Koloski-Ostrow, personal communication). Sulfur was probably mined in galleries and smelted in small furnaces at the surface, the fumes polluting the air and attacking the stone surfaces of the ancient temples (Rossi-Manaresi and Ghezzo 1978), unplanned effects of human activity on the setting. Sulfur extraction both improved the economy of the region during the Roman period and began the destruction of the monuments. Agrigento's decline after 350 C.E. may partially be due to the decline in the sulfur industry.

Other mineral products from Agrigento were salt from the mines (Pliny XXXI.86; philosopher/naturalist Solinus V.18–19) and possibly iron, since slag was found here as well as at Selinus and Erice (cf. Miletus case study).

Christianity is first noted here at the end of the third century C.E. (Bonacasà Carra 1987), and Agrigento was firmly Christian by the fifth century or so. Little is known architecturally of the Early Christian and Byzantine periods, although some very early churches survive in villages inland from Agrigento, and there are one or two ruined churches (e.g., the older Santa Lucia, not the baroque church of the same name) within the city that are said to date from the early Christian Era. Like their predecessors, the early Christians continued to use the site and its existing buildings for their own purposes. The Demeter sanctuary, about halfway down the eastern side of the city, was transformed into the church of S. Biagio (de Waele 1971, 1985), and a residential area occupied the slope above S. Biagio's east end. In the last period before abandonment, some houses of the Greco-Roman quarter were reused, subdivided, or remodeled to Roman/Byzantine tastes (De Miro, 1957, 1985).

Shrunk to a village and shortening its name to Girgenti, the town of the late Byzantine period retreated, after the fall of the Western Roman Empire, to the defendable western summit, turned the Temple of Athena into a church, and waited for better times. Instead, beginning in the seventh century, raids from the Arabs and Berbers of North Africa culminated in 825 C.E. with conquest by these belligerent neighbors, who settled down on the western part of Girgenti. Perhaps the short stretches of rampart around the western summit, not yet securely dated, were built at this time.

The no-longer-extant castle (*11 in Fig. 3.4) at the juncture of the two summits may have been begun by the Arabs and was used by the Normans who conquered Sicily in the late eleventh century. The tectonic fault below the castle may have contributed to its destruction. On Rupe Atenea, the early Christians or the Normans built a small church where the temple is thought to have stood; the rock surface of the summit was cut down to form the base of the chapel, and the ancient stone surface of the summit was cut into blocks for the walls. The Normans also established the Cathedral at the crown of the Girgenti ridge. Norman rule of Sicily incorporated Greek, Roman, and Arabic traditions and construction methods into the unique style called Arabo-Norman. The Sicilian multiethnic culture persisted in the thirteenth century, with additional Jewish and Teutonic peoples living at Agrigento. A newly resident group of eastern Greeks rebuilt the Church of Saint Mary of the Greeks over the Temple of Athena in the twelfth century. Like the rest of Sicily, Agrigento was ruled as part of the Kingdom of the Two Sicilies by a dynasty from Aragon intermittently from 1130 to 1861, so that Spanish cultural features, as well as French elements from the Bourbon kings of Spain, were incorporated in the local culture.

Political and demographic changes at Akragas-Agrigento-Girgenti from the medieval through the early modern period are suggested in the strange configuration of the Purgatorio system of passages, the largest in Girgenti. C. Micceché[5] informed us of the system of rooms and passages under the medieval city, of which the *Ipogei del Purgatorio* is the best known (Fig. 3.5). The passages lead to an artificial cave that Croce et al. (1980) thought was used to store water for the medieval castle above it. Micceché noticed that the tool marks in the cave were unlike those in the passages and channels considered to be Greek; rather, the marks are like those of other constructions of the Arab/Berber period (ninth–eleventh centuries). A number of passages lead downhill away from the cave; these are explainable if they were created by stonemasons following veins of building stone, but pointless if the cave were a reservoir. The cave was a quarry for building stone, just as in Naples, where huge quarries underlay the city and were a safe haven from air raids during World War II. The Naples quarries were begun during Greek times and were expanded through the centuries. The Agrigento cavities were probably amplified during the later Middle Ages.

One gallery of this subsystem leads to the doorway next to the Church of the Purgatorio, from which the network gets its name; Figure 3.5 was drawn from just inside this doorway. The outer face of the doorway from the passage to the church plaza is a product of the Bourbon period (eighteenth century); over its door is the sculpture of a sleeping lion symbolically associated with the Bourbons.

Not until after World War II did the town begin to expand, taking again its Roman name. Construction in central Agrigento since World War II has interrupted the underground channels and obliterated many traces of the earlier water and drainage systems. As late as the 1980s, the city had not begun to recover its former population size, but recent growth has pushed it over the number of ancient residents (80,000).

Unlike Delphi, where we have documentary as well as physical evidence of a series of geological events that can be tied plausibly to the known human history of the site (see Delphi chronology, Appendix A), at Agrigento we have much less evidence of either kind. The limited data suggest more frequent landslides than earthquakes.

After the 1966 landslide, Grappelli's commission (1968) revisited some galleries in the southwest sector. Before the development of speleological methods and modern instruments, the passages were extremely hard to study, and they remain so to nonspeleologists because many have collapsed and others have been filled in either naturally by fallen blocks or artificially by accumulated refuse or concrete foundations.[6] Landslides and (possible) earthquakes have been the most dramatic natural events known here.

The most notable fault of the site, running more or less north-south, separates the two upper summits (Micceché, personal communication, 1994).[7] The large Purgatorio cave mentioned previously is located just below the crease between the eastern and western summits, where the north-south fault intersects the ridge. Because of this fault and the summits' distinctive topographies, the two summits have developed independently, and the ravines that formed above the fault have separate names: La Nave to the north and Valle Pipitusario to the south (Ercoli 1994).

The trench now occupied by the railroad, south of the western summit just below the medieval city, suggests a second fault, running approximately east-west. A row of trees along the side of the hill in this area, visible in early photographs, hints at a number of resurgences along this seam between strata. These are likely to be the outflow points of several underground passages, such as the one connected to the spring under the church of Santa Lucia. Since before 1970, this very old church has been closed as a result of the collapse of both building and foundations—a good example of what happens in the weak calcarenite layer.

Other faults are indicated by the two rivers that flank the site. This fault pattern could account for some of the steepness of the Agrigento site (Cotecchia, 1996, his fig. 6a, tectonic fracturing along the southern edges of the ridge of temples; his fig. 21b, neotectonic sketch showing the deep faults and the surface layers of sandy clay and calcarenite).

Landslides have been a recurrent feature of the geomorphologic history of Agrigento because of the hazardous conditions within the calcarenite outcrops of the site, especially in the two summits and along the southern and eastern edges of the ridge of temples (Cotecchia 1996, his fig. 6a). The mesostructural situation determines the evolution of both the clayey slopes at the bases of the ridges and the borders of the calcarenitic plates: intense erosion and the fall of stone blocks. The landslides of 1966 and 1976 and the numerous earlier landslides along each border of the ridges are geomorphologic processes that change the site continually. Friability of the calcarenite has contributed to serious sliding in this century and may have either carried away the earliest settlements on the western part of the western summit or prevented them altogether. In addition, the clay is constantly moving in the valleys and along lower slopes (G. Trasatti, personal communication, 1994).

Serious sliding in the twentieth century occurred on the south face of the western end of Girgenti, in the Addolorata Quarter, which slid down the hill in 1966. The slide happened in the summer dry season, so it cannot be attributed to immediate saturation from rain. The quarter became completely uninhabitable

FIGURE 3.6. Arab quarter at the northwest edge of Girgenti. Rock cuttings into which houses were built before the 1966 landslide. (Photo by Crouch)

and not reparable, with many of the buildings abandoned, leaving the cuttings among the boulders as intriguing evidence of the adaptability of the Arab residents from the ninth century on (Fig.3.6).[8]

Calcarenite also outcrops over clay at the Juno Temple in the lower city, on the southeast corner of the lower ridge. The temple was endangered in 1976 by a small landslide among the natural rocks and along the ancient Greek rampart of this area. The geodynamics of this event are easier to see and to understand, and easier to ameliorate, than the higher 1966 slide. This landslide was an external manifestation, a complicated result, of the underground water table softening the subsoil. The weathered and corroded *passages* are pierced by new building foundations that obstruct the flow of drainage, changing the outlets and the flow paths within the hill (Ercoli 1991). Landslide conditions at Agrigento are annually reactivated by rainwater saturation, leading to "regressive evolution of the lower edge of the calcarenitic plateau underlying the Temples" (Cotecchia, 1996, p. 5 and his figs. 14a and b). The "cumulative rainfall of 1976, continuing for ninety days, has a return period of twenty-four years" (Cotecchia, 1996, his fig. 18); this rainfall was due again in 2000 but did not occur.

Equally important but harder to notice and date are processes that have affected the geology of Agrigento, both natural and human. In the absence of maintenance, nature takes over and disaster can ensue. Natural processes include erosion, dissolution, wind-blown salt mist attacking the stone, and other forms of weathering. Human processes include scarping the hills to increase their steepness for defensibility and failure to repair landslide-damaged sewer lines. In addition to the natural process of rain infiltration that induces sliding by mechanical action, Agrigento faces the dissolution of stone within the hill (Cotecchia 1996: 5). Although it might be assumed that the nature of the stone would affect dissolution processes strongly, Dreybrodt (1990) found that stronger influences were the presence of CO_2, steep hydraulic gradient, and length of channel. Hence the steep slopes and lengthy downhill channels of Agrigento foster dissolution (see also Rossi-Manaresi and Ghezzo 1978).

Deforestation also contributes to the continuous erosion of the natural clayey slopes. Typical Mediterranean bushes have given way to increasing urbanization, which in turn modifies the underground water flow (McPherson 1974). Postponing remediation through a lagging political process may well exacerbate the effects of rainfall and deforestation.

Materials   What were the local materials, and how did differences between them affect final construction? Two hundred plus years of study (after 580 B.C.E.) of the qualities of local stones made it possible for the ancient builders of Agrigento to choose the individual stone types that were adequate for their purposes: to build temples, civic buildings, or houses. The compressive failure stress of this calcarenite is very poor, only 35 kg/cm^2 (Rossi-Manaresi and Ghezzo 1978; Cotecchia 1996, table of stone strengths at Agrigento). The effect of this weakness on the proportions of the ancient buildings compared, for instance, with the thinner proportions of stronger marble temples is discussed in the Selinus chapter.

Not only is the biocalcarenite relatively weak, it must also contend with climatic and other factors that weaken it further. Cotecchia's chart (1996: 8) of the causes of stone decay at Agrigento demonstrates the interconnectedness of these factors. In both monuments and unquarried stone, the winds, alterations of temperature, and rain waters that may be acid or salty are the active agents of decay. The winds not only cause drying or exacerbate heating, but abrade the surfaces with wind-borne sands. Mechanical separation caused by cyclic salt crystallization affects the stones of the monuments: "dissolution of calcitic cement increases enormously the porosity of the stone in depth, therefore increases the stone's permeability to gas and water, and decreases [both] its mechanical resistance" and its strength (Rossi-Manaresi and Ghezzo 1978). The rocks have different degrees and kinds of fracturing, porosity, texture, and a range of cementation characteristics. The ancient solution to the problem of weathering was to coat the surface of the stone with a stucco of marble or limestone dust (Fig. 1.1); in places where the stucco still clings, it protects the surface from pitting.

In the case of stone layers in the ground, mechanical and solutional variations are evident because of the actions of infiltrated surface water. Bioturbation causes additional changes in the stone. Depending on their proximity to clay or sand, rock layers are affected by different local environments.

For construction, builders sent to the quarries detailed lists of the stones necessary for each project, with the proportions of each determined by the purpose for which it was destined (Coulton 1977: 64–67 with ancient and modern bibliography). Although the role of quarrying at Agrigento has not yet been studied, we know of stone quarries on Rupe Atenea and along the edges of the Girgenti quarter that extracted yellow calcite. We may question, however, whether the volume of stone that was necessary for the numerous temples, public buildings, ramparts, and houses could possibly have been obtained from only these two quarries. This question and others require further study of the terrain for quarries, as we and others have done at Selinus (Carapezza et al. 1983; Ercoli 1994), as Hayward (1995) has done at Corinth, and as Vetters (1988, 1998) and Bammer (1998) have done at Ephesus.

The other important building material here was clay. Blue clay, widely visible in and around Agrigento, is easily available in large quantities from the banks of the rivers. Clay was used extensively for roof tiles and terra cotta pipes, as well as for table and cooking ware (Alaimo and Ferla 1981). Sun-dried clay bricks (adobe) could be used for houses and other small buildings. Hardened to stone, clay formed the marl of the lower hill to the north, where the lower train station is located.

Urban Design and Geological Setting　The cultural aspects of urbanization may be misunderstood without comprehension of the physical constraints they master. At Agrigento, some of these goals seem mutually exclusive because of the sloping terrain and wide extent of the site, which impede efficiency by making access from one part of the city to another difficult even today, and because of changing ethnic and political circumstances. We have previously noted the shifting patterns of residence here in response to conquest and abandonment. The topography, the geomorphology, and the hydrogeology of the site also contributed to Agrigento's urban design.

According to De Miro (1988, 1994), the city's plan, made to accommodate various classes of people, was regulated by its geography. He has discerned two lower terraces, the Lower Agora mentioned previously and the Upper Agora at the western center of the long slope of the hill. The central area of residence adjacent to the latter was the core of the city. Large houses for the elite were built near the lower ridge, whereas the upper northwestern part of town was allocated to craftsmen and workers — each on a geographically separate section of the slope connected by flights of steps. Features of this plan persisted over a thousand years from the archaic period of the sixth century B.C.E. into the early Hellenistic rebuilding after 338 B.C.E. under the new ruler Timoleon, and into the Roman period.

Earliest planning decisions about the division of property show up in persistent street patterns of the site (Fig 3.3). Only on Girgenti are the earliest streets either lost under the existing structures or preserved by modern property lines that maintain "organic" medieval or earlier divisions. Schmiedt (1957) and Schmiedt and Griffo (1958) noted from aerial photography that Rupe Athena may have had a grid. During every period, the city utilized its setting on a south-sloping hill, high enough to protect it from north winds. This orientation was an advantage for savings of fuel otherwise necessary against winter cold, but a difficulty during the heat of summer.

The street pattern of the Greco-Roman quarter was a grid of elongated insulae (city blocks), 35 by 280 meters; this is close to the insula size at Selinus, built six or seven decades earlier in the sixth century B.C.E. (De Miro 1988). There were six east-west streets measuring seven and more meters wide and numerous north-south streets that were narrower, about five and a half meters. These were perpendicular to the major street, twelve meters wide, that ran at a gentle slope down from Gate 2 westward to the temples near Gate 5 (Fig. 3.4). This street seems to have been the base line from which the entire residential quarter was laid out. During the Roman period, the main street changed from the first to the second street at the south, between Gates 1 and 6.

One unusual feature of the urban arrangements at Agrigento is that the very large area between the upper and lower ridges is mostly occupied by residences. Governmental buildings were neither at the center of the residential area nor scattered among the houses, but sited along the western edge. Except for the church of San Nicola near the Archaeological Museum, which may replace an earlier sanctuary of the Chthonian Gods, some 450 hectares of the city's center show no signs of either religious buildings or tombs, and even this sanctuary stands toward the western edge. Rather, the early (temple-less) sanctuaries were assigned the upper and lower ridges, stony banks that were unsuitable for agriculture (Di Vita 1996: 294), and later amplified with the handsome temples that are the glory of Agrigento. Where the stone comes to the surface of the ridges was the site of the largest and most expensive buildings, which were most visible and hence most effective in such locations. This was a fine piece of urban design: to bracket the settlement between upper and lower sanctuaries and to set off the regular residential quarter from the irregular string of government buildings along the sharper slope to the west. The very gentle slope at the lower end of the western edge was the convenient location of the commercial agora around which were grouped fountains, stoas, and a gymnasium.

The size and extent of different vistas was closely tied to the topography (Fig. 3.7). The five temples of the lower ridge respond to the one on Rupe Atenea and the one or two on Girgenti, but the Temple of Vulcan/Hephaestus farthest to the west keeps aloof visually and physically from both of the ridges, being separated from the upper group by distance and from the lower group by a stream running in a ravine and by elevation on a promontory taller than the lower ridge. The Temple of Vulcan was visible from the Hera Temple at the east end of the lower ridge, but out of sight from the lower Dioskouroi sanctuary nearest to it across the ravine at the west end. The Hephaestus promontory and its probable early shrine were within easy walking distance of the extramural ceramic workshops of the late sixth century, on the western plain close to Gates 4 and 5, between the early necropolis and the lower city (see Stier et al. 1967, map 1 of Akragas). The promontory commands an excellent view of the entire site, the western hinterlands, and the sea.

The trough at the bottom of the interior slope, between the ridges, today is planted with orchards, and in the past held the luxurious houses of the elite (De Miro 1988, 1974). The modern visitor, staying or dining at the Villa Athena (hotel and restaurant) built among the orchards, can look up from the orange groves to the temple ridge. The longer slope upward to the summits, however, is both in-

FIGURE 3.7. Panoramic view from Temple of Zeus to the western summit (Girgenti). In the foreground is one of the giant Atlas figures that supported the entablature of the Zeus Temple. Notice that the entire slope between the temple ridge in the foreground and the summit is invisible from here. (Photo by Crouch)

terrupted by hills, ridges, and terraces, and cupped slightly as it approaches the upper ridge, so that the general mass of modern apartment buildings is perceived but picking out an individual monument is difficult. In antiquity, perhaps only the roofs of temples or their upper columns and capitals with the roofs they supported would have been visible from within the Greco-Roman quarter, but more easily visible from the lower ridge (Fig. 3.2). It was a long walk, quite steep in some places, from the elite residences of the lower valley to the temples on the summits, but that was appropriate for the annual or semiannual procession to honor the gods. The physical effort pleased the gods as it improved the character and physique of worshippers.

Just as at Priene, discussed subsequently, the highest summit (Rupe Atenea) had a temple, probably to Athena, a fortress, and other ancient buildings (unidentified). The other summit (Girgenti) held at least one other temple to Athena, the chief deity of the mother-city Gela, near a spring that still flows. The foundation, stylobate, and six Doric column bases of this temple still lie under the twelfth-century church of St. Mary of the Greeks. Some authors postulate a second temple on this western summit. Interpolating from other sites, one would expect the cathedral to have supplanted a prominent pagan temple, perhaps, in this case, a temple to Zeus (Kininmouth 1965). Indeed, Polybius (I.v.5, V.lxxxviii–lxxxix) tells us that Zeus was also worshipped on this hill. Some of the original colonists were from the island of Rhodes, where at Rhodes City the Acropolis had a joint temple to Zeus and Athena.

In laying out their city, the Greek city builders chose to intensify the tension between terrain and plan, as we have seen in discussing the underground water-lines, which are strongly linked to the geostructural setting of the site rather than to either the grid of the Greco-Roman streets or the meanderings of the upper street network. The builders also valued the tension between visibility and hiddenness, as we have noted in the placement of temples.

The major roads ascending the hill or linking the gates offered many opportunities for visual connection with the setting, adding a tangible richness to life. Especially the north-south road that connected the lower with the upper ridge, being necessitated by the geology to wind along the sloping site, alternated great vistas of countryside, sea, or summits with closely blocked views where buildings or trees crowded close to the road. From the midpoint of the residential quarter, one catches glimpses of the temples on the ridge below and the sea beyond. From within the typical Greco-Roman courtyard house or even from within the regular neighborhood of such houses, the limited vistas reinforced intimacy, just as do the short streets of the surviving medieval quarter of Girgenti. But at an intersection, or better still in the bigger public spaces of agora, council chamber, courtyard, or sanctuary, more spacious vistas reinforced the sense of belonging to a larger-than-family entity, a municipality or a religious group. Public porticoes and large public roads were interim spaces, contributing to both intimacy and openness.

In antiquity, when all major buildings and the ordinary houses clustered around them were made from the stone and clay of the hill and its encircling riverbanks, the unity of site and setting was enhanced visually by the sameness of colors and textures. These same buildings were usually constructed by resident workmen, though sometimes under the direction of foreign builders, as the example of Phaiax indicates. Pride in one's work increased the attachment to structures that were both useful and beautiful.

The hidden resource of water was revealed at many places in the architectural form of the city. From the springs along the sides of Rupe Atenea down to the fountain and large cistern of the temple of Asklepion below the southern walls, water elements served as points of beauty as well as utility in the urban scheme. In the early days when Agrigento was rich, much of its wealth was spent on temples, civic buildings, aqueducts, and ramparts that improved the quality of life for all the residents. Even today, with a different distribution of buildings, Agrigento is an exhilarating city to visit and a stimulating one in which to live. Pindar in the early fifth century B.C.E. called it "the most beautiful town the mortals have ever built."

Conclusions: Implications of the Geological Setting

Stability problems in Agrigento arise from both geological conditions and human actions. Agrigento still survives, although the natural erosion of the land has resulted in an incessant erosion of the cultural and historical heritage. Some intervention slowly taking place regarding the problem of the underground passages, such as the Purgatorio. Weakening of the calcarenite in the passages increases the danger of collapse that threatens the security of some buildings.[8] Corrosion of the stones of the visible monuments began long ago; in the eighteenth century corrosion was advanced enough to impel conservation efforts. Conservation has attracted the attention of archaeologists and geologists, but not enough political will for complete solution of the problems.

Similarly the visible deforestation and eventual destruction of the soil began in the sixteenth century when the area's population doubled. Growth pressure impelled the people to cut trees to expand agricultural land and provide building materials and fuel for cooking and warming. Erosion followed, which led to long-term soil destruction. By the end of the eighteenth century, the Agrigentans had to import wheat. Between 1817 and 1847, they lost half of the nearby forest, which negatively affected the infiltration and retention of water, and caused further erosion of slopes. The valleys and plains are still rich land (R. J. A. Wilson 1990), but as of 1985 the city planned to import desalinated drinking water from Gela (G. Trasatti, personal communication).

Agrigento is one of only two major Greco-Roman towns where the ancient Hellenistic quarter is largely free of modern buildings. Precise physical description of the historic settlement requires multidisciplinary research with emphasis on archaeology, geology, geotectonics, urban history, physical and social sciences, and analysis of the ancient and modern political context. The city exemplifies the difficulties of preserving our heritage.

Morgantina

Description and History

Morgantina was built inland, toward the center of Sicily, on the Serra Orlando ridge of the Mt. Erei mountain system near its juncture with the Catania plain, which extends 55 km to the sea on the east. Morgantina's site is 578–656 m above sea level, whereas the valleys to the north and south are 300–350 m lower. To the west lie the plateaus of central Sicily. Located 2.5 km NE of Aidone, the ridge had an area of 3 km², with eastern peaks at Cittadella and the higher Farmhouse Hill, separated by a 30% slope from the rest of the ridge extending to Papa Hill and Trigona Hill on the west (Fig. 3.8).

The earliest population (possibly Sikels) lived at Morgantina from the third millennium B.C.E. (Stillwell et al. 1976), on the easternmost hill, Cittadella. Even this settlement was sited with appreciation of the relevant factors of hydrogeology and geomorphology (Schilirò et al. 1996). Early remains of settlement on Cittadella could be Neolithic, whereas early Bronze Age materials have been found both at the San Francisco Bisconti area at the northeastern end of the main ridge and at the later central agora. At Cittadella, other material comes from the late Bronze Age. After a period when Morgantina was unoccupied, King Morges arrived with his people from South Italy perhaps in the twelfth century B.C.E. and, according to legend, gave his name to the site. The Sikel language of his people is similar to Latin (Antonaccio 1997, an integrated account of archaic life at Cittadella).

Cittadella was occupied during the Bronze Age, the Iron Age, the archaic period, the early classical period, and after a gap in occupation, again during the

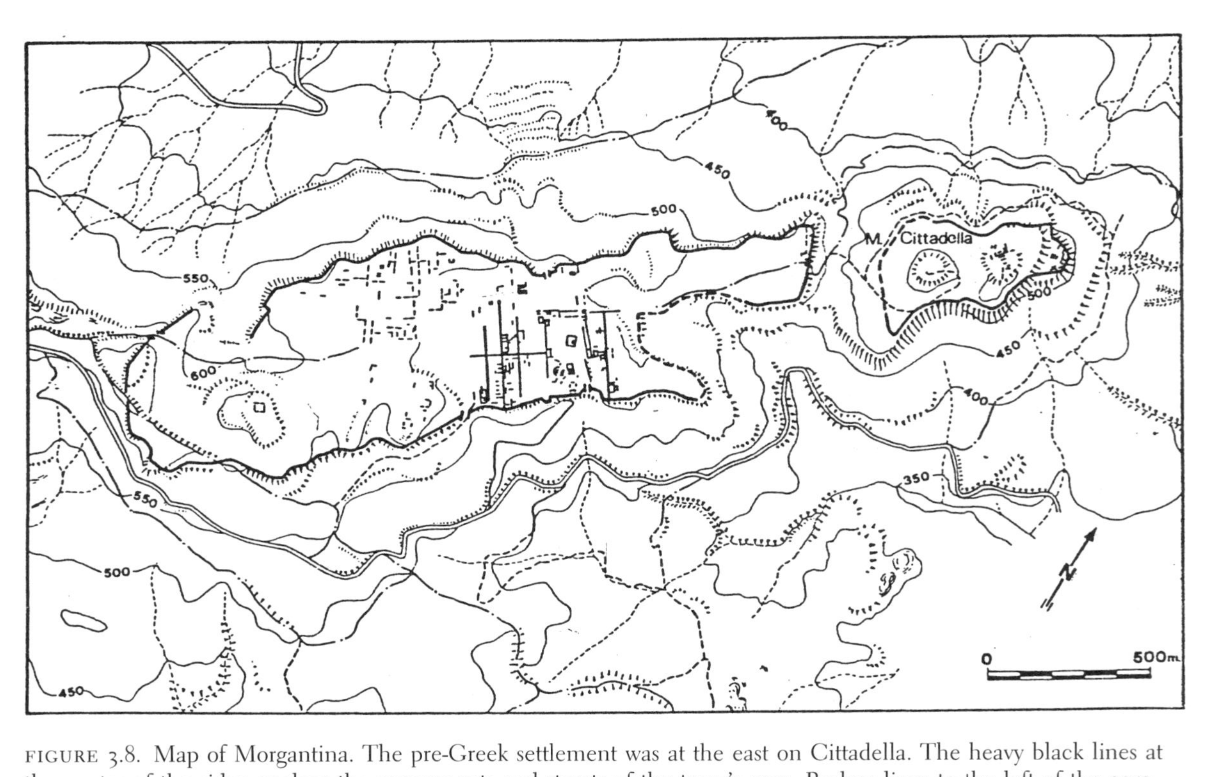

FIGURE 3.8. Map of Morgantina. The pre-Greek settlement was at the east on Cittadella. The heavy black lines at the center of the ridge enclose the monuments and streets of the town's core. Broken lines to the left of the core indicate the extension of the town. South of the town is the modern road (double line) to Aidone, which is off the map to the west. (Courtesy Morgantina Archives, Princeton University)

Hellenistic period, whereas the center and west of the ridge were occupied from the fifth to the first century B.C.E.

Morgantina—although later absorbed into the Syracusan realm—was originally reached by the Greeks via road and river from Leontini. Early Chalcidian settlers of Sicily had met indigenous people at Catania beginning in the seventh century B.C.E. (Vallet 1962; Sjöqvist 1973). From there, they followed the Gornalunga River inland to Leontini. Preferring commercial expansion to conquest, the Chalcidians pressed northwestward, arriving by the end of the sixth century at the site of Morgantina, which "lies practically in the geographical center of the island, at the watershed of the rivers Gela and Gornalunga" (Sjöqvist 1973). They pushed onward to Enna, 25 km or so northwest of Morgantina (Dunbabin 1948: 137; Vallet 1962) by the fifth century to obtain the timber products of the area.

Settlement at Morgantina about 560 B.C.E. involved both Sikels (Raffiotta 1991) and a Greco-Chalcidian group (Lyons 1991); the village may have numbered 1000 people (Schilirò et al 1996). The oldest tombs discovered at Morgantina are from 730–650 B.C.E., but little else was found of the people of that time. The next group of tombs dates from 550–475, the period of the last habitation of Cittadella. Like Enna, Cittadella was profoundly Hellenized during the sixth century, possibly beginning with imports of Ionian wine amphoras of a type found also at Marseilles. Vases from Greece and Greek colonies in eastern Sicily have been found in the tombs, especially from Rhegion, a Chalcidian colony. Both tombs and houses are of the Greek and native types (Lyons 1991), local architecture decorated with Greek antefixes and other details. New tombs in this area are rare from about 450 B.C.E. onward (Lyons 1991, 1996), but existing tombs continued in use by families as long as 200 more years, some containing goods from the sixth to fourth centuries B.C.E.[9]

After the Chalcidians, other Greeks from the south coast of Sicily joined the settlement (Stillwell et al. 1976); these may have been Dorians, coming up the Gela River and bringing their alphabet (Vallet 1962). The settlement overlooked the valuable trade route along the Gornalunga River to the interior and dominated useful agricultural lands. At least partially destroyed in 466 by Hippocrates, Cittadella was taken again by Sikels under Douktios in 459. In the struggle of the indigenous Sikels to unify the island against the Greeks, the ramparts around Cittadella were razed (Diodorus XI 78; Malcolm Bell 1988) and the village destroyed.

In approximately 450 B.C.E., residents moved to a site at the center of the ridge (Fig. 3.9). The new town came under Syracusan control late in the fifth century, enjoying peace until Dionysos of Syracuse died in 368 B.C.E. Some of these settlers were middle class persons from Syracuse (Métraux 1978). Both the houses and the public areas were sited to take advantage of the available water supply and access routes. During the prosperous period of 317–276 B.C.E., the agora was enlarged by quarrying its borders, and after 269 when Hieron came to power in Syracuse (Karlsson 1993; Malcolm Bell 2000, lecture), it was enhanced with shrines, fountains both small and monumental, a theater, granaries, stoas, and the unique Great Steps that form a retaining wall across the center of the agora. The architectural style of the theater and stoas reflect the cultural dominance of Syracuse in this area.

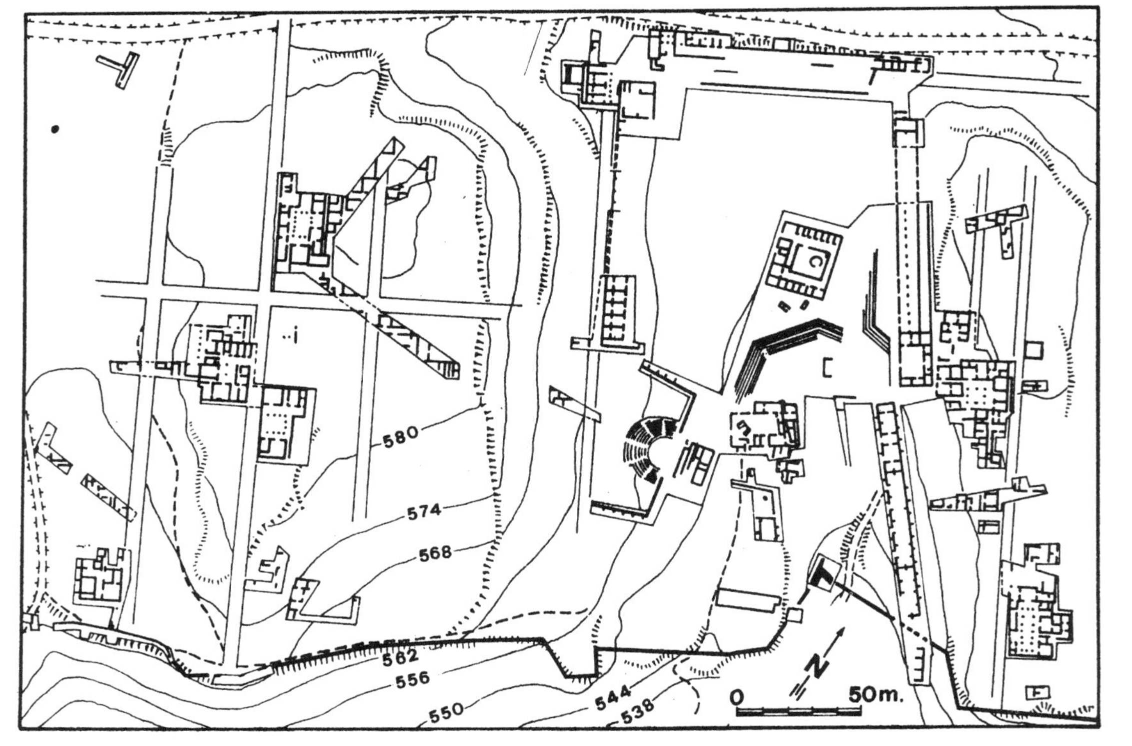

FIGURE 3.9. Map of central Morgantina. The Hellenistic agora, with the theater at the lower left, the never-completed west stoa, the north stoa, and the east stoa and granary occupies most of the right half of the map. House walls in grid patterns occupy the two hills to the east and west, flattened for houses during the Greek period. (Courtesy Morgantina Archives, Princeton University)

After Hieron died in 215 B.C.E., Morgantina was incorporated into the Roman Empire in 211, as part of the takeover of his kingdom. The Romans awarded Morgantina to their mercenary troops, veterans from Spain called the Hispanii—a good example of the Roman practice of settling foreign colonies (Hopkins 1978b: 65, 67 on: exchanges of populations inherent in the Roman slavery and veteran colonies systems). The market building (*macellum*) is one that survives from this late period.

In 104 B.C.E., Morgantina was conquered by rebelling slaves, many of them Syrians, led by Salvio. At about the same time as the rebellion (some time between 106 and 43 B.C.E., according to Goldsberry 1973, citing Orosius V, 6, 2) a noteable eruption of Mt. Etna disturbed the area. The rebellion at Morgantina resulted in a severe reduction in population. By 70 B.C.E., climatic change (Bruno and Nicosia, 1994) and/or human-induced reduction in the water resources of the site required construction of an aqueduct (Allen 1970; Raffiotta 1991) to supply the large agora fountain (Malcolm Bell 1986–87, 1988). Large pipes from this aqueduct have been found along the road from Aidone. During 40–30 B.C.E., fire destroyed a number of houses, the market building, and the north stoa of the agora. General decay went unchecked. Deliberate hacking of statues into bits seems to signal the final death of the town (R. J. A. Wilson 1990; Stone 1993). The place was "no longer a city" (Strabo 6.2.4).

By this time, competition from ample Egyptian grain was destroying the advantage that proximity to the Roman market had given Sicilian farmers. These governmental and economic changes unfortunately were coupled with an environmental change, so that, by the beginning of our era, the site was left deserted. Our new research suggests that the fall of the water table exacerbated the other factors of abandonment.

Physical Analysis

Geomorphology and Climatology   Central Sicily's geomorphology consists of plateaus with deep valleys. Morgantina is situated on the eastern edge of the central plateau, dominating the perennial Gornalunga River, which rises in the hills just north of nearby Aidone. Water from the river supplies irrigation for orchards and vegetable gardens in the valley bottoms; the intermediate slopes produce wheat; and the higher, steeper slopes are used for pasturage and olive trees. As early as the Iron Age, wheat and barley were cultivated hereabouts, as the numerous millstones indicate. An ear of grain was the symbol on the earliest coinage of Morgantina in the mid-fifth century B.C.E.—both wheat and money indicating prosperity. Pace (1935, his fig. 162) reports a large number of fifth-century farm tools found at Morgantina. In addition to grain, the area produced leather, wool, pigs, and cheese for export.

The ridge of the settlement declines from 656 in elevation at Cypress Hill in the west to 578 m at Cittadella, both lower than the nearby town of Aidone at 885 a.s.l. (above sea level); the Aidone peak and the Morgantina ridge are connected by a limestone saddle. Cittadella is up to 400 m wide at the west, tapering over a length of 600 m. to a point at the east. To the west, the main part of the ridge is

1800 m long, and varies from 150 m to 500 m in width; on the ridge, relief is 20–74 m. The small hills that form the relief are triangular for the most part, but Cypress Hill is crescent shaped, bulging to the northeast. Valleys along the ridge follow the northwest joint pattern of the rocks and flank a long central depression, which became the main road. This arrangement of valleys and hills divides the ridge into usable and easily defended units, of which Cittadella at the far end was the earliest occupied and the safest (Judson 1963). Early studies indicate that the climate on Morgantina's ridge has not changed since Paleolithic times (Judson ca. 1958). The typical Mediterranean temperatures of 10–12° C median (33° maximum, 4° minimum) have been measured at nearby Piazza Armarina. There the average maximum temperature is 20.1° C and average minimum is 9.6° C, but the range is 3.7° to 32.8°. Precipitation averages 850 mm at Aidone, but 700–730 mm at Morgantina. Rainfall at Aidone has varied from 442 mm to 1243 mm, with 35% of the precipitation in autumn, 40% in winter, 20% in spring, and only 5% in summer. The wettest months are March and October, whereas the driest is June (Bruno and Sciortino 1994). Climatic variation, postulated by Nilsson (1983) and Pinna (1977), cannot be completely ruled out at Morgantina, especially since local climate variation has been shown to correlate with deforestation, which is well documented for central Sicily.

Geology and Structural Analysis   Morgantina's elongated NE-SW ridge has a nucleus of sand and sandstone (Plio-Pleistocene); however, toward the bottom and along the sides, there are diverse rocky formations (Fig. 3.10).

The stratigraphy at Morgantina is as follows, beginning with the topmost modern layers:

Soils:
a. "Cioccolata," a podsolic brown soil made from plant cover of conifers, oat, and beech.
b. A blackish, humic soil dating from the post-Roman period, found on downslopes and consisting of organic materials mixed with surface sands and silts, as would result from the decomposition of ancient grass, scrub, brush, but not timber; found below the modern plow zone.

Recent lithotypes (Holocenic) visible as slag from mines, talus, and alluvium along the Gornalunga River.

Sand and sandstone (Plio-Pleistocene) of the ridge, agora valley, and the eastern half of the Trigona Valley.

Messinian forms of the Evaporitic Series:
a. Massive, grayish-white limestone of the Calcare de Base Formation and the Gypsum Formation (chalky white limestone below the north gate).
b. Tripoli Formation with rhythmical alternation of calcerous marls and diatomites.

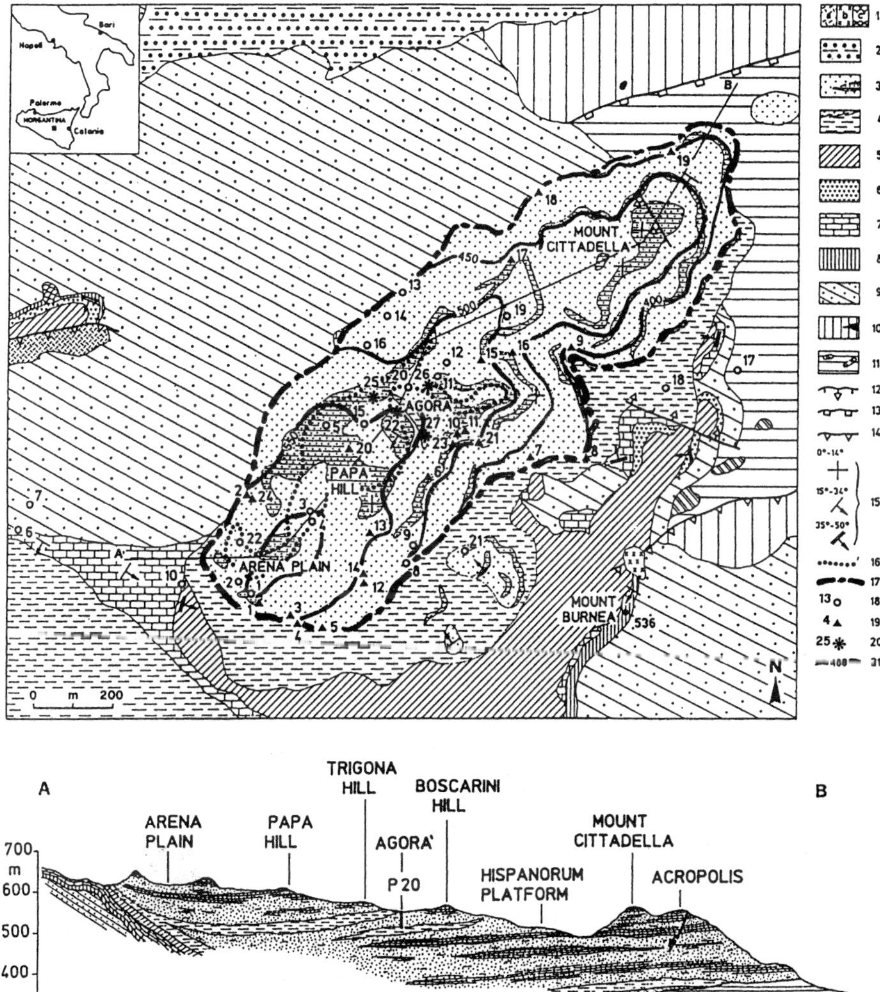

FIGURE 3.10. Map of the hydrogeology of Morgantina and environs: (Bruno and Nicosia 1994). Key to legend: (1a) slag and (1b) talus, both modern; (2) alluvium (Holocene); (3) sands and sandstones of Upper Sands Formation (Plio-Pleistocene); (4) marly-silty clays, Geracello Formation (upper Plio-Pleistocene); (5) marls and marly limestone, Trubi Formation (lower Pliocene); (6) gypsum, Sicilian Evaporitic Series (Messinian); (7) limestones, Calcari di Base, Sicilian Evaporitic Series (Messinian); (8) siliceous marls and diatomites, Tripoli Formation (Messinian); (9) clays and marly-sandy clays, Terravecchia Formation (Tortonian); (10) clays and quartzitic sandstones, Numidic Flysch Formation (Oligocene-Miocene); (11) clays with marly limestones (Polizzi Formation), variegated clays formation (Cretaceous-Eocene); (12) normal faults; (13) overthrust nappes; (14) thrust and reverse faults; (15) strike and dip of strata; (16) hydrogeographic watershed; (17) hydrogeologic watershed; (18) wells; (19) springs; (20) dry springs; (21) isopiestic lines.

Rocky formations:

a. Trangressive marly-sandy clays (Tortonian Terravecchia Formation) of two types:

   i. ivory-white calcareous marls of the lower Pliocene (Trubi Formation)

   ii. blue-gray marly clays of the Geracello Formation (upper Pliocene-Pleistocene)

b. overlain by Upper Sands Formation (Roda 1968)

   i. quartz-calcareous sands with

   ii. frequent quartz sandstone (also called sandy limestone).

The ridge is structurally a syncline, becoming a pericline at the edges (Bruno and Nicosia 1994). This form of the ridge is pre-Pliocenec in age. Faults visible here (NW-SE) are late Pliocene to early Pleistocene. At the Trubi level (lower Pliocene) there is much fracturing. Discontinuities have affected the upper Sandstone Formation too; these occur in families from a few centimeters to 3 m, showing subvertical dips. The principal fracture families strike NW-SE, and secondary ones NE-SW. The two thrusts here present (14 on Fig. 3.10) trend toward the NW in counterposition to the generally SE faults of Sicily (Bruno and Nicosia 1994).

Landslide and collapse at Morgantina involve both the ancient city and the present archaeological site. Most important is the internal crumbling of the syncline, for example, a crumble slide along SE flank of Boscarini Hill to the east of the site. Collapse phenomenon, triggered by tectonic stress, occurs in the sandstone where there is a complex system of joints. Collapse is visible also near the agora, with joints of type K (length greater than 1 m) that plunge at 53° and 232° with dips of 88° and 83° spaced 22 cm and 54 cm apart respectively.

Exceptionally heavy rains can trigger interstitial overpressure, as happened in the winter of 1995, when pressures along the fractured rock caused some landslides near the South Gate. Several dozen meters NW of that area, the wall of the large, late-Roman furnace or kiln (set into the granary east of the agora) fell down, incapable of resisting the thrust of the restraining embankment that overhung the furnace, in spite of preventative work carried out by the Enna Soprintendenza. Southeast of the agora, the slide in the sandstone involved both the lens of marine clay (Geracello Formation) and the material from the excavation dump. In particular, the NE part of the recent landslide involved collapse of the quartzcalcareous sandstone with rock fall.

Judson (1963) thought that there had been no recorded earthquakes at Morgantina in historical time. However, during the enormous earthquake of 1963, the town of Noto Antico southeast of Morgantina was destroyed, and Syracuse was also damaged. Several churches at Aidone were damaged at that time. Morgantina was essentially abandoned then, so no damage was recorded there.

Hydrogeology and Hydrography   Geologists define two aspects of the water resources of the Morgantina area. On the surface is the hydrographic area, divided from the neighboring basin of the Gornalunga River by the surrounding ridgelines. Within is a smaller, subsurface area, the hydrogeologic basin (Fig. 3.10, nos. 16,

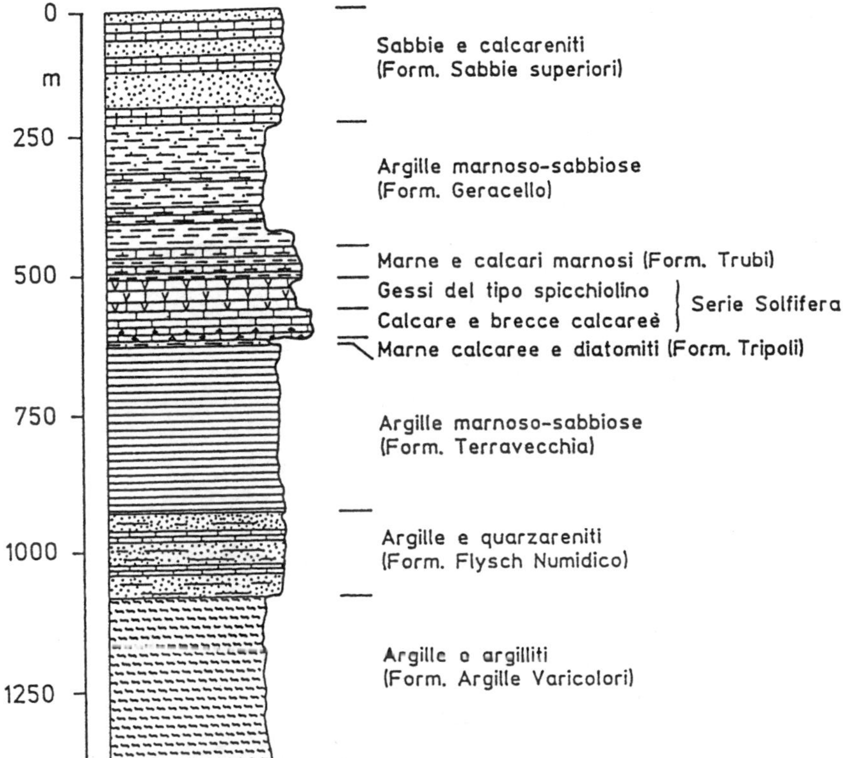

FIGURE 3.11. Geological column, Morgantina, from top to botton (newest to oldest):
Sand and calcarenite, marly-sandy clay, marl and marly limestone, gypsum and breccia
of the sulfur series, marly limestone with diatomite, marly-sandy clays, clay with quarza-
renite, variegated clays. Compare with description of stratigraphy in text (Bruno and Nicosia 1994).

17), which extends west-east from the Arena Plain to Boscarini Hill. This basin
lies vertically between the Upper Sands Formation and the waterproof substratum.
Surface drainage of the Morgantina ridge is NW-SE to NNW-SSE, aligned with
the tectonic fracturing of the rocks.

An important result of our studies is the discovery of a large aquifer inside
the ridge of Morgantina, made possible by the stratigraphy (Fig. 3.11) and the
structure of the arenaceous (sandy) layers of the ridge. That structure is modified
by clay lenses within the rock and supported by the synclinal shape of the substratum and by its impermeability. The aquifer (Fig. 3.12) is thought to consist of
a major water table supplemented by a higher water table suspended in the area
SW of the agora and probably another in Cittadella (Schilirò et al. 1996). Lateral
flow to the north connects the minor water table with the underlying main water
table via a subterranean threshold, over which the water moves under certain

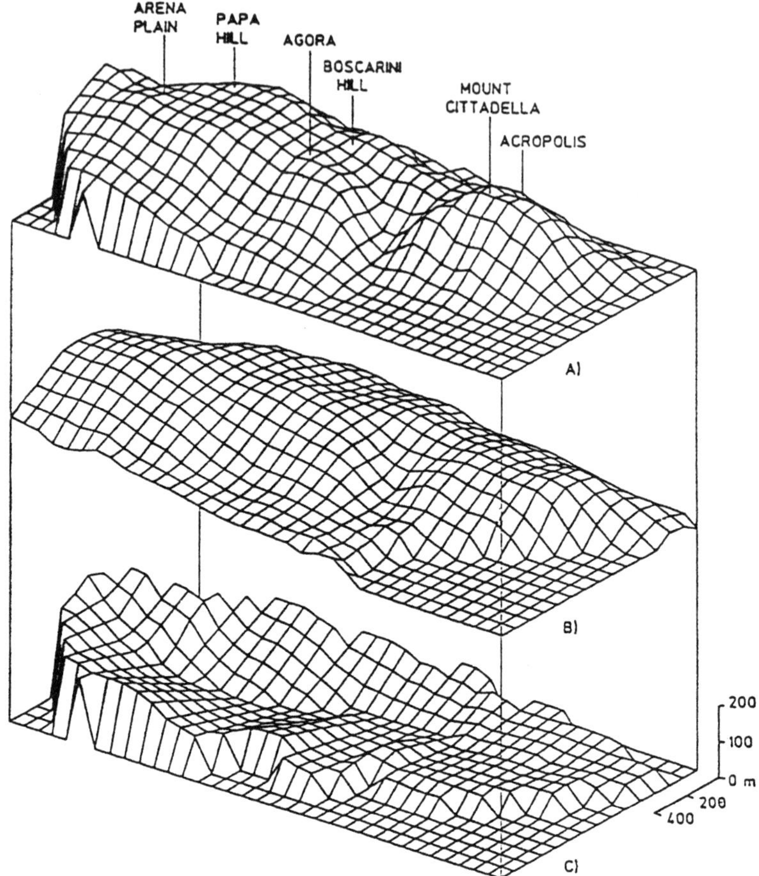

FIGURE 3.12. Three-dimensional reconstruction of the Morgantina aquifer (Bruno and Nicosia 1994, from numerous measurements done on the site). Top to bottom: A. Orography (ground surface); B. Piezometry (top of water table); and C. Impermeable substratum (bottom of aquifer and top of impermeable layer).

climatic conditions such as abundant rainfall. Bruno and Nicosia (1994) did a computer model of the aquifer that shows that the upper water table relates to the underlying main water table near Papa and Trigona Hills. The field investigations turned up erosion that affected the size of the aquifer, indicating historic changes in the morphological setting of the ridge and resulting in a smaller aquifer after the Hellenistic era.

Because of the arrangement of the rocks, the Morgantina water basin is isolated and fed exclusively, now as in the past, by means of rainwater and not by a stream (Atzori et al. 1995). The main recharge area coincides with the highest hydraulic head area (Arena Plain to Boscarini Hill). Hydraulic measurements taken at wells 1–22 and springs 1–29 allow reconstruction of the piezometric surface of the water table. The subterranean watershed slopes to follow the piezometric NE-SW trend. The differences in the piezometric surfaces indicate two different values of permeability on the two sides of the groundwater divide, because groundwater in the northern area rests directly on clays of the Terravecchia Formation and locally on the Evaporitic Series, whereas the southern portion is on clays of the Geracello Formation. In the latter, contact between the sands and clays is not so clear because of the gradual transition from clay to sand with frequent interbedding of the two materials, decreasing permeability south of the ridge.

The large number of springs (Bruno and Nicosia 1994) supplied water copiously during the Hellenistic era (Fig. 3.10). Today's dry springs allow the reconstruction of the ancient piezometric level of the upper water table: Springs 22, 25, 26, 27, and 29 are all located at the edges of the agora, whereas number 28 is on the SE slope of Cittadella and number 25 is much higher at the North Sanctuary of Demeter. The West Stoa's upper back wall was midway in altitude between the North Sanctuary and the springs around the agora, but the water pressure within the hill was able to topple that wall (discussed under Urban Design). Thus the piezometric head at the time of construction (third century B.C.E.) must have been much higher than it is today.

The greatest thickness of the water table is 190 m at Papa and Trigona Hills, but only 100 m at Cittadella (Fig. 3.12). The maximum possible volume of the aquifer is calculated as 2.7×10^8 m^3. A smaller volume, however, is likely because of the variability of the topographic surface and of the water table. The permeability of the enclosing rocks and the porosity of the sands (46% here) also affect the ability of the ridge to retain water (Bruno and Nicosia 1994).

Because of the cupped bottom (synclinal setting) of the aquifer, springs cannot tap all of the water volume. Below 390 (m.a.s.l.), the aquifer is not tappable and therefore 6.44×10^6 m^3 remain unavailable compared with 0.97×10^8 available, from the estimated modern total of $(2.25 \times 10^8 \times 0.46) = 1.035 \times 10^8$ m^3. With the higher piezometric surface in the third century B.C.E., a volume of at least 1.00×10^8 m^3 yr for a population of 10,000 to 30,000 is probable. The population also utilized cisterns for subpotable water (Atzori et al. 1995; Crouch 1993).

An indicator of a water supply crisis is the history of the water for the monumental fountain in the NE corner of the agora. At the beginning, this fountain was supplied from its own spring (26*), but later the supply had to be supplemented with rainwater and eventually supplied by pipelines from the south. Its supply dropped from 41 to 19.5 m^3/yr between the third and first centuries B.C.E. (Schilirò et al. 1996).

Water Quality and Potability   Water along the ridge is visible at 51 points. Twenty-three of these were tested. Chemical analysis of the water shows pronounced interconnection of the two water tables of the ridge (Bruno and Nicosia 1994). A temperature minimum of 15.50°C was noted at springs 2 and 24, whereas maximum temperature was 22°C at well 19. Low temperature reflects a short stay of water in the aquifer, whereas the higher temperature in the deep well 19 indicates decrease of water filtration velocity because of a low hydraulic gradient. The water, with a 6.96 to 8.40 pH concentration, is neutral to weakly alkaline. Electrical conductivity correlates nicely with salinity except in wells such as 7 and 10; springs 7 and 8 have the maximum conductivity, close to the Italian maximum admissible concentration (C.M.A.), whereas wells 10 and 21 are higher (Bruno and Nicosia 1994). The salts in the waters, compared with modern standards of potability, are harder than the modern normal standard. Calcium bicarbonate and a moderate amount of ammonia are found in the water because of human and animal activity on the farms that occupy the recharge areas of the ridge now, and the evaporitic deposits in the substratum contribute sulfates, which also affect the taste and purity of the water.

Materials Analysis   A survey of construction reveals foundations of pebbles, reused stone blocks, reused kiln waste, broken tiles, and baked brick. Walls combined baked-brick lower parts with mud-brick upper walls, reinforced with wooden uprights. Tiled, double-pitch roofs were used as early as the seventh or sixth century B.C.E. on Cittadella. Some plastered interior walls from the sixth century have survived on Cittadella, whereas in the Hellenistic era, on the later site, more attention was paid to house interiors and finer finishes (Barra Bagnasco 1996; see details in Tsakirgis 1984: 326–34 and 1989).

Analysis of construction materials at Morgantina by applied mineralogical and petrographic tests of the pipes, pithoi (very large clay vessels), and roof tiles was carried on by Atzori et al. (1995). Since this is the most elaborate study available of clays at a particular site, these findings are presented here in some detail. Twelve samples of ceramics from the agora were analyzed, as well as eleven clays and two sands from different formations (Figs. 3.13, 3.14, and Table 3.1).[10]

From these petrographic analyses, two distinct groups of pottery, six samples each, were determined, independent of their use as tiles, vessels, or pipes: Group 1 incorporating grog (fired and crushed pottery), also called *chamotte*; and Group 2 without grog but with a high silt-sand fraction containing quartz and feldspars with some pyroxenes and fragments of metamorphic rocks. Group 1 pottery was thicker than Group 2. Grog is a common ingredient in Roman ceramics. Addition of grog to the first group of pottery probably indicates an early form of recycling, since Morgantina was littered with many broken roof tiles and other sherds, dating from the destruction of the town between 215 and 211 B.C.E. The effort of grinding up the terra-cotta to use as grog was less than that of bringing additional new clay from the Gornalunga riverbed so far below. A by-product of the process was the cleaning up of the site.

The groundmass of pottery is optically isotropic or shows weak birefringence because of quartz and mica clasts of small dimensions. Two samples had elevated birefringence, containing microcrystalline calcite and clay, as confirmed by diffractometric data. The colors of samples ranged from dark brown to brick red and yellow ochre, often with different shades between the surface and core caused by redistribution of iron or by formation of calcium and iron silicates during firing (Jacobs 1992). The most abundant inclusions are quartz clasts (fine to coarse) poorly rounded and often metamorphic (wavy extinction). Feldspars in these sam-

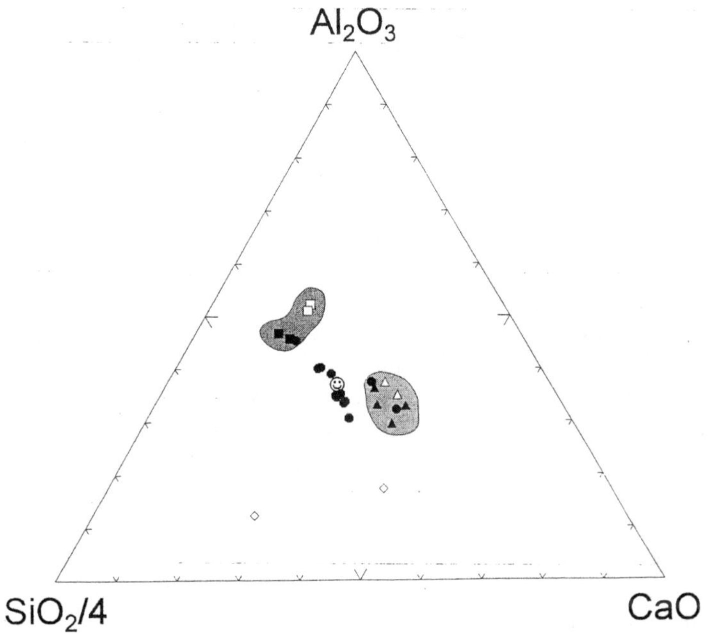

FIGURE 3.13. Ternary diagram of relations of SiO_2, CaO, and Al_2O_3 in terra-cotta and clay samples at Morgantina (Atzori et al. 1996).

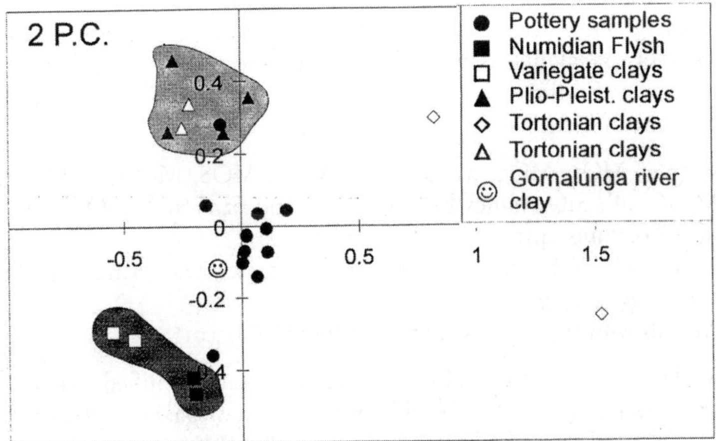

FIGURE 3.14. Multivarious analysis of the clay and terra-cotta, related to the first (SiO_2, TiO_2, Al_2O_3, $Fe_2 O_3$, MgO) and second (CaO, SiO_2) principal components (Atzori et al. 1996).

ples are mostly andesine-oligoclase (plagioclases), sometimes zoned but sometimes microclinic K-feldspars. All samples had minerals[11] that are associated with Mt. Etna or the Aeolian Island volcanics, which were especially active during the last million years B.C.E. (Simkin et al. 1981). The same minerals, however, may be deposits from even older volcanoes in central eastern Sicily. None of the samples had the kind of augite and olivine association found in the alluvial sediment of the Simeto River (Amore et al. 1979) farther downstream from Morgantina.

Two samples include medium-high-grade metamorphic rocks, and one has cummingtonite, indicating sedimentary contribution from the Perloritani Mountains NE of Morgantina (Atzori et al. 1974; Cirrincione 1992) or reelaborations of Terravecchia Formation sediments (Cirrincione et al. 1995). No clasts of volcanic rock as referenced by Cuomo di Caprio (1992) were found in these samples, but some fragments of chert were. Some samples have remains of microorganisms (Globigerinidae, Globorotalidae, and Orbulinae), but these are too poorly preserved to allow dating of the clay. There were many examples of microfracturing (small, local cracks) parallel to the surface of ceramics, especially in the grog group. Porosity was 24.1% to 34.7%, with a mean pore diameter of 0.005–0.538 μ. Grog samples were more porous, but there was no correlation among porosity, chemical characteristics, or mineralogical paragenesis (simultaneous formation).

From X-ray diffraction analysis, it was learned that there were quartz and feldspars in all samples; clay minerals and calcite occur only in two samples, but Ca-silicate occurs in the rest. This indicates that firing temperatures were about 950°C (Capel et al. 1985; Fabbri 1992).

Since analyzed oxides (Table 3.1.) are not usually modified in quantity after burial, they can represent the composition of the raw material. Organic materials Na_2O, K_2O, and P_2O_5 were not examined because they might have been used in the pottery manufacture; likewise, chlorides might have been precipitated after burial of the pottery. The parallel chemical analysis of the pottery and the local clay considered these facts especially. Table 3.1 shows the results of analysis of major elements, recalculated in anhydrous (waterless) mode, for the 12 samples. On the basis of the three main oxides SiO_2, Al_2O_3, and CaO, three groups of pottery have been identified:

1. samples MO2, MO3, MO4, MO5, MO7, MO8, MO211, MO14, and MO16 with SiO_2 values between 60.23 and 62.17 and CaO values between 9.7 and 14.12
2. MO1 and MO6 with low SiO_2 values (54.99, 53.69) and high CaO values (15.7, 19.47)
3. MO 18 with high Al_2O_3 (18.7) and low CaO (7.01)

Raw clays were analyzed also. Two groups were identified, based on their chemistry: One group was the Numidic flysch and variegated clays, rich in Al_2O_3, TiO_2, and FeO and poor in CaO; the other was Plio-Pleistocene and Terravecchia clays with more SiO_2 and CaO but less Al_2O_3, TiO_2, and FeO. Alluvial clays from the Gornalunga River bed, however, have intermediary composition, whereas the Gornalunga sands have very low Al_2O_3 and variable CaO and SiO_2 contents.

TABLE 3.1 Morgantina Oxides[a]

Samples	SiO$_2$	TiO$_2$	Al$_2$O$_3$	FeO*	MnO	MgO	CaO	Na$_2$O	K$_2$O	P$_3$O$_5$
Pottery										
MO 1	54.88	0.84	17.64	6.24	0.07	1.80	15.70	0.60	2.00	0.24
MO 2	62.08	0.74	14.80	5.42	0.07	1.78	12.37	0.64	1.84	0.26
MO 3	61.36	0.69	17.14	5.18	0.06	1.87	10.04	1.13	2.29	0.23
MO 4	61.26	0.71	16.75	5.29	0.07	2.27	9.70	1.04	2.64	0.26
MO 5	62.17	0.68	13.10	5.07	0.06	1.90	14.12	0.66	1.80	0.44
MO 6	53.69	0.77	15.66	5.76	0.07	2.07	19.47	0.50	1.73	0.27
MO 7	60.64	0.78	15.11	5.53	0.07	2.06	12.61	0.76	2.16	0.29
MO 8	60.90	0.73	14.53	5.20	0.06	2.17	13.29	0.82	2.05	0.26
MO 11	61.34	0.74	14.41	5.25	0.06	1.92	13.22	0.77	2.03	0.26
MO 14	60.23	0.72	16.79	5.31	0.11	2.07	11.17	1.03	2.40	0.16
MO 16	61.52	0.77	14.73	5.54	0.07	2.25	12.14	0.73	1.88	0.38
MO 18	61.93	0.89	18.70	6.49	0.12	1.75	7.01	0.62	2.31	0.18
Numidian Flysh clays and Variegate clays										
FN 1	63.02	1.09	18.75	6.49	0.05	1.81	6.53	0.34	1.80	0.12
FN 2	63.97	1.10	18.88	6.37	0.06	1.80	5.52	0.34	1.83	0.12
VV 1	55.77	1.05	22.99	7.41	0.06	2.49	7.06	1.30	1.68	0.20
vv 1	56.82	1.02	22.18	7.15	0.06	2.26	7.03	1.49	1.83	0.16
Plio-Pleistocene clays and Tortonian clays										
PP 1	51.08	0.82	16.37	5.57	0.06	3.07	19.47	0.44	1.81	0.30
pp 1	56.01	0.72	14.00	4.54	0.07	2.47	19.18	0.92	1.90	0.20
PP 2	54.49	0.87	16.96	5.75	0.07	3.18	16.07	0.41	2.00	0.22
pp 2	56.54	0.78	15.47	4.90	0.06	2.65	16.93	0.55	1.92	0.21
TR 1	52.50	0.82	18.26	5.39	0.05	2.67	17.40	0.44	2.28	0.19
tr 1	51.47	0.80	17.35	5.52	0.06	2.50	19.26	0.58	2.24	0.21
Plio-Pleistocene sands										
S 1	83.66	0.04	4.20	0.48	0.03	0.20	9.10	0.95	1.27	0.07
S 2	66.21	0.36	7.68	2.36	0.06	1.13	20.10	0.62	1.29	0.17
Gornalunga River alluvium clay										
Gor 1	60.35	0.95	15.89	6.11	0.08	1.93	12.18	0.42	1.92	0.16

[a]Major oxides, recalculated in anhydrous mode, of selected Morgantina pottery and clays.

Thus the pottery samples of the first group do not appear to have been manufactured using clays from formations outcropping locally. Further, their chemical composition was not altered by the addition of external material, since the petrographical analysis shows that the only tempering agent used was grog, which does not modify the chemical characteristics of the artifact. The chemical analyses do not support the hypothesis that the clays were refined by filtering processes used to separate the coarse fractions. This is evident from the results obtained in the whole local clays on their fractions less than 32 μ. In further contradiction to such a hypothesis, the pottery samples rich in CaO have an Al_2O_3 content similar to or higher than those poor in CaO.

The analysis of the pottery shows a trend of mixing (Langmuir et al. 1977) between a component rich in CaO and one rich in SiO_2; these end members can be identified in the two groups of clay, with a further end member given by the quartz sands. Such different mixtures of the material may be the result of human intervention, although it is not clear what would make the potters of that period undertake such a laborious operation when there is no clear advantage to be gained. A second, more probable hypothesis (Atzori et al. 1995) is that the raw material was taken from the alluviums of the Gornalunga River because this river basin is composed in part of Numidic flysch and variegated clays and in part by Terravecchia Formation clays. This hypothesis was confirmed by the cluster analysis and by principal components multivariate statistical analysis (Fig. 3.14) on the chemical composition of both the pottery and the raw clays.[11] These analyses indicate very close connection between the river clays and most of the pottery from Morgantina (Atzori et al. 1995 cf. Cuomo di Caprio 1992).

Implications for Human Settlement

From before the seventh century B.C.E., the geographical and geological setting here had great impact on the lives of the people. As at Aidone in recent times, the earliest hill town populations relied on grazing animals, especially pigs in the oak forests, on hunting, and on the sale of forest products for their livelihood (compare with Argos). The work of artisans is evident in the spinning, weaving, and metalwork found at Cittadella (Antonaccio 1997), though perhaps raw wool rather than finished weaving was exported.

Unlike many Greco-Roman settlements that were oriented to the compass points, the NW-SE extension of this ridge imposed a plan with the main street running in that same northeasterly direction, for the most part, with the cross streets at right angles, that is, pointed to the southeast. The formation and placement of the small hills along the ridge reinforced this pattern. Streets as well as major buildings and spaces are oriented NW-SE, turning away from the hottest sun and winds from Africa. Most notably, the open space of the agora opens to the southeast. Two of the major gates of the city were sited at opposite ends of this cross valley; the third was at the west, where two small hills flanked the road — a natural place for defense. Roads to the plains below were built down the steep slopes to the north and south, outside their respective gates leading from the agora. Like Enna, Morgantina had access to favorable agricultural land as well. Morgantina was thus sited to take advantage of access routes and water supply.

Many ramparts were built in Sicily under Hieronian patronage. Such walls were a sign of the independence of Greek cities (Karlsson 1993: 38–39 and fig. 1). The pattern of construction is a "masonry chain," a form of buttress built into the wall, made of alternate headers and stretchers arranged vertically, spaced 2.5 to 3.35 m apart, approximately 10 Doric feet of 32.6 cm. This chain is visible, for instance, in the city wall near the San Francesco Bisconti district to the east. Only

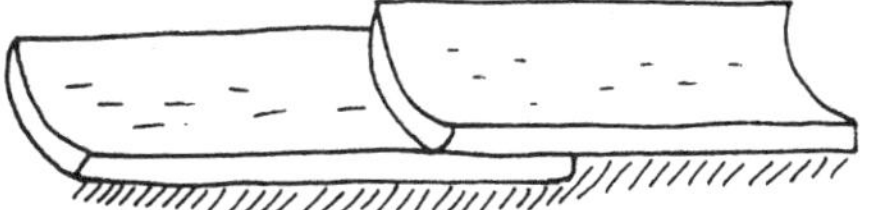

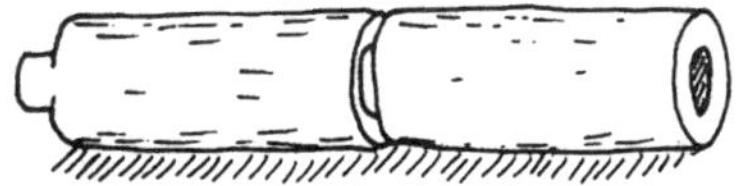

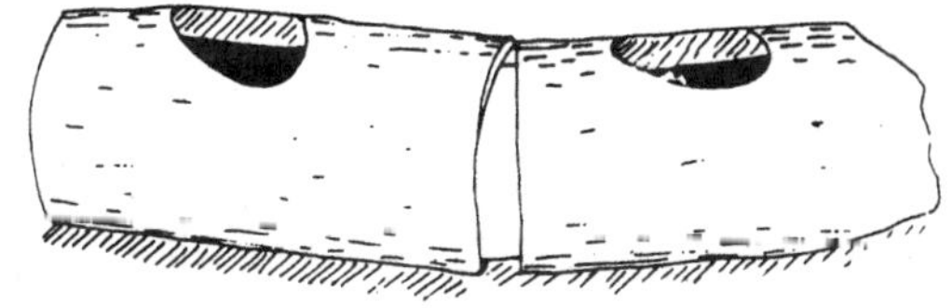

FIGURE 3.15. Types of water system elements at Morgantina. From top to bottom: roof tiles reused as gutters; small terra-cotta pipes with male-female joints; large terra-cotta pipes with male-female joints and so-called "handholes"; and constructed stone drainage channel (Bruno and Renna 1998).

where the ridge was very steep was the rampart omitted; to the north beyond the North Demeter Sanctuary was one such natural slope of sufficient steepness to be self-protecting, like the north slopes at Agrigento. The increased strength and stability provided by the masonry chain pattern might have been at least as important as the status given by the rampart, when used on slopes subject to landslides in the San Francesco area and elsewhere (Schilirò et al. 1996; compare with the stepped wall on the eastern side of the acropolis at Selinus, near the letter B on their fig. 3.18, probably a sheathing against landslides).

Stone for the ramparts of Morgantina was taken from quarries along the line of wall, between the gates. The houses and public buildings of Morgantina were made from local stone taken from quarries within and at the edges of the built-up area, from clay of the site and of the river bed, and from lumber of the nearby hills.

For the most part, Morgantina was well adapted to the geomorphology of its site. In the construction of the West Stoa, however, the Greek builders ran into a problem of local instability similar to that of the two modern landslides mentioned previously. The West Stoa had been planned to have two levels with four meters difference in height, adapted to the existing slope. About 40 lineal meters of its upper wall was never completed because it was being overthrust by pressure from behind it within the hill. The heavy beams and wall blocks of the projected stoa are now fallen or deformed. Behind the ruins of this building, the soil and rock appear today as a cliff over the unfinished back perimeter wall. (Note the back wall and identical cells of shops on the plan in Fig. 3.9 just north of the theater.)

Two causes, both geological, have been advanced for this incompletion: The most commonly presented explanation is instability as a result of the attempt to build a large, heavy building without strong foundations over an excavation about 4 meters deep at the contact between sands and marly clays, where the subsoil was plastic clay. From what we have learned of the geological sophistication of ancient Greek builders, this explanation seems unlikely. Structural geologist Bruno, however, suggests that water pressure within the hill could have thrust the back wall of the West Stoa forward, deforming and eventually pushing down whatever had been erected. The builders added very heavy buttresses, but these were insufficient against the forces within the earth (Schilirò et al. 1996), indicating that the water's head at the time of construction (early third century B.C.E.) must have been notably higher than today, possibly due to the cooler weather and increased rain during the Hellenistic period (see Fig. 7.1). The hydrogeology of the location was not sufficiently understood by the ancient builders. Alternately, extended heavy rains may have coincided with the period of construction.

The East Stoa on the other side of the agora is fortunately grounded on an outcropping of sandstone, so its relatively light construction has been sufficient to manage reduced stresses compared with the situation of the West Stoa.

We close this study of Morgantina with a brief account of the water supply and sewer systems (earlier work: Crouch 1984, 1993), after reexamination of the site with geologists Bruno and Nicosia (1994), and Schilirò, Bruno, Cannata, and Renna (1996), and engineer Sciortino (1993). The functioning of the supply and drainage networks relates to the equilibrium of water supply and water demand (Bruno and Renna 1998).

Morgantina residents used the necessary water architectonically and decoratively, as Greeks had begun to do in the sixth century B.C.E. throughout the Mediterranean area. Even in this modest hill town, there was ingenious use of artifacts and technological solutions, adapted to hydrogeological features. For instance, the changes to the fountains at the northwestern and northeastern corners of the agora show the application of developing technical solutions required because of over-exploitation of the water table and unfavorable climate (Bruno and Nicosia 1994). These are the elements of the system at Morgantina (Fig. 3.15):

1. Drinking water was from springs and from rainwater in private cisterns used especially during summer (Bruno and Renna 1998; excavation notebooks in the Morgantina Room at Princeton University describe these cisterns). Additional supplies were from on-site springs and from a long-distance waterline from which individual pipes were found along the road from Aidone (Allen 1970; Crouch 1984).
2. Pipelines within the site led to and from reservoirs and from springs.
3. Public fountains were placed along the main SW-NE road and in the agora.
4. Water distribution in both the agora and the residential areas was via tubes of modular terra-cotta elements or smaller diameter lead tubes.
5. The terra-cotta pipes for distribution were made with "male-female" joints and varied diameters, depending on function and date. Larger pipes had so-called "clean-out holes." Their size is thought to have depended on function. No definitive work on form, function, and date of terra-cotta pipes has as yet been published. (See Ortloff and Crouch 2001.)
6. Wastewater and sewage were carried in large-cross-section rectangular drains of sandstone, which formed collection nets in the agora and residential areas. Sandstone drains had the greatest linear extent of any water system element here. They were carved in rock or laid in earth and built of rectangular blocks without waterproofing, as seepage into the ground was accepted — even encouraged.
7. The settlement had many cisterns but few wells. Twenty meters (60 ft) was the practical limit for the depth of Greek wells, and it may be that the receding water table made wells ineffective in the last days of the settlement. At least one quasi-public fountain or well is known on West Hill, at the House of the Arched Cistern (Crouch 1993, Fig. 17.4), and another well was found in the courtyard of the so-called Apartment House northeast of the agora (Allen 1970; Stone 1993). The sixth century B.C.E. well on Cittadella was 15 m deep (Crouch 1984. Ill. 2).

Aqueducts, fountains, and drains are likely to have been paid for with public money, their size and cost exceeding private budgets, but domestic cisterns or wells were the responsibility of each family (Crouch 1984). The relative extent of the kinds of hydraulic elements here, their relative spatial distribution, and the length of single branches of the network are seen in Figure 3.16 (Schilirò et al. 1996). The terra-cotta tubes now visible on the surface are mostly of NE, SE, and SW orientation, but are missing from the 270–360° quadrant, where little excavation has taken place, and where the steep slope makes preservation of ancient material more problematic.

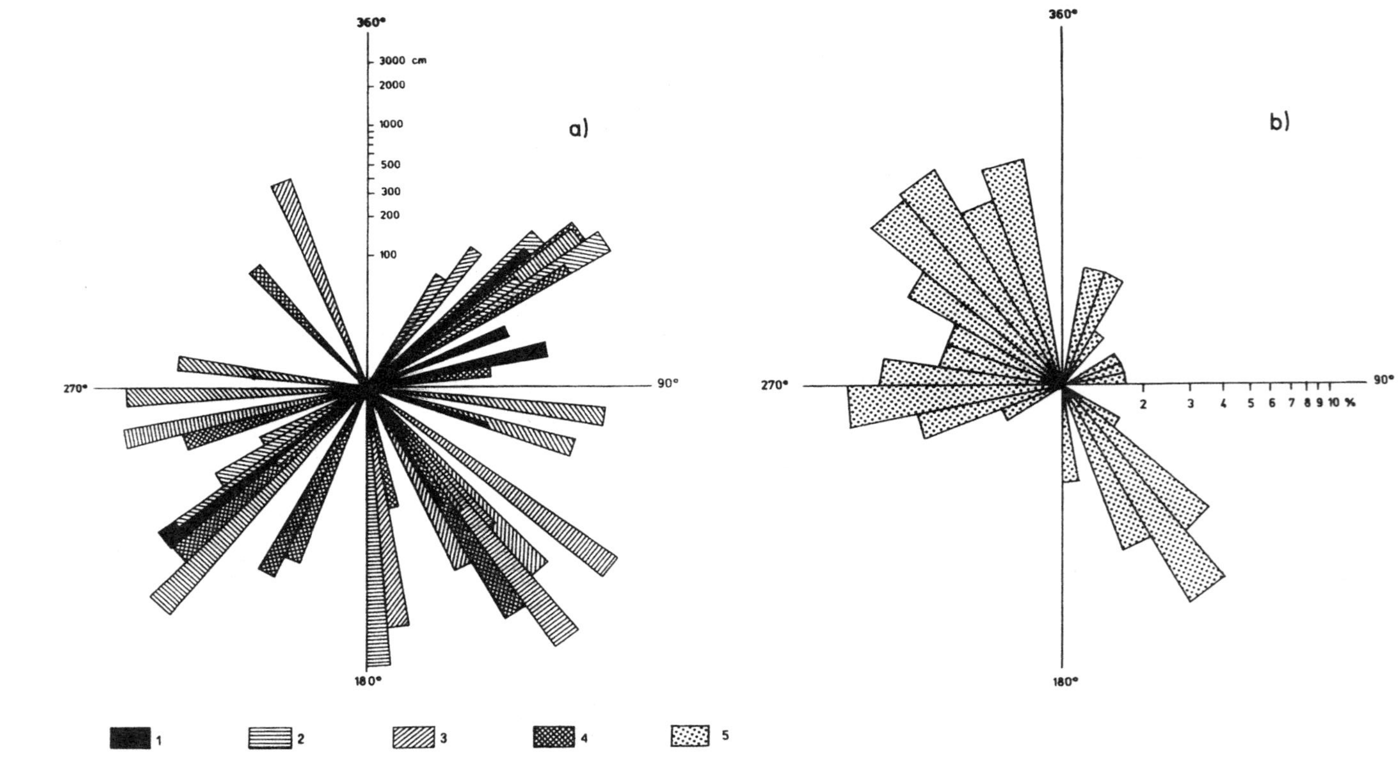

FIGURE 3.16. Orientation diagrams (Schilirò et al. 1996): (a) Orientations and lengths of various pipe and channel elements of the hydraulic network at Morgantina, where the known ruins of the town spread out over 180° from NE to SW; 1, terra-cotta drainage channels; 2, constructed stone drainage channel; 3, large terra-cotta pipes; 4, small terra-cotta pipes. (b) Orientations of discontinuities, 5, in the sandstone outcropping of Morgantina.

Hydrogeological change was a major factor in the desertion of Morgantina since the mid-first century B.C.E. Political and social changes also contributed to this abandonment.

Urban Design: Analysis Based on Geology,
Engineering, and Materials

The mutual impact of urban design and geology at Morgantina is seen most clearly in the history of agora. We note geological constraints on many structures here. Oldest elements are the northeast spring and the Chthonian sanctuary at the center, which probably existed before 450 B.C.E., a shrine to a locally important goddess of earth and sky (Edlund-Berry 1989–90). The central sanctuary was located on an elevated section of the agora beside the road and near the later theater, but not aligned with it. Perhaps as early as the fifth century there was a *bothros* and altars, focusing on connections of the underworld with both the surface world and the realm of the gods in the sky above. In the fourth to third century B.C.E., a more imposing sanctuary was constructed of local stone quarried along the ridge (Judson ca. 1958). The new form had two courtyards and many rooms to serve the worshippers, who were probably common people of several ethnicities. As the city declined, the north and south Demeter sanctuaries were looted and destroyed in the late second or early first century B.C.E., whereas the Chthonian sanctuary alone remained in use. Physical evidence of extended use includes coins from the fifth to the first century B.C.E., but mostly third and second century, and numerous Persephone figures; in addition, over 3100 lamps were found (Edlund-Berry 1989–90).

The second focus of the agora was the spring in the northeast corner. The spring was formalized with a large fountainhouse, probably in the third century B.C.E. when the whole agora area was modernized. Morgantina is the clearest picture we have of Hieronian architecture from Syracuse during the period 250–220 B.C.E. (R. J. A. Wilson 1990): the plan of the agora, the Great Steps, the monumental fountain house, the *bouleuterion*, the renovated theater, and houses on the hills above, such as the one named after its Ganymede mosaic, all reveal the taste and wealth of the town during this period.

The Great Steps were built in the late fourth or early third century B.C.E. (dated by coins). Inserted into the center of the agora as a retaining wall, the steps divided the agora into upper and lower sections. At the right, they were grounded on a sandstone outcrop carved into steps to match the constructed ones; to the left, on a layer of clay; and in the center, on sand and sandstone. Where the clay and sand came together at an oblique angle, the builders provided a large drain that proved insufficient to accommodate all the drainage water. The water in this intermittent stream (Judson ca. 1958) came from surface runoff and from the spring in the cave above the Doric Stoa at the northwest corner of the agora (verified by Bruno). The integration of monumental architecture and municipal drainage is most sophisticated (Crouch 1990).

The theater was also a retaining wall, along the lower west side of the agora. Like the West Stoa, the theater evinces stability problems. The seats themselves were large stone blocks placed against the slippery clay of the slope, but required buttress walls to right and left. Over time, from the slippage of the clay, these had fallen, to be erected again in the late twentieth century by the excavators; earlier repairs came in the second and first centuries B.C.E. In most Greek cities, public meetings were held in the theater, here the overflow from meetings could sit on the Great Steps a few meters away. Possibly (as C. Antonaccio has suggested in a letter), the official meetings of the *ekklesiasterion* (assembly) were held here. Coins from 413 to 211 B.C.E. and other objects (theater moldings from the third and second centuries) indicate the theater was begun in the late fourth century, finished by the end of that century, added to in the third century, abandoned after 211, and reworked in the first century B.C.E. Even older material—early Bronze Age sherds—was found in the fill under the skene building that stood between the seats of the theater and the nearby Chthonian sanctuary. The theater is oriented almost east, ideal for late afternoon and early evening performances, whereas the Great Steps point SE (Fig. 3.9).

To the right and left of the main open space of the agora were long, narrow stoas and granaries. The stoas were braced against the two NW-SE sides of the upper agora, not quite parallel to one another, framing an elongated trapezoidal space. Just such a trapezoidal agora marked the center of Megara Hyblea, from which the Chalcidian traders came to the Morgantina area in the archaic period (Vallet 1962: his fig. 2). The granaries flanked the road that entered the lower agora from the south gate, whereas the stoas were set back against the hills.

In the Greek era, only one major building was left unfinished: the West Stoa, which collapsed before it was finished. The East Stoa survived more completely. The upper end of the agora is bounded by another stoa running SW-NE, thereby shielding the agora from west and northwest winds, as does the hillside behind it sloping upward to the northwest.

Later, as the Greco-Roman social system broke down and the site was deserted during the last century B.C.E. and first C.E., rapid sedimentation combined with blockage of postern gates and drainage slits in the ramparts. Walls became dams, undermined, overpowered, and eventually overthrown by torrential rain. Sedimentation could be rapid in this area; Judson (1963) estimates that the lower part of the agora rose 3 m in 30 years, a result repeated again after the early excavations of the 1950s.

Like Empuries in Spain (Burés Vilasseca 1998), Morgantina lacked a monumental aqueduct, although it did have an underground line from the Aidone area (Allen 1970). Both cities relied on wells, cisterns, and on-site springs. One was the artesian spring above the North Demeter sanctuary. Climbing along the ridge above this sanctuary in March 1985, my feet became soaking wet although there had been no rain for a month. This still-flowing artesian spring in the area could water the ancient sanctuary. Because so many of the springs still visible in the twentieth century are located outside the city walls and somewhat downslope, they would in antiquity have been impossible to tap for piped water to supply the city. Thus it is likely that the ancient residents did not rely on running water for all their needs, instead utilizing cisterns to store rainwater for many domestic and industrial purposes (Crouch 1984, 1993). A cistern of 30 m^3 capacity could be filled by runoff from a roof of 45 m², and refilled several times a year. This would provide potable or subpotable water during most of the year (Judson 1963; Bruno and Renna 1998). Still, the Greeks usually went once or twice a day to the nearest fountain or spring for fresh drinking water.

Houses on the east and west hills all have at least one cistern, some showing repairs as late as the last Roman occupation, when some cisterns were given vaulted brick coverings. They are set in regular blocks between streets not exactly parallel to the stoas of agora, but consistent in their local grid pattern.

Not only was the city fabricated from stones brought from quarries on and near the site, but the buildings stood on specific sites within the local topography. The new town was laid out in the popular Greek colonial pattern of the fifth century B.C.E., with two residential areas on flat-topped hills flanking the lower agora, an example of humans editing the site by excavating stone for the new buildings. Movement of blocks of stone downward to the agora for monumental structures is partially verified by the steep chute between the West Stoa and the theater. Those regular hilltops are, therefore, not natural. The quarrying process that flattened the hills may help to explain why there is little physical evidence, if any, for the houses of the fifth century (Kuznetsov 1999: similar scraping of sites of early buildings in Greek Black Sea colonies).

For moderns, the greatest advantage of a site that has not been built upon by subsequent people is that we can get a clearer idea of what the ancient physical arrangements were. From these we can infer the social arrangements, to some extent. Morgantina was born, flourished, and died during a relatively short period. Its physical and socioeconomic settings changed simultaneously until the site was abandoned. This unique history makes Morgantina a great treasure for the urban historian (cf. Priene).

Geological understanding of the site of Morgantina illuminates some ways that ancient peoples made use of their physical resources. The site exhibits prudent and even elegant use of local materials (stone, clay) for both private and public buildings.

The geological data for interpreting the past is necessarily read from the present situation, but does not constitute a true likeness of past circumstances. Still, if we use data about modern life in the area, with recovered archaeological information, we can arrive at an approximation to the ancient situation—not a precise description. Morgantina of the Greco-Roman period was a place with good resources for the necessities of human life. For a society accustomed to living in town and going out to the fields to work, the site of the town combined sufficient accessibility with excellent defensibility. Comparing Morgantina with the modern hill town of Aidone 2.5 km away and 150–200 m higher, Aidone is more defensible thanks to its high elevation and steep slopes, but harder to supply from the coastal cities, and not as convenient to agricultural lands.

If the water supply in antiquity had been the same as it is now, it would hardly support a population of 30,000 (Crouch 1993) or even 10,000 (Schiliró et al. 1996). Most of the surviving springs at Morgantina now give 1 l/s or less, and many others are dry, yet the geological study shows that in the third century B.C.E. when the population was at a maximum, the aquifer was also at a maximum, and higher springs were fed from it more amply than at present. Bruno and Nicosia (1994) conclude that the causes of desiccation could have been climatic change, but I suggest improper management of water resources when the Hellenized population was replaced by the Hispanii (Spanish veterans). R. J. A. Wilson (1990) stressed the effects on both water supply and microclimate of the progressive deforestation.

Today, such long-range outcomes are still difficult to foresee. Indeed, modern excavations and modern pumping of the aquifer have triggered unforeseen repetition of some ancient hydrogeological phenomena. We can only imagine the difficulties of forecasting geographical, societal, and urban development in a non-scientific society, although the ancients had active oral histories and sophisticated technological traditions to assist them.

Selinus

General Description

The plains of Sicily (Ulzega 1993–94) run to the sea from the central mountainous plateau, with deltas at 10 places around the island. The plain to the west of Selinus was built by a series of smaller streams and to the east includes the delta of the larger river from the Belice Valley (Barone and Elia 1979; Carapezza et al. 1983). Sand dunes extend along the coast from Mazara on the west to Sciana on the east on a base of calcarenite and clay. The sands—some 100 m deep—are mostly carbonate accumulation from sea and wind action, with some lucastrine deposit. Large sand dunes lie west of the acropolis and small ones to the east; these dunes were poorly cemented during the recent two centuries or so (Ercoli 1989).

Instead of seeing Selinus as it is today (Fig. 3.17), a low headland rising between two flat river valleys, imagine it in the eighth and seventh centuries B.C.E. as a peninsula (C. Cavallari and S. Cavallari 1872) pushing into the sea and flanked by two wide river mouths that made prefect landing places for long-distance traders, who brought this western Sicilian site into the orbit of Aegean civilization. The rivers also provided fresh water and made an easy access route into the hinterland (Di Vita 1996; Mertens 1996; de la Genière 1982).

A wide sea terrace was broken into sections by the rivers Modione (ancient Selinus) to the west and Cottone (ancient Cothon), both now completely channelized (Ercoli and Valore 1990) (Fig. 3.18). The central area has been loosely termed an acropolis, although it rises only some 45 meters above the sea (Pugliese-Carratelli 1983; de la Genière 1982; Tusa 1967). The acropolis became the monumental core of this Greek city, although it was approachable by land only from the north. Stratigraphy of the acropolis is as follows (based on Amadori et al. 1992):

> Sands: Some 100 m deep, sands are mostly carbonate accumulation from sea and wind action, with some lucastrine deposits. Large dunes lie west of the acropolis and small ones to the east; these are recently cemented.

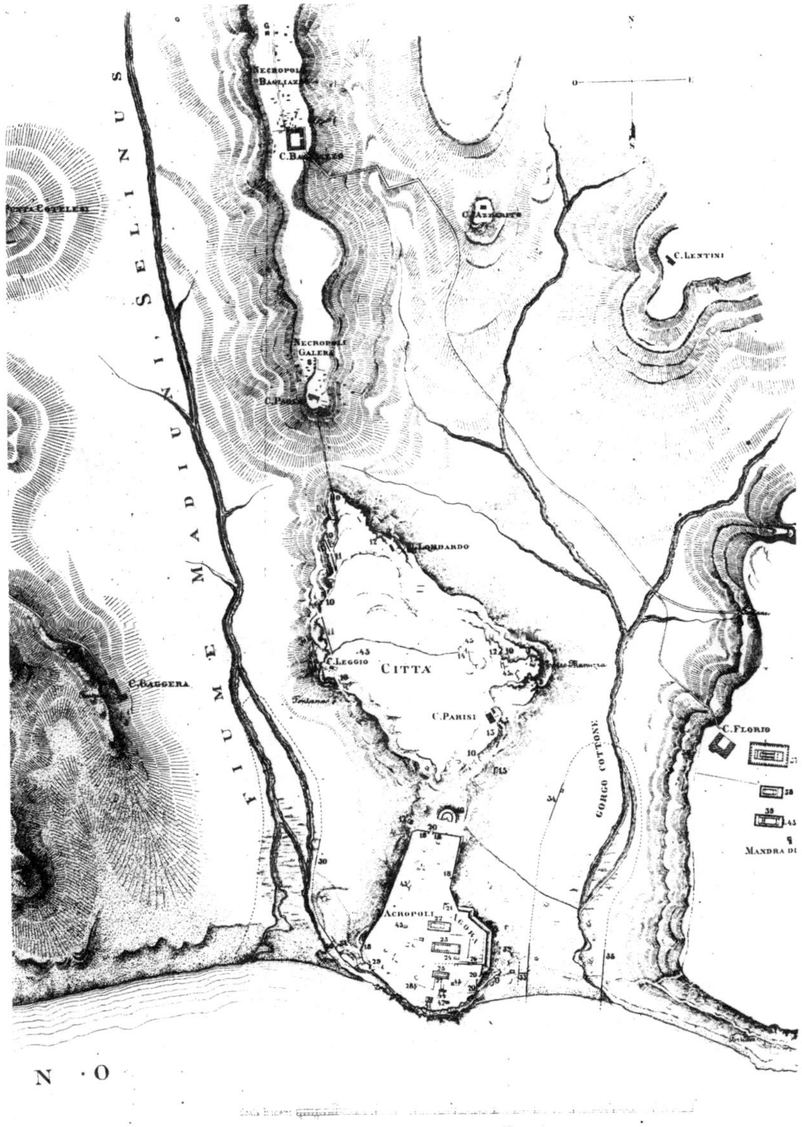

FIG. 1 – SELINUNTE. RILIEVO DI C. E S. CAVALLARI, DA *Bull. Comm. Ant. e B. Arti in Sicilia*, V, 1872, TAV. I.

FIGURE 3.17. Site map of Selinus and environs. Left (west) to right (east): Gaggera spring area with remains of temples; Citta (Manuzza) with acropolis and three temples to the south; and three more temples on east hill (with Casa (house of) Floris). The site is divided E-W into three sections by the two rivers. At the lower center between Citta and the acropolis is the semicircular bastion of the north gate, connected to the perimeter wall along the right side of the acropolis. To the north above Citta is a long, narrow ridge where cemeteries and quarries occupied the southernmost section, and quarries and springs the area farther north (C. Cavallari and S. Cavallari 1872).

71

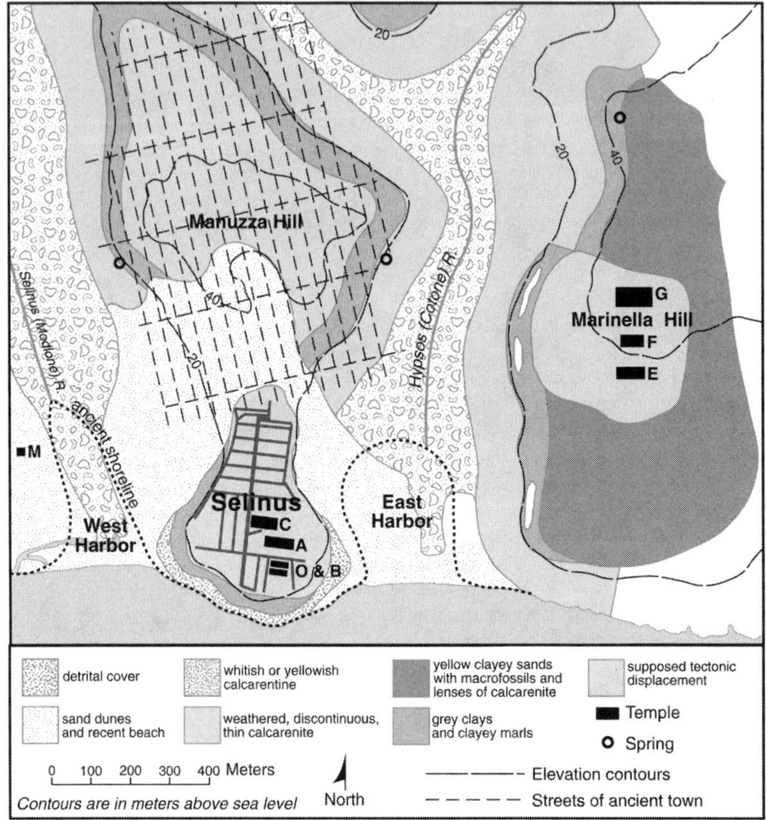

FIGURE 3.18. Streets and buildings located by archaeological evidence, harbors determined from aerial photography and excavation, and temples sited from extant remains all superimposed on a lithological map. (Redrawn by Langford from Amadori et al. 1992a, by permission of *Bollotino Di Geofisica*; and from Crouch 1993 and Ercoli and Crouch 1998, by permission)

Alluvium

Marsala calcarenite formation, 2–15m, thickest on Manuzza; whitish on East Hill with sandy clay above and below; ochre elsewhere; with dolines in the gypsum.

Clayey sand, with calcareous and fossil fragments.

Clay shading to marl.

The clay and sand of the south slope of the acropolis (Fig. 3.19) has eroded gradually, undermining the upper stone layer, so that fallen blocks of the calcarenite form a barrier toward the sea. This is the same collapse process that occurred at Agrigento, exacerbated here by wave action.

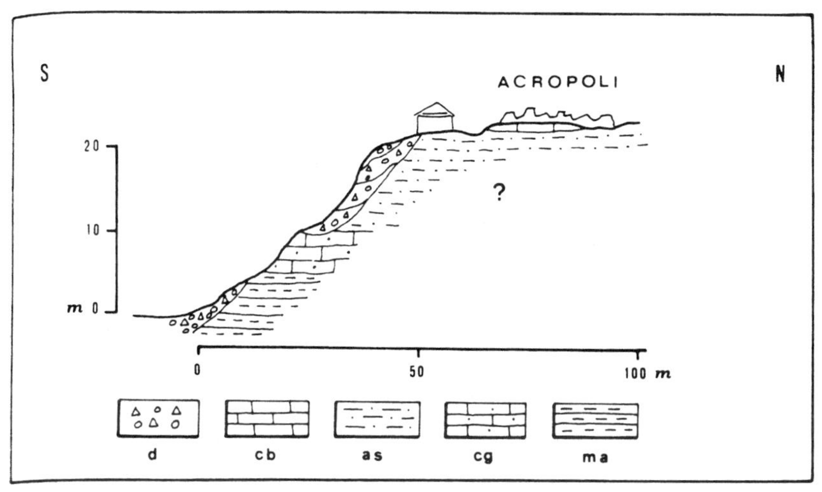

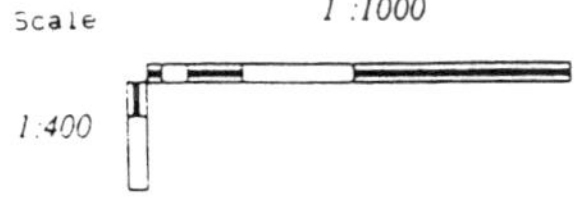

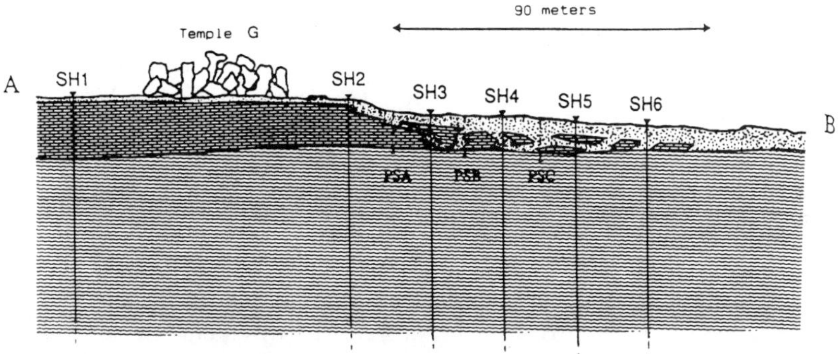

FIGURE 3.19. Geology sections at Selinus: Top: NS acropolis section, with the sea to the south, at left; (d) detritus subject to landslide; (cb) whitish calcarenite; (as) clayey sand; (cg) yellow calcarenite with calcareous modules and macrofauna; (ma) clayey marl. Bottom: NS section of east hill with south to the left. SH 1, 2, 3, 4, 5, 6 are sites of geophysical soundings. The ruins of Temple G overlie a thick lens of calcarenite, which thins to the right (north), (LaPenna et al. 1992, by permission from *Bollotino Di Geofisica*).

A layer of calcarenite underlies the temples of East Hill (Fig. 3.19), the urban district of Manuzza, and part of the acropolis—an area of 2.5 km² total. At the center of East Hill, the calcarenite is whitish between two layers of sandy clay, but ochre-colored elsewhere at the site. The calcarenite layer is thickest (2–15m) on Manuzza. Calcarenite has strong potential for water resources, with aquifers in the extensive stony outcrops, and there is additional water in the sand dunes (Barone and Elia 1979).

Human History

In several ways, the geology of Selinus has been closely related to its human history. Before the Greek traders and subsequent colonists arrived, the site was the home of indigenous peoples, possibly the Elamites (Elymi) (Tusa 1962; Antonaccio 1997), who also lived in the area north of Selinus, and were to some extent Hellenized by early contact and trade goods (Ferla et al. 1984). Of all the Sicilian cities, Selinus had the widest range of ethnic interactions during the longest period of time. This was due to the unrelenting pressure of the Carthaginians to the south, the Elamites and other native Sicilians to the north and west, long-distance trade that brought eastern Greeks, Latins, and probably Sardinians, Corsicans, and Spaniards to their ports, a continuous stream of immigrants, and the mixed Greek and eastern Sicilian heritage of the colonizers. North Africans in the seventh century B.C.E., for instance, built houses with Punic walling techniques on all three Selinus hills (Martin 1980–81). These peoples used the necropoli of Manicaliungo and Timpone Nero, 5–6 km east and west of the acropolis, respectively.

Greek Selinus seems to have begun as a trading post, with some activity as early as the ninth century (Sakellariou 1996). Dorian Greeks first traded at Selinus, then colonized it, as they did Syracuse, Gela, and Akragas. The Greek colonists of Selinus came from the eastern Sicilian city of Megara Hyblea, beginning in 650 B.C.E. (Di Vita 1979; Tusa 1962; Barone and Elia 1979). The *oikis* (founder or leader) of the settlers of Selinus, named Pammilus, came from the older Megara (a Greek city between Athens and Corinth), which had a long history of geographical constraint and consequent colonizing efforts (Di Vita 1984a).

The colonists of Selinus were looking for a convenient site for trade with North Africa and western Sicily, and were delighted to find, in addition, ample agricultural land in a part of Sicily not already claimed by other Greek settlers (De Angelis 1994, ch. 6). Since Megara Hyblea had markedly similar topography to Selinus—a peninsula between two rivers, with ample plains forming the hinterland—it seems that the colonizers were following the Greek tradition of choosing terrain where they would feel at home but at the same time could increase their opportunities and where their agricultural and commercial techniques would be applicable. For 250 years at Selinus, the "dream-city directed by right and law, by discipline and order, by equality and concord" came close to realization (Ehrenberg 1937; Garnsey and Morris 1989); this is particularly evident in the city plan.

In the seventh century B.C.E., Selinus ruled a city-state area some four or five times as large as that of Megara Hyblea (De Angelis 1994). As Greco (1996) reminds us, the survival of the colony depended on control of enough agricultural land for subsistence and for export to earn needed goods not produced at the site. "Enough" could be as little as 9.5 km^2 per 5000 people (Sanders 1984: 251), although prudence called for dominance of as much land as possible to provide for population growth. Penetration of Greeks northward from Selinus in the first 25 years of the seventh century is indicated by an inscription in Selinuntine Greek dialect, found in 1957 C.E. near Poggioreale in the hills some 25–30 km up the Belice Valley (Tusa 1962). Residents of Selinus wanted to expand westward also, but they were blocked by the Punic and Elamites who had already claimed that area (Barone and Elia 1979).

The settlement history of Selinus related strongly to the morphology of the terrain, with settlers occupying one essential location after another (Fig. 3.17). In the seventh century B.C.E., contemporary with the city foundation, the preexisting sanctuary of Demeter/Malophoros first received a Greek altar (Le Dinahet 1984). The sanctuary was given a courtyard at the end of the seventh century and a replacement temple at the beginning of the sixth century. In the same century, the road from Malophoros to the acropolis area, later known as Street F, crossed the acropolis parallel to the coast just north of the sacred zone. F Street extended also to the East Hill, while the equally important east-west street, now called O Street, ran parallel to it at the junction of the acropolis and Manuzza (Mertens 1988–89).

The first Greek settlers of Selinus lived relatively peacefully with the Elamites at the north edge of Manuzza. Excavations show that mixed Greek and indigenous houses from the late seventh and early sixth centuries were soon followed by a purely Greek layer (de la Genière 1982; Rallo 1984). Earliest vestiges of a purely Greek cluster come from the eastern harbor area.

Harbors at this early time, and through much of the Greek era, made use of beaches and river mouths (De Wit et al. 1988). The ships were pulled up on the beach until they were ready to sail again. Early trade and settlement do not seem to have been hampered by malaria (cf. Burés Vilasseca 1998), which may, however, have contributed to the abandonment of the site in the late third century B.C.E. Sedimentation of the valleys has covered additional evidence of early settlement.

Within 50 years, the original trading settlement became a Greek city of the traditional type (Di Vita 1967). The first fortification by the settlers may date from the mid-sixth century just before construction of Temple C (Tusa 1982). During the first half of the sixth century, the settlers decided on a plan that arranged the streets *per strigas* (Latin for "by furrows"; cf. Crouch 1993: "bar and stripes") (Fig. 3.18). The acropolis and Manuzza had elongated blocks close to the insulae of Agrigento in size, whereas from aerial photographs the East and West Hills (unexcavated) seem to be more irregular (Di Vita 1984 a&b; Schmiedt 1957; Schmiedt and Griffo 1958).

The earliest river walls and sewers date to approximately the same time as the new street pattern. By the sixth century, river walls in the east valley were in place to control flooding. Systems of sewers carried the wastes from the residences and businesses of the hill slopes toward the river walls and probably into the river.

Streets were 6.3 m wide and stone paved (as seen in Mertens 1989–91 excavations). (Compare with the river walls of Argos [Gaki-Papanastasiou 1994] and of Akragas, not yet formally studied.)

During the first half of the sixth century, the settlers began to monumentalize the acropolis and East Hill with stoas as well as temples (Tusa 1967; Di Vita 1984a says 580–70). Urban planning is basically the organization of space, whereas architecture is the physical realization of the habitat (Mertens 1996). The Selinuntines were ready to undertake monumental construction when they could meld the "abstract concepts, mental blueprints, functional notions, and first-hand experience of various technological solutions" (Mertens 1996) that they brought to the site with their newly mastered information about the local materials and the indigenous means of utilizing those materials. An important step toward monumental architecture was the decision to cut standard blocks (ashlar) and lay them in regular courses.

A major episode of violent destruction shook Selinus shortly before the middle of the sixth century B.C.E. (Di Vita 1996). An earthquake would certainly have been strong impetus for redevelopment of the site. The Selinuntines razed many old buildings between 550 and 490 B.C.E. (Barone and Elia 1979) to build a more monumental city. By this time Selinus was rich, with a large population (Di Vita 1967) drawn to the site for trade opportunities with North Africa and Western Sicily and by the agricultural potential of the rich plains. Selinus was coining money and producing authors, poets, orators, sculptors, painters, musicians, Pythagorean philosophers, athletes, and charioteers (Barone and Elia 1979).

A rampart constructed between 510 and 480 B.C.E. (Di Vita 1967; Tusa 1967) used large stone blocks, especially on the west side of the acropolis. A desire to awe the indigenous peoples may have contributed to the decision to build this wall. Perhaps this was when the sides of the acropolis were first sheathed in stone to guard against landslides, which were often linked to earthquakes (Carrozzo et al. 1992).

The same desire to express the wealth and strength of the city stimulated construction of temples at many places on the site. The Selinuntines sited their temples to claim territory and invoke the help and protection of their gods. It would be useful to have a "regional geography of cult distribution — especially the mostly-unstudied shrines of the lower classes — in order to understand the growth of ancient cities," (Alcock 1993: 173) because all too often, "analyses of . . . ritual behavior ignore the issue of *where* they occurred. . . . Active human participation, in the form of processions between the core of a community and its outlying territory, underscores this creation of a ritually defined social space. . . . The organization of sacred space acted to create and perpetuate the new social environment" (Alcock 1993: 202, 213; see also Polignac 1998). In 1972, J. Kuper drew attention to locations notable for the "condensation of values," the locus for important transactions of social life: visible and physical evidence for domination by the new settlers which acted as a substitute for verbal arguments.[12] The temples and the houses at the core of Selinus interacted with those on East and West Hills and with the groups of houses and secular buildings on Manuzza and the acropolis.

The monuments and vernacular architecture of the city reached maximum extent and number at the end of the sixth century and the beginning of the fifth century B.C.E. (Ercoli 1989), when Manuzza was completely filled with houses, shops, and agora, and had some quarries and necropoli along its flanks (Di Vita 1984a&b). There were also houses on the northern part of the acropolis, and the hillslopes around the northern ends of the two harbors were thick with houses, shops, and warehouses (Fougères and Hulot 1910, plate 3).

At the beginning of the sixth century, Temple C was built on the acropolis at the center of the site, soon followed by its neighbor, Temple D.[13] The top of East Hill was claimed as a site for three large sanctuaries: the earliest version of Temple E and then Temple F (560 on), and in the last half of the sixth century, Temple G, a colossal temple, rivaling the Zeus Temple at Agrigento and taking a whole generation to build (Le Dinahet 1984 or 1979). At the same time (sixth to fifth century), to the west of the city, an open-air altar of Zeus Meilichios (Le Dinahet 1984) was built on West Hill, where the pre-Greek Malophoros (Demeter) (Ercoli 1989) was joined by Zeus Meilichios (Le Dinahet 1984) and Hecate near the important spring of Gaggera. Zeus's grand sanctuary and colossal fountain occupied the slope, complete with plaza, great stairs, altar, temple, water conduit, and enclosing wall. These three sets of buildings were sited to take advantage of the elevational differences for maximum visual impact from the city proper. (See the following discussion on development of roads and quarries to build these temples.)

A century later, 500–460 B.C.E., Temple E on East Hill was rebuilt (Di Vita 1967; Le Dinahet 1979). O Street was equipped with gutters and a gently arced paved surface to facilitate drainage, all dating from the second quarter of the fifth century (Di Vita 1996). On the acropolis, Temple O and Temple A date to 490–80 (de la Genière and Theodorescu 1980–81; Mertens 1989). Finally, Temple B and its small altar on the acropolis were possibly dedicated to Poseidon but completed in the fourth century (Tusa 1979–82)

Both practical and sociocultural problems were solved in the building of these temples. The temples on East Hill were carefully placed with regard to the underlying geology (Fig. 3.19 bottom). Only under the row of three large temples is there a lens-shaped calcarenite layer 3–5 m thick, providing a much stronger and more stable base than the clayey soils of the rest of the hill (Amadori et al. 1992). This placement was not only an aesthetic success, but functioned as propitiation against future earthquake movements. Geologists Amadori et al. (1992) found evidence of an ancient trench, dug and then filled, beyond the northern edge of Temple G, which suggests to me that the ancient builders were careful to discern whether they were building on soil or stone. A fault running east-west, detected north of Temple G, lowered the south part of East Hill (Di Vita 1967) (Fig. 3.20).

The formalization of the city with public buildings and spaces continued with the construction of the agora, an area of open space at center of public life, on Manuzza during the first part of the fifth century (Rallo 1984; Le Dinahet 1979). Like the trapezoidal agora at the mother city of Megara, this locus of governmental and commercial buildings was a hinge where two street patterns came together

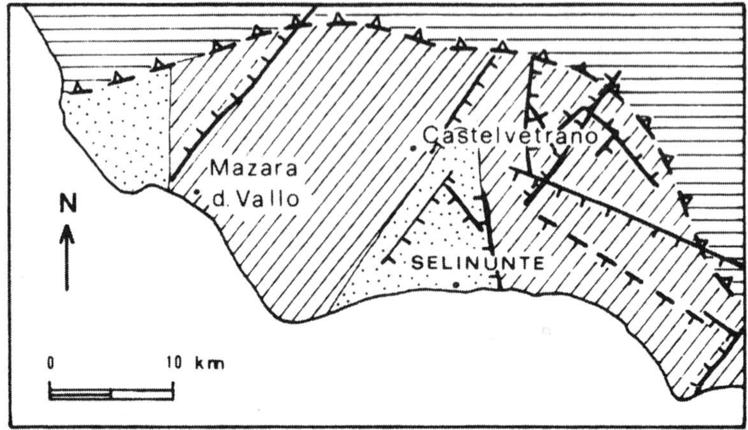

FIGURE 3.20. Faults in the Selinus area. (a) continuous and intense lowering in the Pliocene and part of the early Plieistocene, with most defomation by normal faults; (b) nearly continuous and moderate lowering in the Pliocene and early Pleistocene, followed by some uplift since middle Pleistocene, with defomation mainly by normal faults; (c) intense lowering in the Pliocene and very early Pleistoocene with intense uplifting since then and moderate deformation by folds and reverse faults with some normal faults; (1) thrust; (2) normal fault on downthrown side (Quaternary and older?); (3) normal fault of Pliocene and Quaternary; (4) normal fault (Pliocene and early Pleistocene). (LaPenna et al. 1992, by permission from *Bollotino Di Geofisica*)

(de la Genière 1982; Mertens's plan in Pugliese-Carratelli 1996: 281). For several centuries, Selinus was the greatest grain market of Sicily (Le Dinahet 1979; Garnsey 1983; Foxhall and Forbes 1982), earning the wealth that generated all these buildings.

After 450 B.C.E. when Selinus won its second war against Segesta, a second era of splendid construction began. New public buildings of the classical period include a commercial stoa that was built in precinct C-D on the acropolis, on a terrace over the steep eastern slope above the port. The stoa was similar to the early fifth-century stoa at Argos in Greece (de la Genière 1982; Di Vita 1979).

Although no stone theater has been found, a possible natural *cavea* has been located in the curve of the southeast slope of Manuzza (Le Dinahet 1979).

A significant mid-fifth-century project was river drainage works and new river walls. The philosopher-politician Empedocles of Agrigento (490–430 B.C.E.) built and paid for these public works (Le Dinahet 1979) to help Selinus cope with a river-related epidemic; the ancient descriptions do not allow us to determine whether this epidemic was the first incidence of malaria at the site, or possibly cholera or typhoid. Empedocles was honored in Temple B, sited toward the south side of the acropolis (Mertens 1996), possibly on the site of the founder's house. Thanks to Empedocles' efforts, Selinus at the end of the fifth century had a population of 60,000, new evidence reported by Mertens (1988–89) having doubled the earlier estimate of the Greek population.

By the middle of the fifth century, the second walling of the acropolis was finished in a technique and material different from the earlier one and along a different line than before (Mertens 1988–89). This effort went for naught when, in 426 B.C.E., an earthquake that affected Sicily caused damage here (Amadori et al. 1992) and weakened Selinus for the successful Carthaginian invasion of 409, which caused great destruction in the city (Collin-Bouffier 1994). I suspect that a general principle of ancient warfare was to invade your enemy whenever he was damaged by earthquake, providing your city was far enough away to have itself avoided the damage. Settlement from North Africa was repeated at the end of the century when an early seventh-century necropolis for cremated remains on the southeast edge of Manuzza was destroyed by later Punic houses (Di Vita 1996).

Almost immediately after 409 B.C.E., however, Selinus came under the rule of the Syracusan aristocrat Hermocrates (Di Vita 1967; Mertens 1988–89). Hermocrates, concerned with fortification of an area he could defend with the troops under his command rather than the protection of a large population, repaired the old temples of the acropolis and the ramparts but did not hesitate to truncate all of the east-west streets of the acropolis where his rampart crossed them (de la Genière 1982). The stepped wall along the east side of the acropolis may date from this period. As a defensive rampart it makes little sense because the steps increase access to the upper wall, but strength and stability could be important on slopes subject to landslides, and the wall increases the beauty of a major entrance to the acropolis.

Intensified residential use of the acropolis required a new agora in the southern sector of the acropolis, within the former C-D sanctuary area (Le Dinahet 1979). The east-west design axis south of this precinct became a traffic artery (de la Genière 1982; de la Genière and Theodorescu 1980–81). Hermocrates did not rebuild the ruined harbor district along the Cottone River; that the harbors could be ruined implies they had earlier been formalized with stone edges and auxiliary buildings (Mertens 1988–89).

Selinus was abandoned by Syracuse in 405 B.C.E. (Diodorus XIII.134.1 and XVI.17.5), beginning a period of intermittent warfare for the next two centuries, with the city sometimes independent, sometimes under Carthage or reconquered by Syracuse. In 383 B.C.E., some Selinuntines returned from Carthage bringing with them new Punic settlers. Their dense occupation of the southwest quarter of the acropolis continued into Hellenistic times, disturbing sixth and fifth-century remains of occupation (Martin 1980–81). All through these difficult times, the sanctuary at Malophoros persisted in use for the indigenous peoples of Sicana-Elima and for the Phoenician population.[14]

This series of conflicts escalated into the First Punic War between Rome and Carthage, with Selinus being a frequent prize (Le Dinahet 1979; Diodorus XXII 10.2; Mertens 1988–89; Nenci 1979). All gates of the city show traces of burning from one or more of these attacks (Diodorus XXIV.1.1), and at some point the postern that connected Street F with the street to East Hill was walled up (Fig. 3.21). The Selinuntines held out until perhaps 241 (Tusa 1967). With the Third Punic War, Selinus was attacked by the second Hannibal and his army between 215 and 210. The population was expelled and enslaved, and Selinus became a ruin. Perhaps the scourge of malaria contributed to the ruin as the mix of fresh and salt water became the right environment for malaria-bearing mosquitoes.

When the city was abandoned, some surviving inhabitants moved to Thermae Selinuntinae, a spa about 10 miles southeast of the ancient site; others moved northeast to Mazara del Vallo (R. J. A. Wilson 1990: i). An earthquake in eastern Sicily and Calabria affected Selinus in 17 C.E. (Le Dinahet 1979). A later, basin-wide earthquake of 361–65 C.E. affected much of the central and eastern Mediterranean (Amadori et al. 1992), but it is not known whether the Selinus site also

FIGURE 3.21. Wall east of the acropolis at Selinus, with closed postern gate. (Photo by Crouch)

suffered as a result. A late village built in the ruins was destroyed in the twelfth century C.E. by earthquake. Motion transmitted through the stone lens under the East Hill temples—a motion that differed the waves' behavior in the surrounding clay—contributed to their overthrow (Amadori et al. 1992).

Geological Resources, Processes, and Events

The great natural resources of the site were water, fertile soil, and stone and clay for building. The two natural harbors and their export trade made these resources and local products available beyond the area of the city-state.

Water Supply

Although the plains of Selinus are watered by two rivers, these are intermittent because in the Mediterranean climate the rivers run full in winter, carrying the annual rains, but in summer they are dry, or nearly so. Today, the western or Modione River does have water in summer, but it is the treated wastewater from the upstream town of Castelvetrano. The whole basin measures 113.53 km^2 (Giannotti et al. 1972) and includes a number of springs (Fig. 3.22) that in ancient times contributed to the supply of drinking water.

Most notable of the springs in the area of Selinus is Gaggera (M in Fig. 3.18 and 1 on Fig. 3.22) on the eastern slope of West Hill. Indeed, it was probably this spring that drew the first indigenous inhabitants and stimulated the development of the Malophoros shrines (Crouch 1993; Dall'Aglio 1970; Dall'Aglio Tedesco 1968). The Gaggera spring emerges opposite Manuzza from the seam between Calabrian calcarenite and the sediments of the Sicilian formation, which alternate with white calcarenite and clayey silt, covered with sand dunes. Today, only three other springs survive in the urban core, on the east and west slopes of Manuzza and on East Hill north of the temples (Peschlow-Bindokat 1992).

As early as the sixth century B.C.E., long-distance waterlines were known at numerous Greek sites (summarized in Crouch 1993). Springs to the north of Selinus, at Staglio and beyond, all flow to the Modione River and are doubled by the underground aquifer flowing in the 5-m-deep slab of calcarenite to the depth of the first impermeable clay layer. Geologists Dall'Aglio and Tedesco (1968) noted that the water of the lower aquifer mixes with that of the Staglio spring and flows at about 50 l/s (liters per second). They found, 150 m south of the Staglio spring, a line of shafts of which the last *pozzo*, 37 m deep, ends in a stratum of 11 m of gray silty clay, then grey calcarenite; artesian action of more than 1.5 m yields a flow of 4 l/s (Dall'Aglio and Tedesco, 1968, their fig. 11) which was tapped for the city supply. Ancient builders, like the modern ones, utilized this route for an aqueduct that brought the water to the East Hill of Selinus (Salinus 1884–85; Ercoli 1997).

At the town of Partanna (an Arabic name), one element from the ancient aqueduct survives: a circular tank from the fifth or fourth century, B.C.E. with lead pipes and very large terra-cotta pipes. The aqueduct is related to a line running along the east side of Manuzza past Torre di Manuzza and Casa Parisi to the

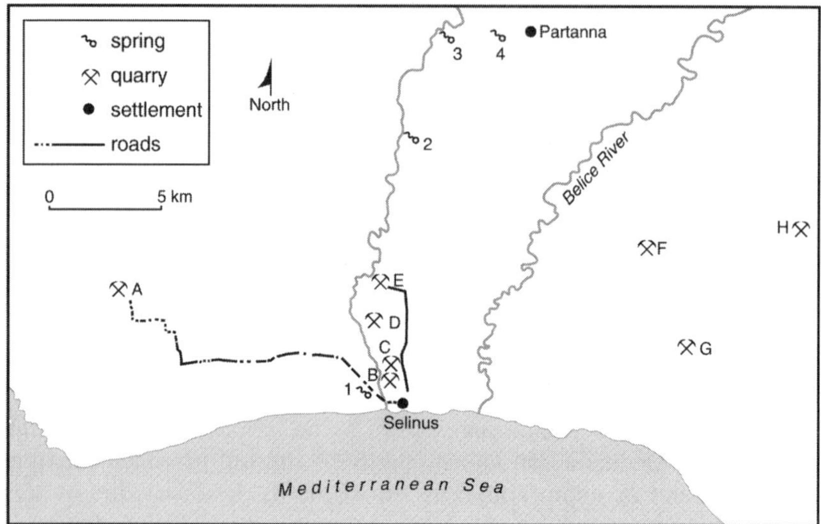

FIGURE 3.22. Regional map with springs, quarries, and roads, from a late nine-teenth-century military map. Major quarries are at A, Cusa to the west; B and C, the earliest quarries Sabato and Buffa, respectively, at the edges of the city; and the first quarry D, called Landaro outside the city. E marks the quarries at Barone; and F, G, and H refer to the last set brought into production at Mirabile, Menfi, and Misilbesi, respectively. The quake-prone Belice River valley lies east of Seli-nus, dividing it from Agrigento's territory. Compare the two major roads with those in Figures 3.23 and 3.25. (Ercoli 1998, redrawn by Langford)

Acropolis. Today, water from other springs of the Gorgo Cottone runs through a newer aqueduct to the hills of the city (Le Dinahet 1979: 48, n. 52). In the case of the Bigini water (Fig. 3.23), the aqueduct paralleled the road from the quarries that were developed in the same area—the Barone quarry or Cave di Barone (Nenci 1979: 1419), also known as "Bugilifer" or "Baglio" ("beam"). The Bigini springs produced 2.9 l/s of water at the end of the nineteenth century, gathered into a round reservoir, the Vasca Selinuntina, 10 m in diameter (Salinas 1884–85) that still survives. The lower part of the reservoir wall is of ashlar masonry almost identical in color and technique to the sixth- and fifth-century river walls in the Cottone Valley. From here, a buried waterline (of which some traces survive as cuttings in the calcarenite), with a slope of 1–6% as was normal for ancient aq-ueducts, led to the city (Peschlow-Bindokat 1992).

Running water in fountains within Selinus was usually reserved for drinking. The rest of domestic water needs were supplied from cisterns. Although Le Di-nahet (1979) reports only two particular houses south of E Street near the main north-south street as having cisterns, I found that every house had either a cistern or a well (see "Water System Elements of the Selinus Acropolis," in Crouch 1993. Fig. 12.5). Wells and shafts are known east of Temple B and in houses along E Street in the north part of the acropolis. Within the city, little channels carried

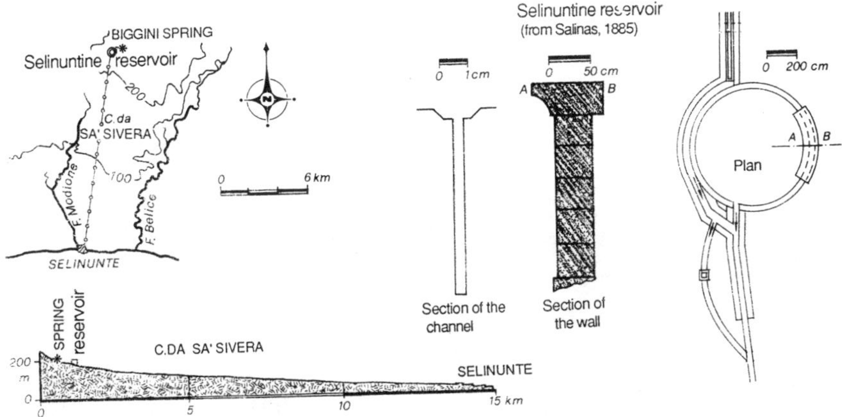

FIGURE 3.23. Bigini water elements, in an area rich in both water and stone (from Salinas's 1884–85 study). The Selinuntine reservoir, in plan at right, survives and is still used. The gentle slope of the land made shipping water and stone to the city equally easy. The profile from the Bigini springs (near the Barone Quarry) to the city is shown as a dotted line on the plan and in profile below. Compare with Figure 3.25.

the water to places of use; for instance, a sixth-century water channel was found under the later stoa of temenos C and D. Other channels provided drainage.

The ample water supply of the *chora* (hinterland) of Selinus was a blessing for the growing of crops and for human requirements—drinking, cooking, cleaning, and crafts such as pottery. At the same time, the streams of water continued to change the terrain as they had for millennia, first creating the two river-mouth harbors and then filling them up with silt, so that by the end of the third century B.C.E. at the latest, the harbors were nearly gone (Fig. 3.24). Ionian cities had similar sedimentation problems. Since the geological history of Sicily is incomplete, we cannot say whether a rising of the interior mountains similar to the upthrust in the hills inland from Ephesus contributed to this silting process.

Quarries and Roads   Constraints of slope and steepness were similar for the delivery of both water and building stone to the city from the outlying quarries, so often associated with springs.

In building any ancient temple or public monument, the builders were constrained to get enough stone of the same good strata to complete the whole building. It would be useful to have a study of how much stone a large temple requires, so that we could estimate how many temples could be built from one quarry.[15] "Three quarries [at Selinus] with different qualities of stone were developed for different uses on the Acropolis" (Collin-Bouffier 1994). On Manuzza alone and convenient for the earliest construction, there were quarries to the north and northwest of the residential area, such as the one at Casa Sabato on the west side (Carrozzo et al. 1992. Fig. 7; Rallo 1984; Mertens 1988–89; Ercoli 1989; Nenci

FIGURE 3.24. Sand bars at the mouth of the eastern river at Selinus, a typical result of the process of sedimentation. (Photo by Crouch)

1979) (Fig. 3.22). The rock of the slopes of Manuzza was quarried for materials for ordinary buildings such as houses (Mertens 1988–89). Perhaps the best stone quarried on Manuzza was a white calcarenite used for roads and ramps (Ercoli 1997).

In the sixth century B.C.E., when the Selinuntines decided to build monumental structures, which demanded large quantities of strong stone, they had to leave the city core and look for quarriable stone in the hinterland. Perhaps the oldest quarry in the hinterland was the Latomie Landaro (Latumiae), which has the shortest delivery route from any of the quarries to the site, whether to the East Hill or along the main north-south street of the acropolis and F Street, and is said to have been used for the earliest construction (Nenci 1979). The rock was coarse in structure and texture with elevated and discontinuous amounts of quartz and iron, which varied the colors and patina, that required plastering; basically, it was a stone one could not trust. It was similar to the sandstone from Manuzza that was used as foundation blocks and in road building. Although the quarry was convenient, it was very hard to get good blocks for sculpture or metopes from Latomie Landaro. Still, the Latomie Landaro and Cave di Cusa are most consistent in mineralogy, with a quartz composition of 5% to 30% (Carapezza et al. 1983).

The development of Latomie Landaro gave impetus to the reconstruction of the roads that had to bear the weight of the stone during delivery to the acropolis, namely, the north-south street and F street. Both were carefully paved to bear the heavy weight of the architectural elements: a solid bed of beaten marl with a thick fill of stones, built during 570–56 B.C.E., according to sherds found in the fill (de la Genière 1982). In contrast, south of the juncture of the two streets, the extension of the north-south road was lightly paved during the sixth century as a pedestrian way, although it was important visually. We may expect to find comparable preparation for heavy transport when the major streets on East Hill are investigated (Burford 1960; Ercoli 1997; Wurch-Kozeli 1988).

In the search for better stone, the builders looked for quarries farther away from the city core. Only a kilometer or so north of the Landaro quarry, the Cave di Barone (Figs. 3.17, 3.22) lies about 6 km north of Selinus. Its stone was already utilized in the sixth century for important buildings such as the large columns of Temple G and again later for the columns of Temples A and O (Nenci 1979). This quarry is very close to the springs at Bigini.

Another major quarry for the monuments was Cave di Cusa (Fig. 3.22) (in Arabic Cave di Ramesear; Nenci 1979), which lies at a distance of nearly 15 km to the west-northwest, west of modern Campobello and higher than Selinus. The distance of the quarry from the city indicates Selinus's dominance over the chora or hinterland. The diverse geology of this quarry provided good-quality stone for construction of Temples C, D, E, F, and G (Nenci 1979). The stone's consistent grain, color, and cementation provided reliable building elements, but because of discontinuities (Carapezza et al. 1983), which can cause cracking or sudden changes of color and texture, and especially because of the presence of quartz, the stone was less useful for metopes and sculptures. A horizontal quarry, Cave di Cusa offered a great expanse of rock, 1300 m long, 70 m wide, 7 m deep, approximately 60,000 m^3 of material. Nenci (1979) has discussed in great detail the process of quarrying here. The technique of work was unlike marble quarrying because it was done without scaffolding and without stairways. [See Durkin and Lister (1983) and Herz and Waelkens (1988) for a discussion of quarrying in Roman quarries.] Still visible here are a number of partially excavated column drums whose size evoked the exclamation "stupefying!" from Carapezza et al. (1983).

From Cave di Cusa to the city (A in Fig. 3.22), the transportation challenges were somewhat different from those of the northern quarries, with obstacles of marshes and sand dunes. The road at first runs southeast in an area where it is rock out through but then runs east-west parallel to the sea. Maps, produced by the Instituto Geografica Militare in 1880 before land redistribution that altered the road pattern in the area, show a road just *above* the dunes west of the city (Ercoli 1997), rather than on the more southern route in, or seaward of, the dunes, as suggested by Peschlow-Bindokat. Geologists Amadori et al. (1992) found evidence of an ancient trench, dug and then filled, beyond the northern edge of Temple G, which suggests that the ancient builders were careful to discern whether they were building on soil or stone. The evidence — visible remnants of old roads, slope, distance, difficulties of terrain, and availability of water to combat worker fatigue and to lubricate the rollers — converges, confirming that the ancient road was that identified by Ercoli (Figs. 3.22, 3.25). This route is both 1.6 km shorter and flatter in the central stretch than Peschlow-Bindokat's route. Fifty percent of this road is traceable on an 1880 map. Reaching West Hill, the road ran down the east slope to the Modione River valley, across the river on a bridge, up over the ridge of the acropolis, down into the second valley, across another river, and finally up the

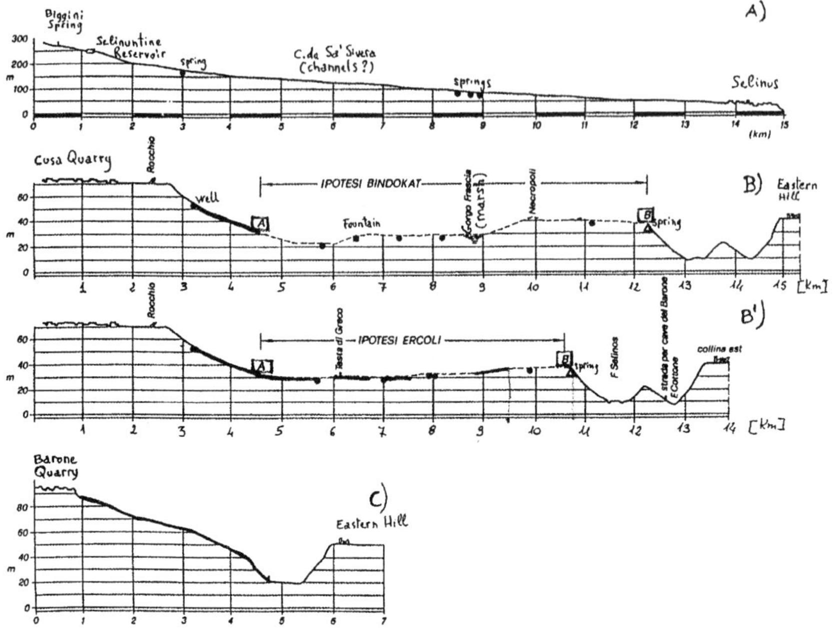

FIGURE 3.25. Profiles of roads from quarries to Selinus. The upper profile (A) from Bigini repeats that in Figure 3.23, which is also the route of the waterline. Compare with the route from Cusa quarry to East Hill (B) proposed by Peschlow-Bindokat and the alternate proposed by Ercoli (B'). The fourth profile (C) is the much shorter and steeper route from the Barone Quarry to East Hill. Heavy lines along the top of each profile indicate remaining physical evidence for the route (Ercoli 1997; Ercoli and Crouch 1998).

west slope of East Hill. Roman remains of a bridge are still visible (Siviglia 1979), part of their coastal road around the island.

Another set of important quarries includes those of Cave di Portella di Misilbesi and Cave di Menfi (Fig. 3.22), both beyond the Cottone River valley east of the Belice River valley F, G, and H on Fig. 3.22. The first quarry lies at a distance of 20 km (Carapezza et al. 1983) and the second 30.2 km (Ercoli 1997) northeast of Selinus in the territory of Agrigento. Cave di Misilbesi has many varieties of calcarenite (Carapezza et al. 1983: 39 noted perfect chemistry, mineral composition, texture, and internal structure for fine construction). For metopes the sculptors needed compact rock of homogenous mineralogy, which allowed for "lightness and sweetness in the nude female bodies" (Amadori et al. 1992; Carapezza et al. 1983: 36), such as the honey-colored stone from the *Menfi* quarry. To the naked eye, indeed, the metopes seem different in texture from the rest of the stone of the temples, which is probably how the ancient builders distinguished between stones of such physico-chemical similarity. The fine-grained biocalcarenite with incorporated fossils used for metopes has only microscopic differences from the stone selected for other building elements. Carapezza et al. (1983) studied the stone by microscope, X-ray diffractometer, spectrometry, chemistry, and chemico-physical tests. Statistically significant findings led them to the conclusion that the stone from this quarry is closest to that of the sculptures.

The superior quality of this stone made it essential to overcome the route's problems in transporting heavy stones (Ercoli 1997), but no ancient road from these three quarries to Selinus has been rediscovered. The terrain along this way is difficult, with many ravines and small rivers to cross. Scholars view the Modione River and other streams only as obstacles in the transport of materials to Selinus (Nenci 1979), except that Pace (1938) wondered whether the ancients brought blocks by river to the site during the winter floods. We will see at Ephesus that the local rivers there were indeed used for the transport of stone.

Selinus thus utilized the stone of at least seven quarries in her three centuries of active construction. The builders accommodated their designs to the capacities of the stones, achieving a distinct local style. Calcarenite lacks the inherent strength of the marbles of the Greek mainland, so an aesthetic of strong, bulky forms arose to exploit the local stone's structural qualities. The stone from the major Selinus quarries comes in white, gray, yellow, rosy, brown, and varicolored calcarenite. The stones exhibit various grades of cementation, some with fossils; grains are fine to coarse; rock is hard and compact with layers of clay-sand-silt and therefore is friable; most are 30% quartz and 70% carbonate minerals. The highly variable homogeneity and assortment of minerals mean that the calcarenite is inconsistent in quality.

Design and History Related to Geology

In several ways the geology of Selinus has been closely related to its human history. The rivers were conveyors for trade to the interior but also carried the mountains to the sea. For a time the process was in stasis, so that from the seventh through the fourth century B.C.E. the Selinuntines benefited from the earlier results of the process and defended themselves fairly well from the continuing geographical change. The river walls, protection against winter floods, are perhaps the most striking evidence of their ability to live in balance with nature. Eventually, however, natural processes combined with political events to bring ruin and desertion to the city. For success, traders and architects require peace, and they mostly had peace during the first 250 years of the settlement's history. But beginning about 400 B.C.E., the problems of defense in the lives of the Selinuntines replaced the challenges of building monumental architecture.

The silting up of the valleys and the growing problem of malaria likely contributed to the demise of the city (Figs. 3.18, 3.24). The Greeks knew malaria in three forms, two benign and one malignant. The disease had been known in the Mycenaean era, then disappeared, reappearing in the fifth century. Many think that outbreaks of malaria could explain the abandonment in the Roman era of sites like Paestum in Italy and Selinus in Sicily that had flourished in the Greek era (Collin-Bouffier 1994). Thirgood (1981: 64) suggests that Carthage introduced malaria into Sicily in the fourth century B.C.E. In the first century, Strabo (III.5.12) reported that the disease was a new factor, known since the end of the third century.

No city from Sicily to the Black Sea could boast a finer combination of vista, site, and splendid architecture—except perhaps the nearest neighbor, Agrigento. Even the curtain wall around the acropolis and the elaborate structures of the North Gate, both probably from about 400–300 B.C.E. (Mertens 1996), show elegant handling of the local calcarenite.

Just as they were resourceful about the opportunities and problems of the rivers, so the builders of Selinus were resourceful about using the building stone found nearby, making their city one of the great showplaces of the Greek world. Bringing one quarry after another into production as demanded by the construction schedule, they produced twelve temples in less than 200 years. Their local tradition was to incorporate Doric and Ionic decorative elements into temples designed for religious processions and other rites (Mertens 1996); these temples were elongated and had larger aisles than contemporary temples in mainland Greece. Since the stone was not as strong as Pentellic marble, the designers compensated by increasing dimensions, giving the temples a massive appearance that I think is not due solely to experimentation or provinciality.

Unfortunately, the intense marine aerosol (particles suspended in the wind from the sea) at Selinus has led to degradation of the stone. Because the surface of the coarse stone was subject to weathering, builders protected columns with a stucco made of limestone dust; where the stucco clings, the surface is still protected (Fig. 1.1). To preserve the metope carvings, it has been necessary to remove them from the site to museums (Carapezza et al. 1983: 40).

Less exacting were the standards for building materials for houses. During the period 625–590 B.C.E., houses had foundations of small stones mortared with clay. Walls were clay reinforced with matting (a kind of wattle-and-daub), and roofs were tile for permanency, fireproofing, and storm protection. As the city prospered during the period 550–490 B.C.E., people demolished the old buildings and built on the acropolis and its slopes more elegant new houses of ashlar laid in equal courses (Le Dinahet 1979).

The population of Selinus remained sparse in Roman and Byzantine times. Deities continued to be invoked for protection against natural disasters and human enemies: Some evidence of a Christian cult at Gaggera is known from a carved monogram of Constantine found there (Le Dinahet 1979). An earthquake or some such natural disaster in the sixth century C.E., mentioned in the Byzantine sources (Ercoli 1989), caused major damage (de la Genière 1982). In the sixth century C.E., Selinus was used as an occasional base for Byzantine attempts to reconquer the western area of the Roman Empire. Many acropolis houses continued to be reused then and into the Middle Ages. During the eighth and ninth centuries, a series of earthquakes—the worst one in 852—occurred to the northwest near Castelvetrano (Naselli 1972; Amadori et al. 1992: their fig. 2, a map of the epicenters), apparently destroying the little Byzantine chapel in the ruins of the acropolis (Mertens 1989). The ramparts of Selinus were thrown down, and the fort built over Temples A and O were destroyed, as were the Byzantine houses nearby (Mertens 1988–89, 1989). During the ninth through eleventh centuries, the Arab rulers of

Sicily utilized the quarries of the site, using the much smaller river mouths as harbors. On East Hill, a twelfth-century earthquake ruined Temple E and damaged the other two temples, raining down capitals and beams on the recently built houses along their steps and into the area between the columns and the cella wall. Aragonese rulers from Spain (thirteenth–nineteenth centuries) reused the quarries for their grand architecture, shipping stone out of the still smaller harbors. The site then sheltered hermits and a small Christian community (Ercoli 1989), as indicated by ceramic finds of thirteenth- and fourteenth-century sherds (Naselli 1972; Mertens 1989). Even later, in the eighteenth century, materials from the ruins were used to build neighboring houses (Ercoli 1989), and some columns were taken to Paris for reuse.

Sicily is now classed as having intermediate seismicity, especially to the northwest of Selinus, with some intense zones such as the Belice Valley immediately to the east, where a major earthquake took place as recently as 1968 (Amadori et al. 1992; Bosi et al. 1973). Ancient Partanna north-northeast of Selinus, for instance, was largely ruined in the latter earthquake (Kininmouth 1972). Some fierce gales of the early twentieth century laid bare the remnants of the ancient harbor walls (Carapezza et al. 1983). The site is now visited by only a few tourists.

Syracuse

Setting and History

The political and cultural history of Syracuse cannot be understood in isolation from its physical setting. In the following account, three topics will exemplify the interlocking of human and geological history: the geographical setting illuminated by the history of the ports and the plains to the west, the hydrogeology, and the regional and site geology.

The southeast corner of Sicily was occupied from 1425 B.C.E. by Sikels, first at the site of Syracuse and all around the bay by 1000 B.C.E. (Fabricius 1932). Greek traders arrived before the mid-eighth century B.C.E., setting up their trading post on the teardrop-shaped island they called Ortygia (Italian: Ortigia) (Fig. 3.26). The early traders brought jewelry and bronzes, as well as wine and oil in plain and decorated amphoras. From the island base, they explored the trade routes along rivers to the interior (Collin-Bouffier 1987). The Syracuse area, or Syracusana, had five rivers, all originating in the Hyblean Mountains, the Anapos being the longest. These rivers were navigable in the river craft that the Greeks used, making possible easy expansion into the hinterland. So significant were the rivers that they were scenes of battles to control both the streams and the lands they watered; one such example is the long struggle between Selinus and Segesta (Diodorus XIII. 43–44, XXI. 82.3).

Colonists followed the traders to Ortygia. Greek colonists looked for earth and rock formations that reminded them of "home" (Crouch 1993; Collin-Bouffier 1987: 677: "Les colons d'archais trouvant à Syracuse une situation geologique comparable à Corinth"). Practically speaking, existing technologies for construction, water management, agriculture, and pottery making would work better if the new

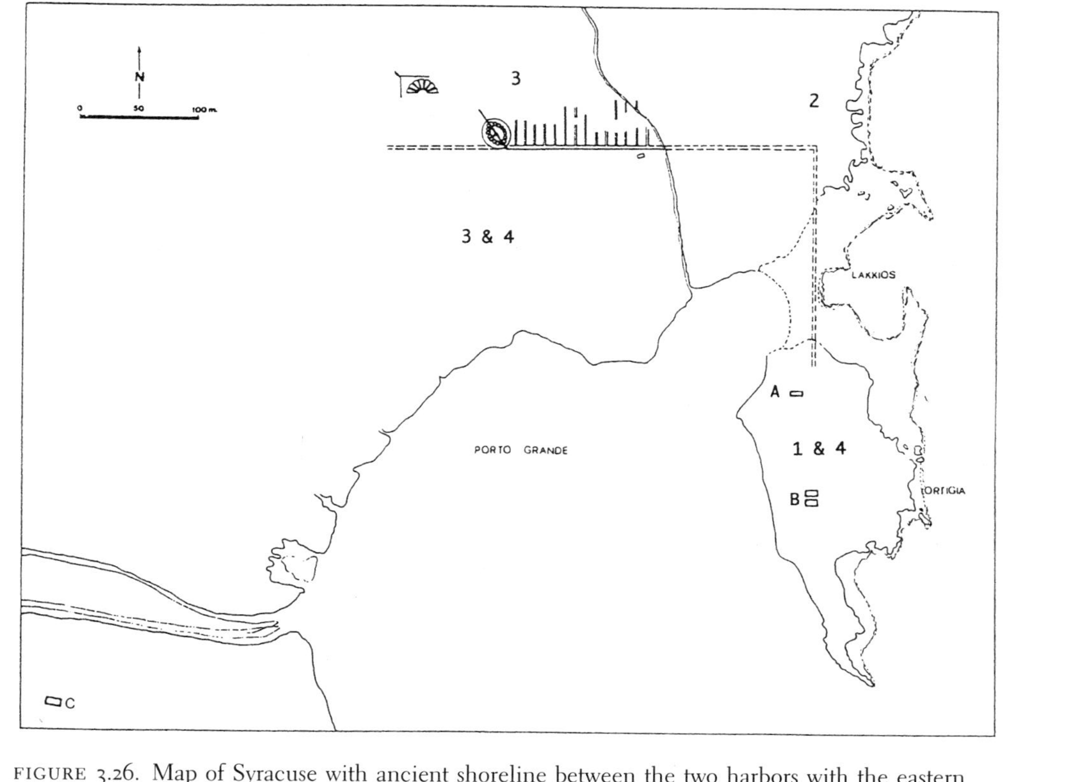

FIGURE 3.26. Map of Syracuse with ancient shoreline between the two harbors with the eastern one marked Lakkios; the great bay is incorrectly marked Porto Grande, a title that should be reserved for the harbor between A and the little river from the north. The mole from Ortygia island at the lower right to the mainland at the upper right is shown as a double dashed line continuing as the major street of the early expansion of the city. Greek and Roman settlements and major temples: (1) first Greek settlement; (2) early Greek expansion; (3) later Greek expansion in the theater district; (4) Roman settlement; (A) Temple of Apollo and (B) Temple of Athena and the Ionic temple, all on Ortygia; (C) Temple of Zeus in the plains at the lower left (after Crouch 1992)   ancient coastline (according to Gargallo 1979)

site closely resembled that for which the technology had been developed. The manufacture of pottery at both Corinth and Syracuse relied on suitable and accessible beds of clay (Qa of Fig. 3.27) (Aureli et al. 1997). The stones also were similar to those the colonists knew how to work. The stratigraphy of Syracuse is as follows:

A. Beaches and alluvial sediment

Mc. White-gray limestone of the Mount Crimiti Formation constitutes the Epipole plateau (1–2 × 3 km) and alternates with clay in quasi-horizontal layers; often karstic, with springs along the contact with the underlying volcanic layer (Cv). Seen at the Euryelos Castle, the Theater, the Amphitheater, and Scala Greca. A lower terrace, terminating north of Ortygia, contains the San Giovanni catacomb.

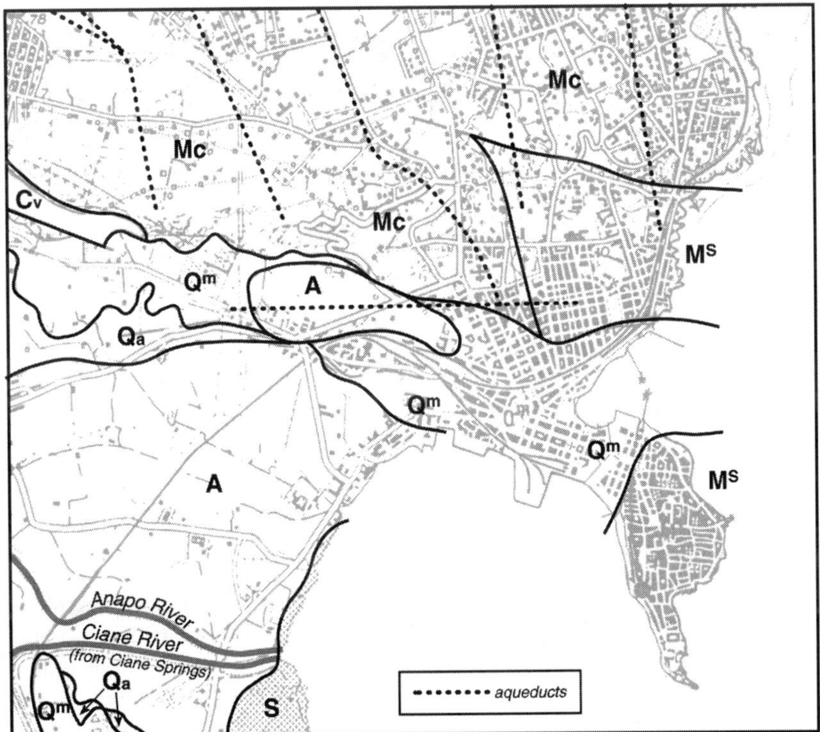

FIGURE 3.27. Geological map of modern Syracuse: (Mc) greenish grey to white, Mount Climiti calcarenite, often karstified; (M^S) friable calcarenitic marls of the Mount Carrubba Formation; (Cv) purple volcanic undersea stones; (Q^M) yellowish sand and calcarenites with crossed stratification; (A) alluvium; (Qa) marly gray-blue clay; (S) swamp. The aqueducts bring water from the mountains to the west. (Redrawn by Langford from Aureli 1994 with aqueducts superimposed from Crouch 1993)

Ms. Friable Quaternary calcarenites of the Mount Carrubba formation, found on Ortygia and north of the Archaeological Museum, are from marine transgressions. Locus of catacombs such as Vigna Cassia.

Calcareous series makes up the bottom of the grabens and outcrops in the horsts.

Qc. Fossiliferous calcarenites and sand, massive or with crossed stratification, flanking the Anapos River and extending under the bay.

Qm. Recent (500 years) alluvial deposit of yellowish sand and calcarenities with crossed stratification, forming the isthmus connecting Ortygia today with the mainland and also forming beaches and alluvial sediments along the rivers and canals.

Qa. Marly, gray-blue, middle Pleistocene clay, running along the Anapos River, resting on more recent calcarenities, sands, and alluvial sediments; in some places located behind Qc, away from the river.

Cv. Volcanic undersea extrusions appearing between Mc and Qm, and at higher levels of Mc. Substrata of volcanic rock, marl, and clay.

Like the preceding traders, the colonists preferred a peninsula connected by a narrow isthmus to the mainland or an off-shore island, not only for safety against attack from the indigenous peoples, but also so they could have a sheltered harbor no matter which way the wind blew. We will see the same preference at Miletus and Ephesus.

The *oekist* Archios, of the ruling Bacchiads at Corinth, and his group arrived at Syracuse in 733 B.C.E., probably preceded by traders. The colonists were Corinthians of several classes and land-hungry farmers from Tenea in the hill country south of the Corinth plain (Strabo VIII 8.6.22). Their thirst for trade goods and land was encouraged by a Delphic oracle, who told them where to look for a good site:

> Somewhere in the misty field of the sea where Ortygia lies by Threnakia,
> Alheios' bubbling mouth intermingles with Arethusa's streaming water-springs.
>
> quoted by Pausanias (IV.4.3)

Archaic Greek society understood that founders of cities needed legitimization as arbitrators of the social order. They frequently applied to Delphi for instructions and approval, drawing on the oracle's practical wisdom about founding a city and about proper urban behavior at the new site.

Close control of land — the basis of survival and wealth — for the benefit of a few founders took place in some colonies. During the first period of aristocratic rule, sale of land was probably permitted but rare. Rationing of land was bound to cause unrest among nonelite settlers, leading to an extended period of tension (Roebuck 1979: 1928, n. 10): "The causes of social crisis in the seventh century [in Sicily, lay] in the pressure of a growing population on a virtually closed system of land tenure." The tension over land continued into the late archaic period (sixth century), as witnessed by the second revolution in Syracuse when the *demos* (people) and the serfs ousted the Gamoroi ("land-sharers") oligarchy. In the Greek homeland, the same kinds of problems were causing unrest and contributing to the willingness to colonize. Some Greek colonies set aside a reserve of land for later settlers.[16]

The new Greek city had courtyard houses, the basic type in Sicily, and these are still evident in the former Jewish quarter of southeast Ortygia (Zampilli 1979). Houses generally had wells, as the freshwater table was easily reached by piercing the few meters of calcarenite. The dead were provided with necropoli on the mainland. Built in 734 B.C.E., a necropolis on Fusco Hill had more than 700 graves; it went out of use at the end of the first century B.C.E. (Frederiksson 1999), and a second was opened much closer to the later Santa Lucia district to the east (Fabricius 1932).

Although no public buildings have survived from the seventh or sixth centuries, by 648 B.C.E. (Coarelli and Torelli 1984) quarries (*latomie*) came into use, indicating that monumental construction had begun. The settlers built three major temples on Ortygia during the sixth and fifth centuries B.C.E., one to Apollo or to Apollo and his sister Artemis, one to Athena (over an eighth-century sanctuary), and the third, an Ionian temple standing next to the Athena Temple, of unknown dedication (Ciurcina ca. 1985). The three temples were strung out along the north-south axis of the teardrop-shaped island.

In the plains southwest of the city, the Temple of Zeus Olympias was built in 491 B.C.E.; open plains were often chosen as sites of temples to Zeus. Perhaps the 492 battle against Gela, which had taken place in those plains (Collin-Bouffier 1987), reminded the Syracusans of the importance of wetlands to their pastoral economy and the necessity for the gods' protection.

Though swampy, the area was valuable, especially in the northeast sector of the plain. The people derived many benefits from the marshy prairie, among them grazing in spring, fishing, especially for eels; and harvesting reeds and canes for building. The construction of the Zeus temple was a way of declaring Syracusan dominance of the plain.

These wetlands also played an important part in the defense of the city not only before the long walls were built at the beginning of the fourth century, but also during later warfare. An invading army had to camp on the western shore of the bay southwest of the urban core of Syracuse, because there was no other place near the city with an unlimited water supply and facilities for a great fleet. There was an anchorage, much less sheltered, to the north of Syracuse but no water sources along the shore before Megara Hyblaea, about 10 miles north along the coast. The level plain southwest of Syracuse had two water sources: the long Anapos River, which flows even in summer, and the short Cyane River (today the Fiume Cardinale), which runs near the temple of Zeus Olympias. Since the Anapos River occupied a shallow bed, it was greatly affected by floods and tended to form natural lagoons. This poor drainage led to stagnant water that made the area susceptible to malaria and other illnesses. Marsh water is not intrinsically stagnant but becomes unhealthily so from the heat of the sun (Villard 1994) and is worsened by poor drainage. Evidence from Ephesus and Miletus suggests that stagnant water that is a mixture of fresh and salt water is especially likely to breed malaria mosquitoes.[17] Villard concludes from examination of the accounts in ancient literature that the epidemics required a combination of geography, heat, and a very large number of soldiers. This combination, he wrote, "chastises the enemies of Syracuse" (page 338; see also Schubring 1865).

Other epidemics—typhus or smallpox—during the sieges of 396 and 212 B.C.E. afflicted invading Carthaginians, according to Diodorus (XIV.71.1) and Livy (XXV,26.8). In 212, the Carthaginians camping on the shore of the bay suffered fully from the unhealthy place, but the Romans on the heights of the plateau of Epipolai ("high field")[18] within the outer walls of Syracuse escaped much of the epidemic; they had already been there two years and were acclimated to the air and the water (Schubring 1865; Villard 1994), but they also had better drainage and better access to clean food and water. Efforts to control the drainage and improve health by building channels have continued from antiquity until the present (Aureli and De Maio 1994).

After the defeat of Athens and the war with Carthage triggered by that defeat, the new ruler Dionysos built new defenses for the increased population. In 402 B.C.E. he set the city's farthest wall much farther out to the northwest connected to a great new fortress at Euryalos, where an isthmus leads to a higher plateau to the west. Euryalos is the grandest fortress surviving anywhere in the Greek world from this period. The new rampart incorporated the plateau into the city, taking in the theater and the quarries, and thereby adding another 215 hectares (Drögemüller 1969) to the city. In 397 a southern wall was built from the fort over to the area of the theater. The northern wall was repaired in 356, and monumental gates were built, for example, the Hexapylon ("Six Gates") near the Scala Greca (Greek stairs), where the road ran north to Megara Hyblea. These two long walls made Epipolai not only a reserve area of agricultural land, but also a fortified area into which people from Syracusan villages could flee with their families and cattle in case of war. Materials came from quarries later rediscovered in this area, such as the one at Cave di Santa Panagia north of the Latomie Casale and just southeast of the Scala Greca (Fabricius 1932).

Although Syracuse never filled the entire area within her defensive ramparts, during the Greek centuries she expanded gradually onto the mainland. The population in the early fifth century was more or less 45,000, of which 12,000 lived on Ortygia. By 413, Syracuse had become a major city of the Greek world (the second city, according to Berger 1991 and Drögemüller 1969) after the rulers forced the inhabitants of several other towns to move to Syracuse. By 440, Syracuse had no rival in Sicily; even Plato was happy to live here. New quarters and cemeteries, as well as new aqueducts and public buildings, developed on the mainland (Fabricius 1932, Drögemüller 1969). Older areas, such as the quarter above the Small Harbor (*Lakkios*) and the northern agora, were rebuilt. The population in 300 B.C.E. was probably closer to Diodorus's figure of 80,000 (XIII.134.1; XIII.43–44, 63.3; XIV.7.1, 70.5, 71.1, 421; XIV.42) than the 200,000 inferred by Fabricius (1932).

From 304 to 215 Syracuse was a significant intellectual and artistic center, having contacts with Greece, Asia Minor, Egypt, Carthage, and Spain. Tyrants such as Hieron reinforced their glory by building or rebuilding monuments like the theater in 238. He also sent architects to redesign the agora of Morgantina (Berger 1991; R. J. A. Wilson 1990, Coarelli and Torelli 1984).

After the Roman conquest of 211–201 B.C.E., Syracuse was smaller, about 325 hectares (Drögemüller 1969), although it was now the provincial capital of Sicily. The Greek residential area north of the east-west street that ended at the amphitheater was abandoned, but the area to the south of that street continued to be occupied. In the late first century B.C.E. civil war between Octavian and Pompey, Syracuse was nearly destroyed and abandoned (Coarelli and Torelli 1984). Octavian Augustus reconstructed the town in 21 B.C.E. as a veterans' colony, extending it as far as the outskirts of Acradina on the mainland (Strabo VI.5.22). In the first century C.E., the Emperor Caligula repaired temples and ramparts, while Claudius improved the water supply and added a bath on Ortygia. The Roman bath near the present railroad station was made of blocks from the Dionysian wall (Fabricius 1932). An amphitheater near the Altar of Hieron used the local white limestone rather than the black basalt used in the amphitheater at Catania (R. A. Wilson 1979); the area of the amphitheater may have been a quarry earlier.[19]

Eastern Sicily was well located for trade with the imperial capital at Rome, the eastern Mediterranean, and Africa. The Roman economy of Syracuse was based on timber, grain, lamps, textiles, wine, cattle, hides, wool, saffron, oil, fruits, fish, sulfur, and transshipment of alum from the Aeolian Isles just off the north coast (R. J. A. Wilson 1990). With ample beds of clay and nearby forest for fuel, Syracuse produced the widest variety of Roman lamps (found in the catacombs), a "long but not distinguished production" from the late third century B.C.E. to the ninth century C.E., and also exportable brick and tile (R. J. A. Wilson 1990: 268–70).

Roman Sicily was a "haven of almost unbroken tranquility" (Wilson 1990) from 211 B.C.E. through the fifth century C.E., with only three armed conflicts (a slave uprising led by Spartacus in the early first century B.C.E., a Frankish raid in 278 C.E., and Vandal incursions beginning in 440). Syracuse suffered, however, from natural disasters in the fourth and fifth centuries C.E. A devastating earthquake on July 21, 365, which was probably centered in the sea near Crete (see Table 1.1) generated a tsunami that damaged Syracuse, and even the palace at Piazza Armerina in the center of the island was shaken.

Geology

Syracuse's favorable situation is obvious by its geography, petrography, and hydrogeology. Examples of the geographical advantages may be seen in the two ports and in the marsh lands, which were both productive and defensive. We will discuss lithography in terms of the quarries for building stone, and hydrogeology as it pertains to water supply.

Ports   Aureli defines the bay of Syracuse as a graben (a downthrown, crustal block bordered lengthwise by normal faults), running immediately northwest of Ortygia (compare the shorelines in Fig. 3.30). The graben continues north of Syracuse along the coast leading to the Augusta peninsula (Aureli and De Maio 1994, Aureli et al. 1997, Gargallo 1970; Drögemüller 1967). In the mid-eighth century B.C.E., Ortygia was a rocky island oriented north-south, lying immediately off shore. Today, the island has become a peninsula running north-south, attached to the mainland by a NW-SE isthmus. Change of sea level here is only slightly due to eustatic variations, but mainly to tectonic submergence of the edges of Ortygia and uplift of its center during geologically recent times. The Small Harbor northeast of Ortygia shows 3 m of submergence, and this submergence continues all around the island (Gargallo 1970; Kapitan 67–68; Voza 1976–77). In the Siracusana the most notable evidence of submergence is at the Plemmyrion peninsula to the south of the bay, where landings for handling quarried yellow "sesame seed" calcarenite for archaic monuments are now submerged, and furnaces for burning limestone for mortar are now at four m below sea level. It is not always easy to differentiate tectonic submergence from transgression of the sea. Such active transgression is revealed in erosive and corrosive forms in the shelly limestone of the coastal areas. All along this SE coast of Sicily, "where the coast meets outlets of the larger water courses, especially those having greater erosive capacity, the sea penetrates for short distances inland, forming a 'ria' type of submerged coast" (Basile et al. 1988, p. 151).

As early as the sixth century B.C.E., a man-made mole or bridge mentioned by poet Ibycus (cited by Gargallo 1970) connected Ortygia with the mainland to the north. The original use of Ortygia as a defensive outpost was thus sacrificed to the convenience of access to markets and homes on the mainland. The mole or bridge's relation to the two ports (Fig. 3.26) was described by Cicero in the first century B.C.E./C.E. (*II Verres*, Book IV of the Second Pleading) who said the ports were

> almost enclosed within the walls, and in the sight of the whole city; harbors which have different entrances but which meet together. . . . By their union a part of the town, which is called the island, being separated from the rest by a narrow arm of the sea, is again joined to and connected with the other by a bridge.

Vestiges of the ancient harbor works, such as encircling walls, foundations of warehouses, and remains of the old mole or bridge leading north, were located in the 1880s (Cavallari and Holm 1883: 28–29). These ruins lie along the coast, east of the presumed limits of Small Harbor (Drögemüller 1969). Cavallari located foundations of the arsenal for military ships and the naval yard along the north of Small Harbor, south of Santa Lucia, as described by Thucydides. Further evidence of the early arrangements has been discovered under the Santa Lucia area, where wells and cisterns contain sherds of Corinthian pottery from the archaic period (Fabricius 1932; Bongiovanni 1996). The present Piazza Santa Lucia may have been an open area accommodating the traffic from the isthmus (Bongiovanni, personal communication). In the first century B.C.E. Diodorus had ascribed harbor installations to the ruler Dionysos of the early fourth century (Zampilli 1979: 42). Since trade was an important component of the Syracusan economy, piers, warehouses, and other provisions for berthing ships and handling goods were necessary. The subsidence of Ortygia's shorelines has put many remains beyond the reach of excavation. When the sea level changed as much as 3 meters, Small Harbor's shoreline was drastically changed, and the size of Ortygia diminished.[20]

The ancient mole or bridge in time became an obstruction, partly covered by the sea, and was finally destroyed by earthquakes in the sixteenth century, probably in the great quake of 1542. The earthquake together with the currents and backwash from the rivers created the sandbar between the harbors, which gradually encased the mole in solid land. Similarly, Lade Island was enveloped by the plain at Miletus as was Ayasoluk Island at Ephesus/Seljuk. Holy Roman Emperor Charles V used the new land between Ortygia and the mainland as the site of a Renaissance fortress. The streets dating only from the last 400 years connect across the present isthmus. Small Harbor — Porto Piccolo — now visible to the east of the modern isthmus, incorporates the eastern part of the ancient Great Harbor and the western part of the ancient Small Harbor. Schubring (1865) pointed out that the needs of the Romans to reach their Praetorium (near the present railroad station) and the Bath of Daphne (in the same area as the potteries, west of the isthmus) made necessary either an access road from the northern end of the ancient mole or bridge or a more direct route to the northwest where the present isthmus lies (Bongiovanni interview 1994; Coarelli and Torelli 1984).

The name "Great Harbor" belongs not to the whole bay stretching west from Ortygia over to the plain where the Zeus Temple stood, but rather to the more limited but still spacious area at the northeastern edge of the bay. Both bay and harbor were altered prehistorically by actions of the Anapos River, which carved out a great channel or canyon in the sea bottom. A less dramatic but later and more visible result of erosion was the deposit into the bay of debris from the ancient creek lying between the Casale Latomie and the site of the present Archaeological Museum. The deposits were carried by coastal currents along the shore eastward, gradually filling in the waterway between the island and the mainland (Aureli et al. 1997). Great Harbor to the west of the ancient isthmus was intact until the mid-sixteenth century C.E.

Lithography and Quarries   Syracuse was like Corinth in having quasi-horizontal layers of limestone and clay that sloped gently from WNW to ESE, across a plateau 1–2 km × 3 km (Fabricius 1932) to the sea, where it terminated at the island of Ortygia. The groundwater could come to the surface along the seams between the layers of stone and clay. Yet, unlike at Corinth, the limestones at Syracuse consisted of a sparsely karstified upper layer over a harder lower layer. At Corinth, some sandstones and cemented conglomerates alternate with clay to produce karstic water resources. This loose parallel suggests that the similarities sought by the surprisingly sophisticated colonists were those of geological process rather than of lithology (Aureli et al. 1997). (Stratigraphy is set out on pages 91–92 and in Fig. 3.27.)

With such a variety of available stones, relatively little had to be imported for construction projects. Most vivid, perhaps, of the imported stones was the pink breccia with white veins brought from Taormina for Hieron II's bath (Wilson 1990). Coarelli and Torelli (1984) estimate that at least 4,700,000 m³ of stone was extracted from Syracuse quarries in classical and Hellenistic times. About 850,000 m³ of building stone came from the Paradiso quarry; the same amount from Cappuccini; 700,000 from Intagliatella and S. Venera; 450,000 from Casale; and 1,850,000 from other small quarries. Some quarries were open pits, some partly underground, and some mixed in form. They were developed according to the quality of stone and its proximity to building sites. The theater and amphitheater may earlier have been quarries. The Cappuccini quarry gave material for the Tyche district to the north. Other northern quarries were Casale and Santa Venera. Outlying quarries included the Cave di Santa Panagia, some 130 m deep and 1600 m wide, lying near the north shore amid layers of basalt and clay (Drögemüller 1969), and the Buffalero at the west of Epipole. The latter produced stone for the ramparts and the buildings at Castle Euryalos (Cavallari and Holm 1883).

The largest of the quarries was the Paradiso, 45 m deep, yielding the best white limestone with fine grain for monuments (Coarelli and Torelli 1984) (Fig. 3.28). Although no longer accessible to tourists because of collapse damage from twentieth-century earthquakes, this great void is still impressive when viewed from the edges of the site. It incorporates the 23-m-long, S-shaped "Ear of Dionysos," where prisoners were supposedly kept after the Athenian defeat.

At Syracuse, stability is a major environmental problem, affecting both the artificial cliff faces in the latomie and the aqueducts and catacombs. Both are negatively impacted by earthquakes, and the latomie especially by the weathering increased by air pollution. In an attempt to distinguish natural from human actions, Ercoli and Speciale sampled a cliff face of 3000 m² on the east side of the Latomie Paradiso. This quarry was begun in the seventh century B.C.E. by the Greeks, who utilized discontinuities for ease of quarrying. Substantial alteration of the ground surface has been evident in city maps and views since the seventeenth century. Since 1800, for instance, 1–10 m of soil has been deposited on and around the fallen blocks in the bottoms of the quarries. Ercoli and Speciale (1988b) found three kinds of changes in the latomie: morphogenic processes of thousands of years; long-term consequences of human actions (hundreds of years); and abrupt recent environmental changes (Fig. 3.29). Morphogenic developments include the discontinuity patterns that separate the rock mass into prismatic and slablike blocks, which facilitated the excavation of the quarry. Later failures of the overhangs are human-induced morphogenic actions. These failures interact with others that are structurally controlled, such as collapse in earthquakes of over-hanging walls or ceiling rock. The stability of blocks depends on the rock's tensile and shear strength, block interlocking, and the pattern of discontinuities; if any one of these factors deminishes, stability is threatened.

Even within one quarry, such as Paradiso, the lithological sequence in the calcarenite varies in color (white, light brown), in grain (fine to coarse), in porosity and cavities, and by weight and strength; mechanical properties of the stone are related to its weight. Although the rock everywhere in this quarry is rather disintegrated, both on the surface and in depth, Ercoli and Speciale (1988b) discerned three lithotypes, independent of color, texture, and cementation: one type of low porosity, another of homogenous, very high porosity, and a third having uneven, large cavities. The large variation of physical and mechanical qualities depends

FIGURE 3.28. Quarry (latomia) in Syracuse. Cut into karstified limestone, the quarry collects water draining from the plateau. The stone of the piers splits off, forming reverse stair steps. This photo was taken in 1977. Earthquakes have caused ceilings and piers to fall in, so tourists are no longer allowed to enter the area. (Photo by Crouch)

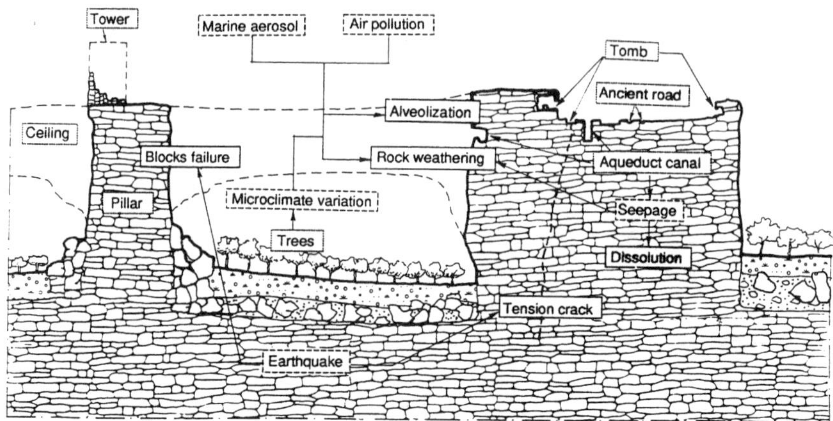

FIGURE 3.29. Stone disintegration mechanisms in the Syracuse quarries: Contributing natural factors (in rectangles with dashed borders) are marine aerosol, air pollution, microclimate variation, seepage, and earthquakes. Effects are block failure, alveolization (cavities), rock weathering, dissolution, and tension cracks. Human constructions are the tower, tombs, ancient road, and two aqueduct canals. The trees are either volunteers or an orchard. The "pillar" is actually a remaining pier from quarrying, meant originally to continue supporting the "roof," that is, the ground surface now fallen in, leaving piles of large rocks at either side of the pier (Ercoli 1988b).

on the sizes and shapes of particles in the rock and their relationships, as well as the heterogeneity and degree of alteration or weathering. Discontinuities here are characterized by three subvertical sets and by the bedding planes. The subvertical discontinuities are normal (perpendicular) to the stratigraphy. Some persist almost to the top of the quarry wall. Several meters apart, they are irregularly spaced, and one extends to at least 15 m below the bottom of the quarry. In these conditions, the prismatic and slablike blocks spall off from the bottom, forming upside down "stairs." When the pillar is no longer supported at the bottom, even a small seismic event may be enough to bring crashing down the pillar and the part of the roof it had supported (Ercoli and Speciale 1988b) (Fig. 3.29).

One surprising conclusion of Ercoli and Speciale's study is that there is more modern damage to rough surfaces where the cliff face has experienced spontaneous failure than to smooth surfaces left by ancient quarrying. This occurs because water in the rocks evaporates and recrystalizes forming a hard crust over powder; then when the crust flakes off, the powder blows or washes away, creating omega-shaped cavities in the cliff. Until late in the nineteenth century, the ancient aqueduct near the top of the latomie still brought enough water to run some mills and to water gardens in and near the quarry. Leaking water from this aqueduct rotted the stone below. Water and vegetation can increase failure when rain infiltrates cracks and increases capillary pressure, and when plant roots prevent interlocking of the stone blocks. Weathering is hastened also by fumes from the chemical industry and by smog from vehicle exhaust.

The quarries and catacombs also sustain damage from tectonic action. The Siracusana has three fault systems, running E–W, WNW–ESE, and NW–SE, recently moving according to normal mechanisms (Fig. 3.30). A fault runs from Sortino to Syracuse along the Bongiovanni and Climiti Mountains (Mirisola 1996). The main seismically active structures relate to the Maltese escarpment, a submarine, steplike fault at the east edge of the plateau, running from NNW-SSE to N-S. This sector is confined by the Etnan volcanic region to the north. Consequently, strain energy release in the compressed area usually occurs by large, violent seismic crises followed by pauses, with a return period of 70 to 500 years. Seismic shocks are usually of magnitude 7 or more, but since 1693, they have been increasingly frequent and of lower magnitude, indicating that a seismic creep is releasing some energy. Earthquakes damaged the major pottery manufacturing area in the third quarter of the first century C.E.; at the same time in Tripolitania in North Africa, an earthquake occurred (R. J. A. Wilson 1990) that could have been felt in Sicily if it were caused by slippage of the African plate. We have already mentioned the areawide earthquake of 365 C.E. In the past thousand years, the greatest local earthquakes have occurred in 1140, 1169, 1542, 1693, 1757, and 1846. In all, there have been 16 earthquakes in Syracuse and its hinterland in the last 1000 years (Boschi et al. 1998). The 1693 quake, largest and most deadly of those recorded, had a magnitude of 7.7–7.8 (Ercoli and Speciale 1988b) or 9 (Boschi et al. 1998), causing serious building damage and human injury, leaving 6000 dead in Syracuse and 60,000 dead in Eastern Sicily (Boschi et al. 1998; Fig. 1, map of epicenters). The Temple of Athena on Ortygia was so damaged that it needed a new facade; in 1728 a new facade in the baroque style was added. The fortress at Euryalos and the wall of Dionysos were also damaged, as were the Hexapylon (gates) on the north side of the city (Fabricius 1932). The epicenter of that earthquake was in the bay between Augusta and Catania, to the north of Syracuse (Ercoli and Speciale 1988a & b). Huge waves attacked the necropoli along the coast. Torrents carried tons of debris over slopes and plains (Fabricius 1932).

Hydrogeology   The hydrogeology of Syracusana follows the local stratigraphy. Water accumulates from brief and fierce winter rains of up to 700 mm. The Iblei Mountain range to the west is a rain barrier, receiving water in late fall and winter. Springs at an altitude of 100 m on Mt. Lauro (a basalt mountain) are the origin of Anapos River, one of the longest in Sicily. The substrata of volcanic rock, marl, and clay is impermeable to the water in the important phreatic nappe (groundwater in horizontally thrust rock) in the overlying limestones. The permeable rocks, a succession of layers between impermeable ones, give rise to many springs along the upper third of the Anapos River, watering the subsoil. In the middle third of its course, the river runs through hard, karstified, sandy limestone inclining to the east, with springs along its seam with the lower impermeable layer. An aqueduct, supposedly Greek, runs for many kilometers in carved passages at the outer edge of this layer, west of the city.[21] Water from rivers and springs is stored in the alluvium (deposited silt and sands) (Collin-Bouffier 1987; Drögemüller 1969). Near the city, the river runs along the foot of Mt. Crimiti to the sea; here,

TECTONIC MAP

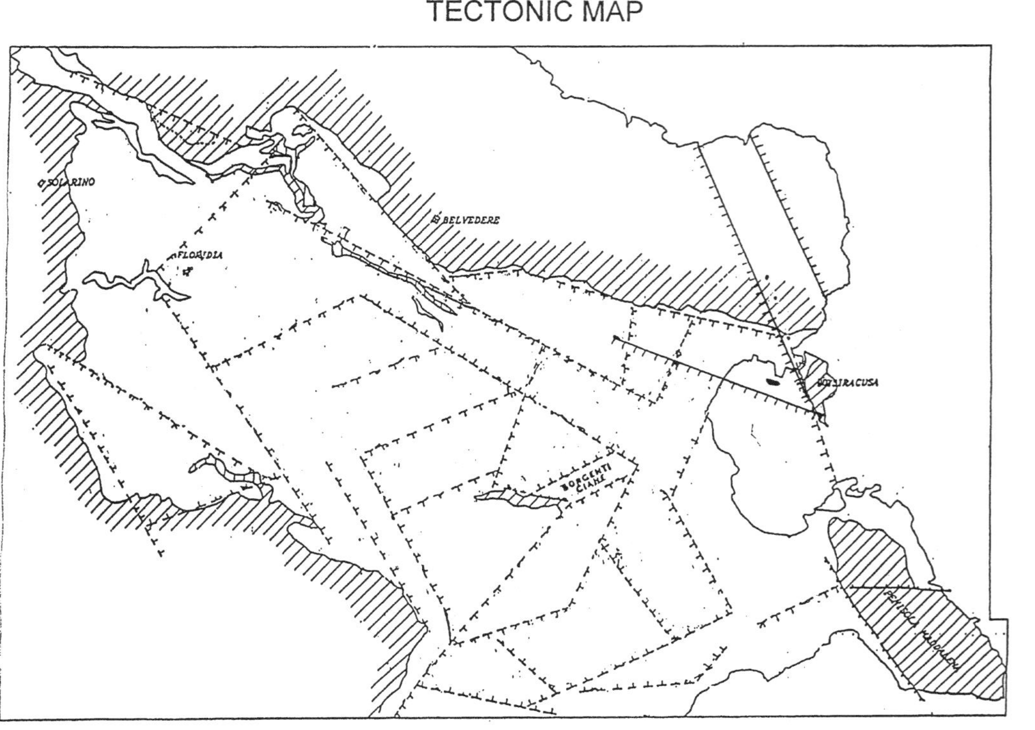

FIGURE 3.30. Fault map of Syracuse area. Ortygia protrudes into a NNW running graben; the fault along the island's west side intersects a second fault running WNW, and the spring of Arethuse rises along the "pipe" formed at the intersection. Another important fault lies above the second one and forms the southern edge of the limestone cliffs into which the quarries (latomie) are cut. The horst of Ciane/Cyane, whence arise the springs called Pisma and Pismotta, is labeled Borgenti Ciane and appears as a limestone island in the alluvial plains. At the lower right the headland of Pemisola Middalena is the area where yellow calcarenite quarries were located; today, some of their shipping facilities are as much as 4 m under the sea (from Aureli et al. 1997).

the lower valley is impermeable, with some volcanic intrusions (Mirisola 1996). The local calcareous series makes up the bottom of the grabens and outcrops in horsts (upthrown blocks between two faults). This series and the calcarenites and sands of the lower Pleistocene and the volcanic stones constitute a hydraulic continuity, which has been called "water-bearing B." This unconfined aquifer runs under the whole area. Additionally, in some sectors, clay of the middle Pleistocene resting on geologically recent calcarenites, sands, and alluvial sediments confines another aquifer, called "water-bearing A" (Aureli et al. 1997; Aureli and De Maio 1994).

The hydrogeological potential of these formations is exemplified in the outflow of two sets of notable springs. At the center of the alluvial plains west of Syracuse, drained for agriculture about 1915, the artesian springs called Pisma and Pismotta rise at 2 m above sea level, 5 km from the present coast (Aureli et al. 1997), to feed the River Cyane [Sorgenti Ciane (Cyane) in Fig. 3.30]. These springs flow from the center of a horst. A series of faults permitted the calcareous rocks to rise from the clay. Under pressure, the water flows quasi-horizontally between rock and clay, then upward through the horst to the surface. In 1930 the flow from these springs was 1260 l/s, but by 1988, because of pumping, it is not over 700 l/s. A late Hellenistic shrine stood on a small hill not far from the springs (R. J. A. Wilson 1990), probably located on one of the stony outcrops of the horst. Legend connected the spring's nymph Cyane with Pluto, saying that the god transformed her into a spring because she objected to his kidnapping of Persephone (Coarelli and Torelli 1984).

The other notable set of springs are the freshwater eruptions into the bay, culminating dramatically in the spring of Arethusa[22] on the western edge of Ortygia (Fig. 3.31). Arethusa's spring, some meters inland from the shore, had already been arranged as a fountain by the fifth century B.C.E. In the early fourth century, Plato walked in the gardens of Dionysos the Elder near the spring (Letters, II, 313a–b, III,319a, VII,347a, VIII,349c). In the first century B.C.E., Cicero described the spring in his *Verres* oration (IV, 53, 118). More recently, before Napoleon sailed to Egypt in 1798, his ships took on fresh water for the voyage from the Arethusa fountain, according to a letter he sent to the English Parliament (Aureli et al. 1997; Ciurcina, n.d. [ca. 1985]: 35). Today this spring wells up in a basin set in an esplanade dating from about 1864 (R. J. A. Wilson 1990), at sea level or perhaps one meter above. In 1930 Arethusa flowed at the rate of 400 l/s. Near Arethusa, in the salt water of the bay, is the "Eye of Sillica," one of several eruptions of fresh water into the sea.

The Arethusa spring and the springs of the bay are more complicated than the other extant springs at Syracuse, appearing along the contact of the calcarenitic layer Mc with the underlying Cv (volcanic) layer (Figs. 3.27 and 3.31). The new geological explanation of the harbor springs (Aureli et al. 1997) is that they occur where two faults intersect, creating a natural "pipe." One fault system lies in the Mount Climiti formation (Mc) and tends NW-SE, where it is masked by Quaternary clay both on land and under the harbor. The other fault system tends approximately NE-SW and intercepts the first; it is associated with the calcarenitic marls of the Mount Carrubba Formation (Ms), which are semipermeable on the

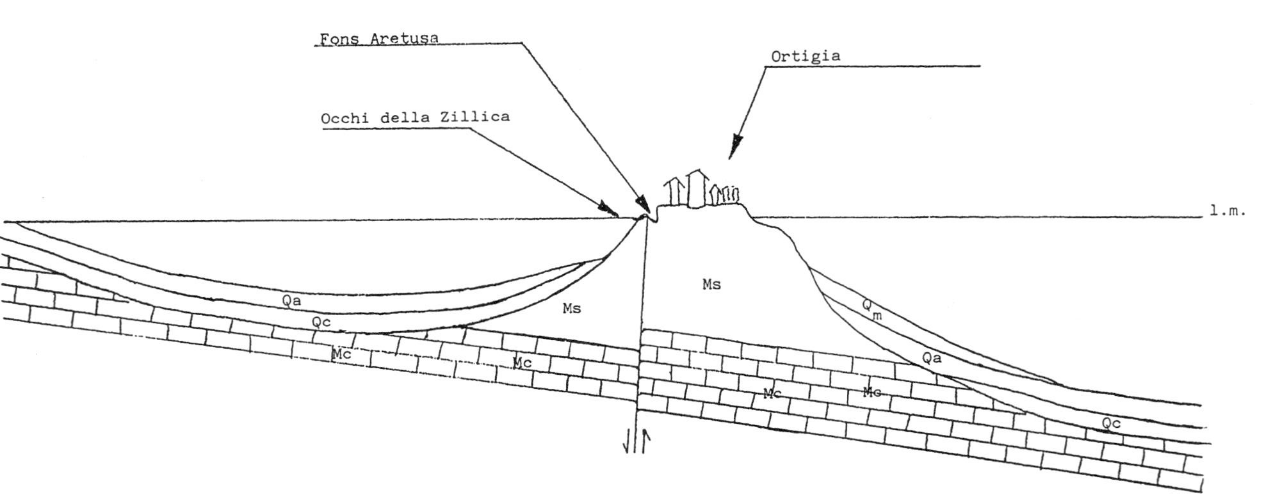

FIGURE 3.31. Section of Arethusa spring. Geological layers are as described in the text and in Figure 3.27. The settlement on Ortygia is indicated by the buildings. On the left (west) side, the spring is denoted in Italian as Fons Arethusa, and the other springs in the harbor as Occhi della Zillica. These springs occur where faults intersect, indicated by the half-arrows at the bottom of the section (Aureli et al. 1997).

Quaternary clay. These two systems of faults intersect at the west edge of Ortygia and again farther out in the sea. The fresh water from the hydrologic basin inland, having passed (under pressure) beneath the clay and under the sea, rises to the surface by means of natural "pipes" at the intersection of the faults, both in the Arethusa basin and farther out in the sea where the Qc layer terminates. An earthquake in the twelfth century, however, drastically changed the flow of a number of the springs in the Great and Little Harbors (Schubring, 1865; Maugeri 1994–95; Aureli et al. 1997).

Natural drainage within the city is north to south and southeast, from the upper edge of Epipolai to the ancient urbanized zone. The water table of the city area can mostly be reached by wells of 20 m in depth or less, which was easily managed with ancient Greek technology (Aureli and De Maio 1994; Crouch 1993).

More could be written about the springs, wells, and aqueducts on the plateau and in the lower areas of the city (Collin-Bouffier 1987; Tolle-Kastenbein 1991; R. J. A. Wilson 1990; Crouch 1993). The geology determined where water was available. Sources of water in turn formed the nucleus of clusters of habitations, workshops, commerce, and public buildings.

Interlocking Human and Geological History

The underlying geology and the visible geography constrained ancient city builders as they do modern engineers. Locations of the ports determined the earliest Greek urban organization at Syracuse, and physical changes in the ports continue to result from interaction of human decisions with geological processes. The complete displacement of the isthmus from N-S to WNW-ESE still complicates the traffic pattern in that area. In every part of the city, streets and neighborhoods had to adapt to appearance or disappearance of water sources, and available building materials. The routes of major streets are largely determined by ancient physical features such as faults, karstified areas, collapsed quarries.

From early Greek times, the colonists produced ceramics, and potteries continued in operation to the end of the Roman Empire. The pottery industry was set up in three clay areas. One area was the belt of clay at the foot of the karstic caverns, forming a sort of bluff across the center of the site in an arc between San Giovanni and Santa Lucia. A second area lay along the ancient river running southeast between the Latomie Casale and the other latomies to the east. The third, a blue clay area, ran along the south edge of the earliest suburb, just where the little river originally emptied into the Great Harbor (Fig. 3.26); ruins of ancient workshops (probably Greek) have been found in this area, indicating exploitation of the resource for pottery. At this location, they had convenient access to the nearby port for shipping (Drögemüller 1969; Agnello 1954). As the clay beds were emptied, some resulting voids were used first as shrines to deities favored by potters; shrines found in niches of very early catacombs suggest that some of these openings began as horizontal extraction pits for the industry (Crouch 1993).

Pottery works require abundant water and provision for water storage. Consequently, many elements of water systems have been found in this zone, including deep wells to the groundwater, which now appear as circular openings from the surface down into the catacombs; circular conduits and drains related to carefully waterproofed square or rectangular cisterns; conical and bell-shaped cisterns that became family tombs with recesses cut all around their edges; and the original aqueduct galleries. Cisterns and wells of early periods,[23] the surface openings of which later became skylights of the catacombs, later were cut off below by ne-cropolis galleries (Giesheimer 1989).

Possibly the catacombs for burial of the large Greco-Roman population (R. J. A. Wilson 1990; Coarelli and Torelli 1984) began to be constructed (in the Ms layer of Fig. 3.27) in the third century B.C.E., although no reliable dating exists for such an early period. In the third century C.E., from 220 on, the catacombs of Santa Lucia, Santa Maria di Gesu, and Vigna Cassia were in use by Christians.[24] The grottoes and tombs outside the Scala Greca on the old northern road to Catania, cut into the limestone terrace (Fabricius 1932), are also difficult to date. Not distinctively Christian, catacombs were used by pagans, Jews, and other groups as well.

The catacomb under the church of San Giovanni (Fig. 3.32) has many wells and at least seven cisterns up to 10 m in diameter. The three types of water elements in this catacomb are wells at intervals of 6.4–7 m except for two that are 8.5 m apart; bell-shaped (not rectangular) cisterns; and the aqueduct running NE to SW, which divides the catacomb into two areas. The round funerary rooms now called Marina, Adelfia, and Sarcofagi—once possibly bell-shaped cisterns— are about 18 m apart in a planned arrangement (Maugeri 1994–95). The aqueduct became the major axis and is still visible at the east and west ends "in perfect condition" (Giesheimer 1989). At the east end, the aqueduct turns SW away from the large gallery; after another 6 m are the remains of what was probably a chamber for dividing the stream of water. As the height of the gallery varies, the aqueduct is visible in the ceiling of some passages and along the side of others. Following the ancient Greek aqueduct along a vein of stone, the Roman and early Christian builders safely enlarged the passage. The depth of the aqueduct gave a constant slope, ensuring a uniform flow in the earlier stage and regularity for the later gallery. The southern sector of the San Giovanni catacomb slopes downward to avoid penetrating the open air and in some places is truncated to avoid its eastern neighbor, the Catacomb Predio Maltese. The northern sector slopes upward, but no further excavation took place after the builders ran into hard, cemented breccia, which could not be vaulted and where there were no aqueduct remains (Giesh-eimer 1989). The neighboring Vigna Cassia Catacomb, also under a church, was dug along an ancient aqueduct and incorporates cisterns. Its southern extension ends at a rectangular shaft with an audible flow of water in the bottom (Maugeri 1994–95; Ciurcina ca. 1985).

Did the Greeks deliberately utilize discontinuities or faults as the axes of aqueducts? Geologist Maugeri gives these arguments: On the one hand, such an aqueduct was easy to dig because the rock is weak along faults, making it easier for builders to follow the predetermined direction. On the other hand, it is difficult to maintain the roof because of the same weakness of the rock. Builders must waterproof the floor or use pipes. If the fault is active, it will destroy the aqueduct by its movements (R. Ruggeri 1996). In the Vigna Cassia, the fault's main direction

FIGURE 3.32. Plan of San Giovanni catacomb. The main axis running WEW to ENE was based on a preexisting Greek aqueduct, which has left carved channels in the floor, here indicated by a heavy line to the east and west. The rotundas dedicated to tombs of important families may have earlier been large cisterns. Smaller round and rectangular openings to the surface may have earlier served as wells. Excavation seems to have followed veins of easily carved stone (Griesheimer 1989).

is not the same as the aqueduct's direction. Further, the catacomb is carved *in* the rock, but water came from seams *between* two kinds of rock. Maugeri (1994–95) concluded that some aqueducts were dug along faults, depending on local circumstances. Ercoli's findings at Agrigento suggest a similar use of natural channels (Aureli et al. 1997).

The catacombs are a striking feature of the urban areas of Syracuse. Unlike those of Rome, which are mostly situated at the fringes of the ancient city, those at Syracuse were sited for the most part under the main centers of population during most of the city's history, taking advantage of the easily dug, friable calcarenites of the Mount Carrubba Formation (the Ms layer). This stone occurs on Ortygia and in the area near the new Archaeology Museum, which was the site of very early suburbs. The builders were not confined to one lithotype, however. The catacombs of San Giovanni and Vigna Cassia, built only 200 m apart along the tunnels of classical aqueducts, occupied different geological beds. The Vigna Cassia is cut in quaternary sediments from marine transgressions (that is, calcarenites), whereas San Giovanni lies in a lower terrace of the Epipole plateau limestone. The grottoes (later, a cemetery) above the theater are in the Mount Climiti limestone (Mc) layer and have karstic springs, which were later amplified with waters from the Galermi and other aqueducts. Differences of typology affect the chronology of the catacombs. The San Giovanni and Vigna Cassia catacombs are separated by a fault running EW in the Acradina district. This fault contributed to the difference in level along the north side of the district and made easier the exploitation of the classical and Hellenistic quarries, where rock was cut into compact, resistant strata, 3–4 m underground (Giesheimer 1989). Even in Ortygia there is evidence of much earlier aqueducts cut into the soft rock.

The irregularity of the Vigna Cassia and Santa Lucia catacombs suggests an early date. Simple galleries may be third century C.E. whereas rotundas seem to be mostly fourth century, with the most recent from 452 C.E. At the end of the fourth century, the catacombs reached maximum extension, and by the beginning of the fifth century they were fully occupied (Giesheimer 1989) and ceased to be actively maintained.

Two reasons account for the preservation of the underground passages of Syracuse. Earlier, the channels were the branches of aqueducts. With the water flowing only in pipes or in smaller floor channels, the passages were quite dry, which contributed to their preservation. Later, when the aqueducts were enlarged into catacombs, the sites' increased sanctity fostered maintenance and discouraged human destruction. The catacombs are still under the protection of the Vatican rather than the Archaeology Service or the Municipality (Ercoli and Speciale 1988a&b; Aureli et al. 1997; Coarelli and Torelli 1984).

Conclusions

Examining Syracuse in terms of the interaction between geological factors and human history increases our knowledge of the city as an entity. At certain key points in the development of Syracuse, elements of the geography and changes in the geology have strongly affected the city. After initially choosing the site for its geologically based situation and resources, settlers next took into consideration the access provided by rivers into the hinterland and the ports that made feasible connections with the rest of the island and the Mediterranean world. The geological events that shaped the city were specifically the earthquakes of the first, fourth, twelfth, sixteenth, and late seventeenth centuries C.E., and the submergence of the edges of Ortygia by 3–4 m at an unknown date, but probably after the Roman period (possibly related to the 1542 earthquake that wrecked the bridge from Ortygia to the north). Then erosion and deposition along the north shore of the bay and in the Large Harbor completely changed the pattern of the two ports. Cracks in quarries and catacombs appeared with the 1693 earthquake, and a great tsunami damaged coastal necropoli. Compared to this, the quiet changes of the erosion of hillsides, silting up of swamps, and action of salt air on the stone of monuments could easily pass unnoticed.

At Syracuse, an investigation of the way geological features were altered by humans is just beginning with the study of port development, the investigation of quarries that were important in the life and construction of the city, and the reconstruction of irrigation and drainage systems that altered the rural landscape. The following would enhance the archaeological and historical understanding of Syracuse and of urban development in the Greco-Roman world:

1. Comparing Syracuse with its mother city, Corinth.
2. Mapping the underground waterlines and other cavities of Ortygia and integrating this knowledge with the archaeological and geological information about the island and with the history of the city. This project was undertaken by the municipality of Syracuse in the 1990s.
3. Studying other natural features affected by human intervention, including karst channels, shafts, and springs used in the urban water system. Determining which resurgences at Syracuse are true springs and which are the outfalls of aqueducts would be useful.

Although the human environment generally consists of both natural and man-made features, to understand the history of the city and the relationship of the built form to the geological base we must distinguish natural from man-made resurgences. A joint project of hydraulic engineers, applied geologists, and archaeologists would go far to elucidate this aspect of the environment.

Reflections on how Syracuse compares with its neighbors, Morgantina, Agrigento, and more distant Selinus, are in Chapter 6, Comparisons of Cities.

Central Greco-Roman Cities

Argos

Argos, situated in the southern peninsula of Greece called the Peloponnese, lies on the northwest side of the Argos Plain, backed by hills to the north and west that are the eastern edge of an extensive region of mountains and intermountain basins (Fig. 4.1). A road runs northward through the valley and over the hills to Nemea and Corinth. Eastward beyond the capricious rivers lie the old Mycenaean cities of Mycenae and Tiryns on their knolls, with the port of Nauplia closing the circuit to the southeast. Beyond Nauplia is the Argolid peninsula with the ancient pilgrimage and health center of Epidauros. (The term "Argolid" as used in the literature sometimes means all the area near Argos and sometimes means only the peninsula south and east of Nauplia. Herein, we will use Argolid for the latter and Argive Plain for the former.) Between Argos and the gulf about 6 km south is the marshy area of Lerna, remnant of a lake that once reached nearly to the outskirts of Argos, while the southeast part of the plain was until recently a series of lagoons (Piérart 1992). To the southwest, skirting the mountains, runs the road to Sparta. The advantages for Argos of being situated at the center of gravity in the triangular plain (Runnels 1995) continued throughout all the periods studied herein.

Argos is unusual among ancient cities because we have ample modern geological investigations of regional structure, morphology, karst geology, and hydrogeology, literary evidence from antiquity, and archaeological data from decades of investigation. These materials contribute to a detailed understanding of how human settlement built on and responded to local resources. We will therefore describe the regional setting of the city before turning to an examination of the urban core.

Argos and Water

Below its mountains, the city of Argos stands on a shelf overlooking a plain of extensive fertile agricultural land that curves around the site from north to

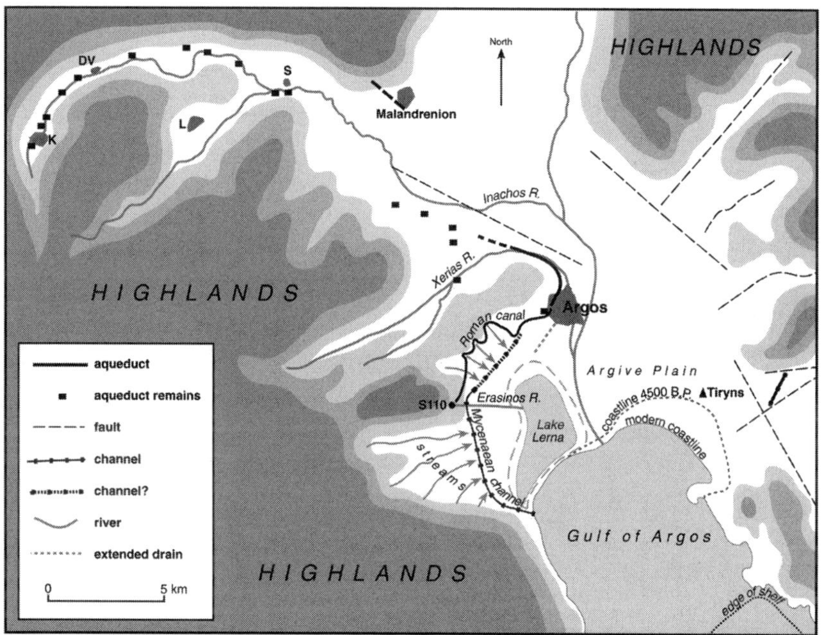

FIGURE 4.1. Map of Argolid plain. Argos lies inland at the foot of a spur of the central massif, on a ridge that protrudes into the Argive Plain. Two rivers (Inachos, from the NW to the gulf, and Xerias, which joins it from the west) drain the mountains between Argos and the Tripoli Basin to its west. Along the northern river are remains of Roman aqueducts, bringing water from the higher karst systems, especially from K (Kefalovrysso) and DV (Douka Vryssi). Southwest of the city ran another Roman aqueduct bringing water from Kephalari Spring (see also Fig. 4.2). At a somewhat lower elevation, a Mycenaean channel (ca. 1200 B.C.E.) intercepted the many streams from the hills and carried their excess water to the Gulf of Argos; a similar channel in date and construction has been found across the valley in the hills east of Tiryns. Kephalari spring and the Erasinos River fed Lake Lerna. The lake was larger in the second millennium B.C.E. than it is now. To the east of the Inachos River, the shoreline of 4500 years ago curved inland nearly to the foot of the Tiryns hill. Still earlier, during the last Ice Age, when sea level was much lower, most of the Gulf of Argos was dry land, with the shoreline or "edge of shelf" at lower right. Dashed lines represent faults where the area is pulling apart tectonically. (Based on maps of Aupert, Knauss, and Zangger; drawn by Langford. Numbers of springs are from Morfis and Zojer 1986.)

southwest.[1] The stratigraphy is as follows, beginning with the topmost modern layers:

> Higher plateau and mountains are Tripoli limestone.
>
> Tripoli plateau sits amid karstic mountains.
>
> (Older) Triassic and Jurassic limestones to the northeast.

Larissa and Aspis hills are blocks of mid-to-late Cretaceous limestone, thrust over younger flysch.

Flysch sediments form the floor of the Deiras valley.

Plain is a graben filled with alluvium.

Alluvial soil in the northern valley permits agriculture.

Marshy southern coastal plain permits grazing.

Under the plain are buried marine deposits.

The city's site, at the seam between the alluvial plain and the mountain range, allows access to underground water from the Tripoli plateau above, as well as access to surface springs and seeps. The plain is a graben partially determined by faults along the foot of the mountains (Gaki-Papanastasiou 1994). One fault runs northward from the area of the Kefalari spring; the Inachos River runs northwest along the middle section of the second fault; and the third fault runs along the foot of the mountains on the eastern side of the plain.

The rounded mountains of Late Cretaceous limestones to the west and earlier Triassic and Jurassic limestones to the northeast contrast with the jagged eastern mountains of flysch. Argos sits in the hollow between the foot of Larissa (276 m) and the southern edge of the Aspis hill (80 m), which together form a promontory into the plain. The two hills are blocks of mid-Cretaceous limestone, thrust over younger flysch sediments that form the floor of the Deiras valley between them. The major processes that affected the developmental history of Argos are karstification in the mountains and transport and deposit of alluvium in the plain, which caused changes of shoreline. Although Greece is generally subject to earthquakes, we have no literary or inscriptional evidence that seismic events were significant at Argos or elsewhere along the western side of the plain during the historical period—the argument from silence. However, Pariente told me that there is physical evidence of a fifth-century earthquake leaving destruction in its wake across the agora south of the Hypostyle Hall. Earthquakes on the eastern fault during Mycenaean times are surmised from data in destruction layers at Midea and Tiryns (Scarborough 1994). In the northern plain Gaki-Papanastasiou (1994) noted some uplift, and Zangger (1993) and Knauss (1996) reported uplift in the Lerna Lake area, from the village of Magoula southeastward to the coast. Ambraseys (1994b) reports a "major tectonic feature" in the sea off the Argos-Lerna coast, which may be related to remains of human structures dating from the fifth or fourth century B.C.E. or earlier (Hall 1995, quoting T. Van Andel, personal communication) 100 m south of the river mouth in the sea. These definitely show that the shoreline there was depressed after the classical period. Yet for about 5000 years, the sea has been relatively stable near its present level and position, the coastline advancing by delta formation along the Inachos River from a mere 250 m southwest of Tiryns to its present location about two km distant (Higgins and Higgins 1996, their fig. 5.4).

Precipitation falling on the mountains around Argos, and penetrating them from surface dolines through solution cavities, resurges later and lower in the form of springs and sometimes as rivers. Karstification of the limestone mountains at the center of the Peloponnesos is thus linked to the water supply for the Argos. The ancients already knew the subtle connections between several dolines on the Tripoli Plain in the western mountains down to Lake Lerna and the perennial Erasinos River in the plain south of Argos (Fig. 4.2). They used painted pine-cones thrown into dolines to check this hypothesis, then they observed the springs below to see where the cones surfaced again (Herodotus VI, 76; Diodorus XV.49.5; Strabo VI.2.9 (C371) and VIII.6.8. (C389); Pausanias II.24,6; checked by Pritchett 1965, and described by Stringfield and Rapp 1977). Pausanias, for instance, reported that

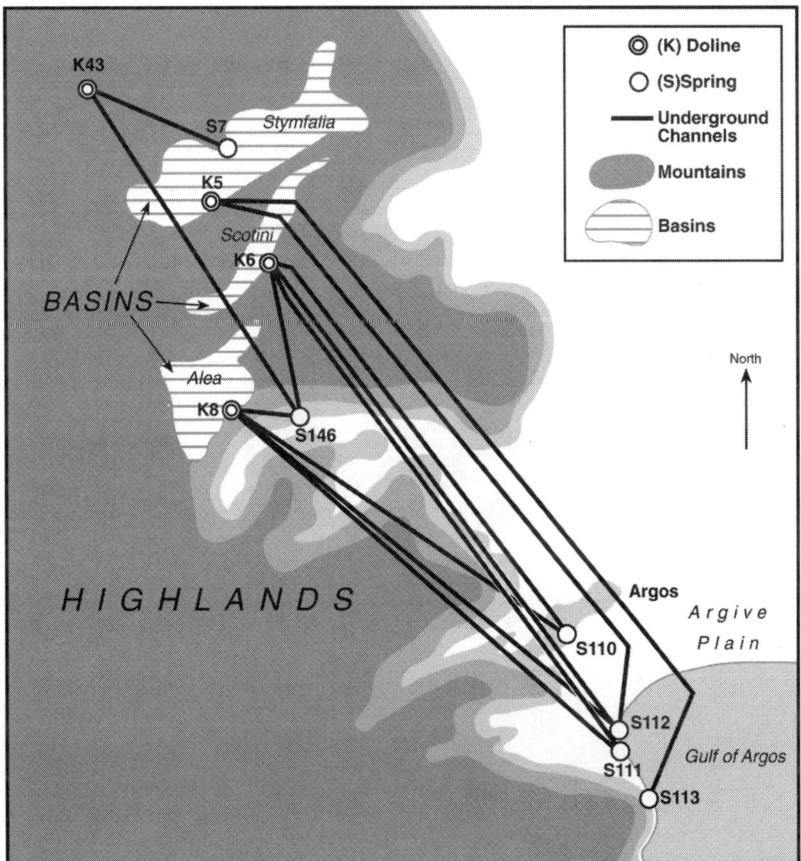

FIGURE 4.2. Underground connections between Tripoli plain and Argolis coast springs. The waters from polji basins of Stymfalia, Scotini, and Alea drain through underground passages (heavy black lines) to the springs (S circles with numbers) on lower levels. DV on Figure 4.1 is the same as S(pring) 146 here. In periods of heavy rain, excess water can cause a reversal of flow so that the water does not simply flow downhill but—temporarily mixed by piston flow—in an uphill direction. (Redrawn by Langford from Müller and Schotterer's fig. 28 and Gospodaric and Leibundgut's fig. 148, both in Morfis and Zojer 1986)

rainfall in the mountains disappears into cracks in the rocks but reappears in the springs below, such as at Diné,[2] a spring to the south of Lerna, associated with the earliest royal family of Argos, the Danäids. Pollution of these subtle connections is a modern problem.

There are three hydrographic basins serving four underground drainage systems from the Tripoli area. The one important to our investigation flows north-northwest to south-southeast beneath the Stymfalia, Scotini, and Alea *poljes* (large, flat depressions with underground drainage through karst networks) to the northern springs of the Argos bay (Morfis and Zojer 1986, figs. 148 and 150). The biggest coastal spring near Argos is Kefalari, about 5 km to the south (Figs. 4.2 and 4.3), drawing on subterranean watershed of 100 km^2 (Scarborough 1994). Although the main sources of this spring are the Alea and Scotini dolines in the Tripoli plateau, a minor connection (operative by piston flow mainly in periods of heavy rain) has been traced from the Stymfalian doline to the Scotini and hence to the Kepfalari spring.

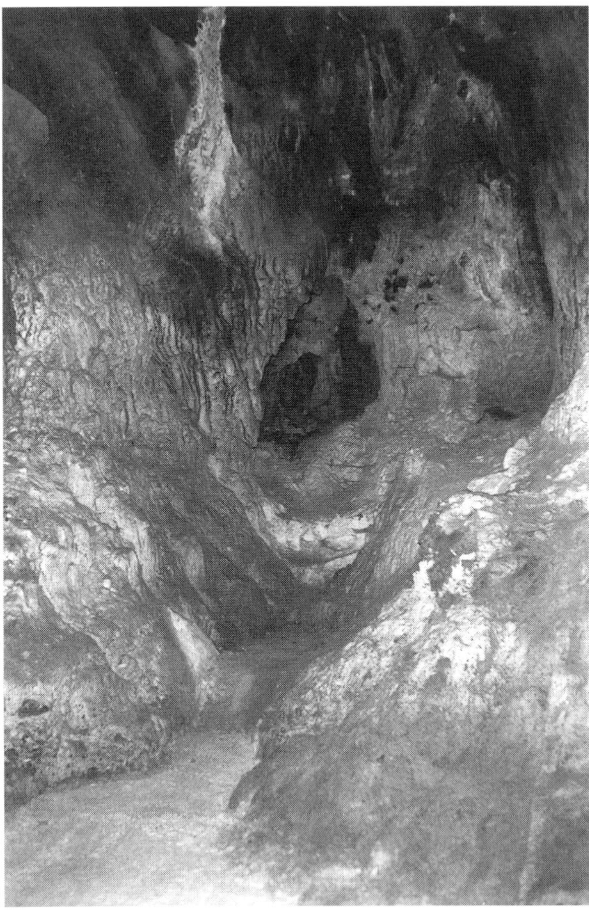

FIGURE 4.3. Karst cave behind the shrine that protects the Kefalari spring south of Argos. (Photo by Crouch)

An early modern understanding of underground connections between the upper plain and the coastal springs is embodied in Marinos (1978; but see the cautions expressed by LeGrand (1973). Some of those connections, now termed "major," were already well understood at that time. However, the intermittent links among the Stymphalia, Scotini, and Alea poljes with outflow at the huge Kephalovryso and Douka Vryssi springs downhill in the Inachos River valley, as well as with the upper and lower springs of Kefalari and Kiveri near the coast, could not be verified until more sophisticated measuring tools such as remote sensing, electromagnetic VLF-resistivity prospection, tracing with salts, spores, phages, or chromide-51 were available. A clearer understanding of the influence of tectonics on underground discharge also contributed to the new understandings published in Morfis and Zojer (1986).

The occurrence of springs at the edge of the sea or in the water at some distance from the shore of the Gulf of Argos is predictable in karst, as was the occurrence of harbor springs at Syracuse. As a result of saltwater and freshwater interacting at the edge of the sea, saltwater can emerge from springs or wells on land and freshwater can emerge at sea.

There are two major kinds of aquifers in the Argos area: those in karstic formations and those in alluvium (Marinos and Plesas 1994). The two interacted at the site of Argos as the urban area grew. There were originally springs and oozes all along the eastern base of the karstified limestone hill of Larissa and, by the Greek period, wells in the alluvial plain on which the city sits (Scarborough 1994). Because of their limited size and the separation from the higher mountains to the west, the Larissa and Aspis hills were not connected to the upper aquifers, so that the yield from their springs was low, probably not over 550 m³/day (Fig. 4.4, sections of Larissa and Aspis). Well-building in the alluvium caused increased draw down (lowering) of the small, local water table in the two hills. This dewatering was a second factor, in addition to population growth, that required off-site water sources and resulted in construction of long-distance waterlines.[3]

When the springs on the east side of Larissa were not enough, the Romans looked for water farther away (Fig. 4.1). It was fairly simple to utilize the water of Kefalari by leading it directly to Argos along a contour-line channel in the second century C.E. (Lehmann 1892); confirmed by Banaka-Dimaki et al. 1998). Remnants of the channel were traced by our group part of the way from Kefalari to where the channel entered the urban area near the Aphrodite Sanctuary high enough to supply the somewhat lower eastern quarters of the city.

The Kefalari line apparently overlapped with another waterline of the same Hadrianic era (ca. 120 C.E.) to supply the core of the city. This second Roman aqueduct tapped other water resources in the west-northwestern hinterland of Argos, which was investigated by archaeologist Aupert (1989). Traces of the Greco-Roman collection and storage system for water near Kephalovryso and Douka Vryssi at the altitudes of 900–1000 m and at distances of some tens of kilometers from Argos in the valley of the Inachos River had been visible until just before Aupert's investigation in the 1980s. He also found traces of ancient piers and channels extant all along the river valley, some still visible in 1995. Local tradition had always related this aqueduct to the Stymphalian Lake by means of the abundant

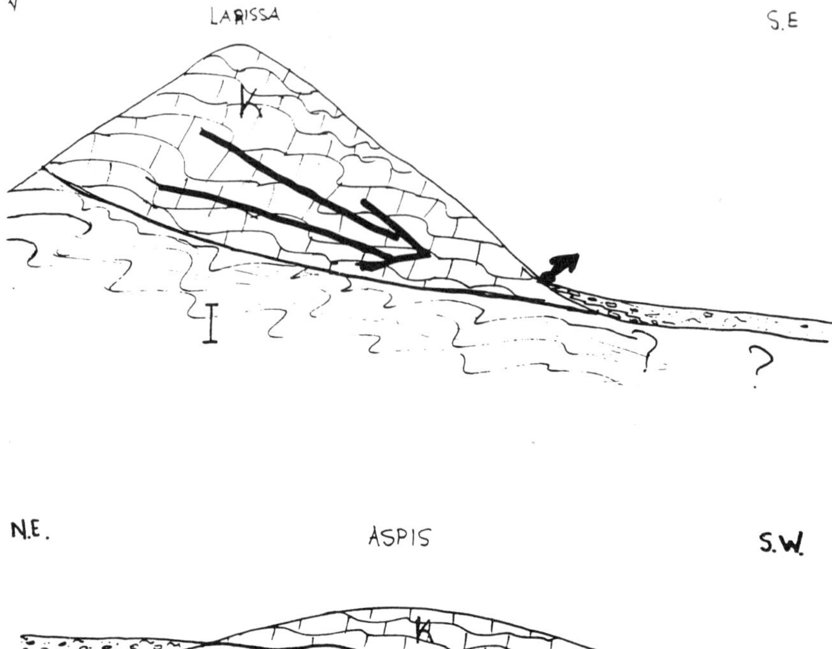

FIGURE 4.4. Sketch of sections through Larissa and Aspis. The two hills are mostly limestone thrust over later flysch. Water flow is indicated by arrows, with the small arrows denoting springs. Valleys are filled with sediments and rock, mainly schist (Marinos and Plesas 1994).

spring at Kephalovryso, located farther upstream than Douka Vryssi. Although Aupert claimed that Lake Stymphalia was not connected to these springs and hence not to the aqueduct he was studying, the technical study of the karst drainage system by Morfis and Zojer (1986) has shown that these springs were fed from all three poljes. Kephalovryso and Douka Vryssi springs have a discharge rate of several tens to several hundreds of m³/hr. Slow passage through the rock regulates the discharge, so that there is significant water even during the dry period of the year. Aupert concluded that this aqueduct supplied the impressive fountain on the east side of Larissa, called the Kriterion, and then was carried on to supply the theater area (Fig. 4.5). A third long-distance waterline is suggested by recent discoveries near Malandréni to the northwest of Argos (Pariente and Touchais 1998).

Several baths and fountains used the abundant water at Argos in the first few centuries of our era. Elements of a distribution system have been found along the streets (Banaka-Dimaki et al. 1998; see map of bath locations in Panayotopoulou,

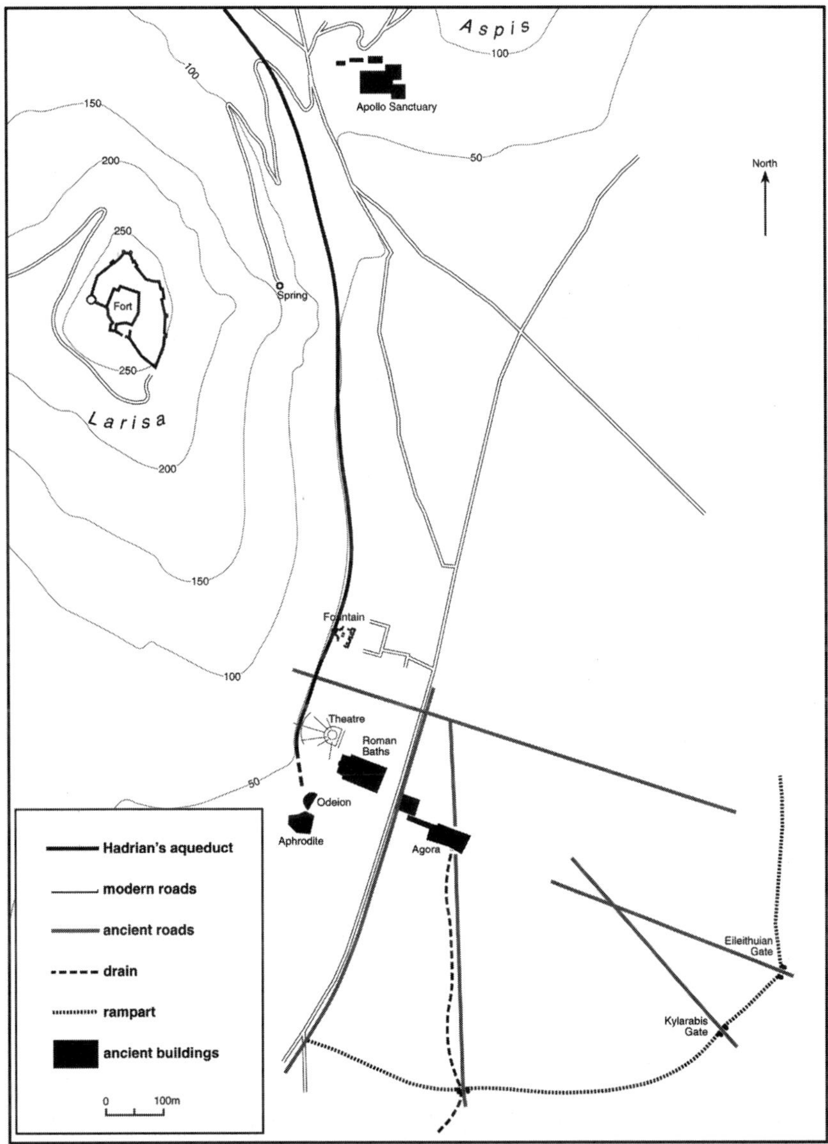

FIGURE 4.5.   Plan of Greco-Roman Argos, with chief geographical features. From north to south, Aspis hill with Apollo Sanctuary. Road leading to the spring where the ancient Hera Sanctuary and later church to the Virgin Mary were located. Larissa mountaintop with fortress used from Mycenaean to Turkish times. North Hadrianic aqueduct. Major modern and ancient roads. Fountain (Kriterion). Theater. Roman bath. Odeion and Aphrodite sanctuary of the archaic-classical core of the city. Agora. Large drains, gates and rampart (based on plans published by the French School of Archaeology in Athens; drawn by Langford).

FIGURE 4.5. (*continued*) Photo of Kriterion fountain. Water once emerged from between the two kinds of rock visible here; the fountain appears to have been developed around an original spring. In addition, the second-century C.E. aqueduct was led to feed the upper part of the fountain, where massive brickwork extends the fountain's walls upward. (Photo by Crouch)

1998: 381, same volume). This contrasts with the centuries-long decline of the cities along the eastern edge of the plain from the fifth century B.C.E. on. Scarborough (1994) suggests that tectonic activity at Mycenae, Tiryns, and Midea in the later Mycenaean era had altered the water resources and that deforestation contributed to the rate of decline of the water table. No such combination of disasters is known at Argos (Van Andel and Zangger 1990b).

At the same time that the Argives were utilizing the aquifers of the underground karst system, they were challenged by surface waters from rainfall and the outfall of springs. South of Argos and west of the Inachos River, a wetland existed variously as a deep and wide lake, which in prehistoric times reached the outskirts of Argos (Zangger 1993) (Fig. 4.6), a marsh, and a lagoon. This landscape was described by Aristotle (*Meteorologia*, 1.14), as follows:

> In the time of the Trojan wars the Argive land was marshy and could only support a small population, whereas the land of Mycenae was in good condition (and for this reason Mycenae was the superior). But now the opposite is the case . . . the land of Mycenae has become . . . dry and barren, while the Argive land that was formerly barren owing to the water has now become fruitful. Now the same process that has taken place in this small district must be supposed to be going on over whole countries and on a large scale.

To control the surface drainage from many small rivers descending from the ring of foothills west of Lake Lerna was a matter of urgency during Mycenaean

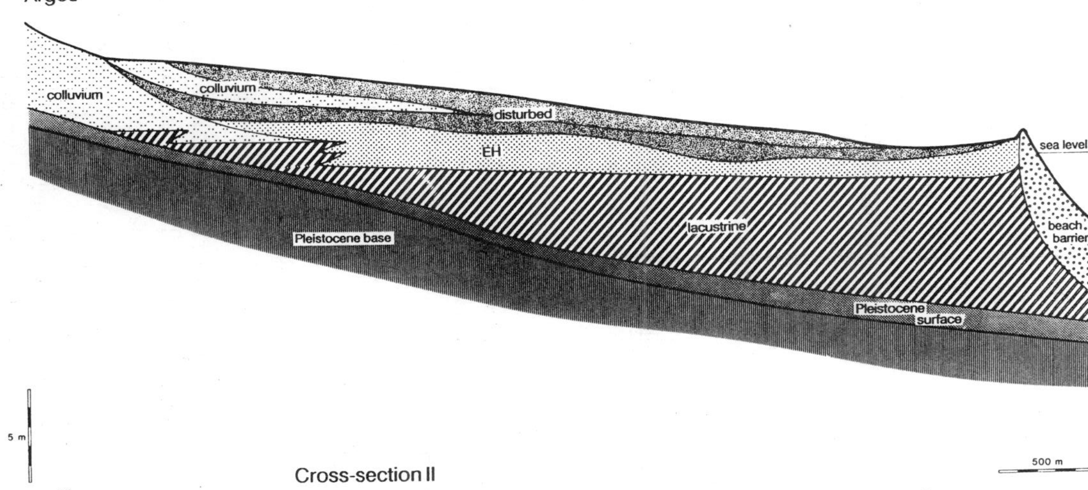

FIGURE 4.6. Section from Argos to the sea. On the sloping Pleistocene base, lacustrine (lake) deposits of the third millennium came close to the edge of the Argos site (at left). Above them at EH is alluvium from the rivers, which filled up the lagoons. Colluvium from the rocky slopes altered the surface at two different periods. Only the upper 2 m or so have been deposited since the Roman period (Zangger 1993, used by permission).

times (second millennium B.C.E.). A collector channel was ingeniously built along a contour line of the hills above the lake (Fig. 4.1). Water from the channel could be directed to the lake in a drought year or to the sea in a wet year (Knauss 1996, his fig. 6). In a dry year, the water of the Inachos River could also be diverted into the wetlands if necessary. Touchais (1998a: his note 44) suggests that, in middle Helladic times (first half of second millennium), a river may have poured through the gap between Larissa and Aspis; later floods from this direction were partly managed by the man-sized drainage tubes built in Roman times in the agora. The Inachos River still floods every 15 years, depositing large amounts of sediments. The classical dike in the bed of the Charadros River, mentioned by Touchais (1998b), is an example of ancient attempts to control the river.

A practical understanding of how preservation of forestlands ensured continuing productivity of springs is evident in the forest preserve above the spring at Lerna, which in the second century C.E. (Strabo XIV.2.6; Pausanias I.21.9, II.37.1, and X.37.5–7) extended down the mountainside to the sea.

In times past, the overflow from the Kefalari and Lerna springs combined to fill Lake Lerna, which then drained to the sea via the Erasinos River. The spring and lake of Lerna are the locale of the Bronze Age myth of Herakles and the Hydra. Engineering historian J. Knauss (1996) connects the myth to the ongoing physical processes here and concludes that the area was subject to alternating drought and torrential rain, which could elevate the surface of the lake as much as 1–2 m in a single day, causing the lake to overflow and destroy people and cattle of the plain; this is possibly the origin of the Hydra myth. An early nineteenth-century traveler (Gell 1810) with a classical education reported:

> The plain of Lerna is at present infested by the same Hydra formerly slain by Hercules and in fact requires the assistance of the hero as much as ever, for the monster . . . desolates the plain with as much ferocity as in ancient times. This formidable scourge is really composed of several very abundant sources, which form the Erasinos, and which in winter the Turks—notwithstanding the many dams they have build in order to prevent it—find it impossible to restrain from flooding the neighboring lands. The water becoming stagnant, produces in summer those fevers and other maladies with which the inhabitants of the plain are annually afflicted.

Immediately west of the Inachos River's mouth, near the modern town of Nea Kios (ancient as mentioned by Herodotus, Thucydides, Strabo, Plutarch, and Pausanias), some wetlands, which were once part of Lerna Lake, persist (Niemi and Finke 1988). As at Syracuse, wetlands were valuable for grazing cattle and may have contributed to the reputation of Argos as a breeder of horses.

Flooding at Argos has not been as thoroughly studied as the mountain karst. On our field trip, Dr. Gaki showed me the river walls built to control flooding and erosion during Greek times, especially at bends in the river. The walls of beautifully cut ashlar masonry are similar to the river walls at Agrigento and Selinus; both may be fifth century B.C.E. The Argive River walls lie below several meters of later alluviation. At one place, we saw a cistern of the Roman teardrop type, at least 3 m deep, cut into the alluvial strata of pebbles and small round stones. The bottom of the cistern lies about 2 m above the ashlar wall. This was on the part of the river curving to the north of Larissa, where piers of the aqueduct also survive. The depth of alluvium above the present river bed and above the ancient walls gives impressive testimony to the ability of these rivers to move debris. The plain silted up, at the average rate of 3.1 mm/yr for 2000 years (Gaki-Papanastasiou 1994), 8 m in 1000 years (Zangger 1993). Debris for erosion by the torrential rains of winter was provided by human actions, though unintentionally (Pope and Van Andel 1984).

As Runnels (1995) points out, if climate were the basic cause of erosion, most of Greece would have been simultaneously affected, but the variation of dates suggests human triggers for the calamities. These actions included certain kinds of slope clearing such as expansion of olive groves onto poorer, steeper sites in the fifth and fourth centuries B.C.E., failure to build and maintain agricultural terraces, and overgrazing (Van Andel and Zangger 1990b). When land was not used at all, such as during periods of complete agricultural collapse, it was safer from erosion than during times of partial economic collapse in which orchards and field crops gave way to herding. The passage of many flocks eventually broke down the terrace walls and then rain flushed the soil down and off the hills, filling in the valleys (Pope and Van Andel 1984). As early as the fourth century B.C.E., Plato wrote of these conditions (*Critias*, III, D–E):

> In consequence of the successive violent deluges . . . there has been a constant movement of soil away from high elevations; and owing to the shelving relief of the coast, this soil, instead of laying down alluvium as it does elsewhere, has been perpetually deposited in the deep sea around the periphery of the country or, in other words, lost. . . . All the rich soil has molted away, leaving a country of skin and bones [so that rainfall] is allowed to flow over the denuded surface [directly] into the sea.

In the Argolid, agricultural decay is especially evident during the Hellenistic period (third–first centuries B.C.E.). Although no major erosive catastrophe is known in the Argive Plain during this period, the area could not completely escape the results of the politically chaotic period of the successors of Alexander the Great and the additional stress of colder and wetter weather. Some modern writers (e.g., Meiggs 1982) insist that Argolid soil is poor, with agriculture suffering from little rain and much heat, but others note the flourishing Mediterranean vegetation. (See the relevant climate data in Figure 7.1 for a cold period coinciding with the Hellenistic era.)

Geology and History

The geological investigation of the Argos area can scarcely be studied today without attention to archaeological findings, which were affected by geological events and processes and in turn provide dating for them. Prominent among these works is the geoarchaeology study of E. Zangger (1993 and earlier articles) dealing with the chronological changes of the plain since 50,000 B.P. (before the present), and with the bordering talus slopes. Carefully using physical sampling methods, including cores dated by human artifacts, he has worked out the locations of the coastlines during human occupation of this region (Fig. 4.1).

Much of the earliest geological and human record of the Argos region has been obliterated naturally, because the sea intruded on and receded from the land by as much as 10 km in response to glacial melting and refreezing between and during the ice ages. After the earliest occupation of Kefalari cave 50,000 years B.P. (Fig. 4.3), there is a long gap in our knowledge until the next rediscovered habitations between 35,000 and 10,500 B.P. A milder climate after 10,000 B.P. encouraged coastal settlements, which were mostly submerged when the sea rose 80 m in height during the next 4000 years. Neolithic pottery has been found at Lerna, Argos, Mycenae, Tiryns, and Porsymna, with Lerna, Mycenae, and Tiryns developing into important settlements during the Bronze Age 3100–2150 B.C.E.

Gaki-Papanastarious's geomorphology dissertation (1994) viewed Argos's regional geology as it developed from seven to two million years ago. Argos has been settled since the end of the Neolithic period, and the oldest architectural remains found here are those of a house of the fourth millennium sited east of Larissa. The place has been continuously occupied since the middle of the third millennium (mid-Bronze Age) (O. Picard 1998). In a study of the geological influence on political boundaries, Niemi and Finke (1988) found that vegetation is noticeably different between the alluvial, agricultural northern valley and the marshy, southern coastal plain for several reasons: the extent and thickness of the buried marine deposit, former transgressions of the sea, land stability as reflected in soil horizons, and overlying alluvial deposits. The altitude-related boundary between the alluvial fans next to the mountains and the flood plain at the center, and differences in the vegetated and bare zones of the coastal area have been identified by Zangger (1993).[4]

Big changes in the shoreline of the Argive Plain took place before and during the Bronze Age, recently enough for oral history and myth to reflect these changes. At Argos the tombs at the foot of the Aspis hill are very old architectural remnants dating from the Middle Helladic (2000–1580) and Mycenaean (1580–1100) periods; these have been reused several times. Small karst caverns here might have been converted easily into tombs; it would be interesting to have them checked by a geologist. Middle Helladic settlement at Argos was concentrated on the Aspis, which served as the acropolis at that time, but which extended to the southwest along the foot of the Larissa hill, utilizing the springs and seeps. The same areas were occupied in Mycenaean times, but a new fortress was added at the top of Larissa (Reynal 1986). Scarborough (1994), working at the scale of human history, investigated the political landscape of the plain during the late Bronze Age (ca. 2000–1000 B.C.E.); the end of this period is the time of the supposed Dorian takeover of the area as reflected in burial practices (Foley 1998).

The unstable political situation of 1400–1200 B.C.E. is evidenced in settlement destruction, stronger fortifications, and new small forts. Voutsake (1995) used mortuary and geologic data to explore the formation of a hierarchical society and a centralized political system in the Argive region during the Bronze Age. Anthropologist Scarborough (1996) thinks that an earthquake, possibly involving volcanic action and tidal waves, severely damaged the eastern cities of the Argive Plain at this time, leading to the downfall of the Mycenaean civilization. As evidence he cites the skeleton found below the east gate at Midea (not a common place for burials), upthrust faults at Midea and Asine as well as an upthrust less than 1 km to the east of Tiryns, and overthrust faults at Mycenae and Argos. The falling water table may have been related to this tectonic evidence, although deforestation also played a part. But Scarborough admits that "precision in the magnitude as well as the timing of these quakes will likely elude us for some time to come" (Scarborough 1994: 33). Geologist Zangger (1993) doubts this interpretation, placing the thrusts, for instance, in the more remote past.

Argos became important only in the early Iron Age (1100–675 B.C.E.) when its residents may have adapted the new iron technology to weapons for conquest of neighbors. Over 600 soundings and excavations have been made in the area, 200 not yet assimilated or published (Pariente and Touchais 1998). During the Geometric era (900–700 B.C.E.), settlement at Argos extended over the area of the present city (Fig. 4.5a), which attained its maximum size then. After a late eighth-century conflict with Asiné to the south in which the Argives destroyed Asiné's Apollo Pytheos sanctuary (Pausanias IV.34.9–10), they constructed a sanctuary to Apollo above the northern road on the side of the Aspis hill. From then on, Apollo was important in the local and foreign politics of Argos (Billot 1992). Early rituals of Apollo on Aspis are dated by sherds from the last half of the eighth century; the god was honored in five shrines in and near Argos, from the eighth to the early fourth century B.C.E. (Billot 1992). The emergence of public space is evident at Argos and other geometric sites (Polignac 1998).

Outside the Aspis, architecture that survived from the early period is scarce and intermittent: An Athena Temple crowned Larissa inside some Greek fortifications (Morgan and Whitelaw 1991). Across the modern road from the agora and extending to the right of the street to the theater, excavators found a primitive image (wood?) of Hermes and Aphrodite, suggesting a possible shrine. Hermes, god of commerce, was closely associated with the agora, and Aphrodite's shrine was at the foot of the Larissa hill; this intermediate site was a logical meeting place for them (Marchetti et al., 1993, 1995; Marchetti 1994). At Argos, there was already a sixth-century theater cut into the slope of Larissa, probably at the site of the rectangular theater steps and not the present larger theater to its north, whose shape was the arc of a circle.

From its slight elevation above the plain and especially from the tops of the Aspis and Larissa hills, Argos dominated the Argive Plain and its cities by the early fifth century B.C.E. According to the fifth-century texts published by Kritzas (1992), Argos's main resources then were agricultural products. Argos is a good example of a city where herding was important, with a traditional pattern of transhumance: herds traveling to the pastures of Arcadia in the summer and returning to the plains of the Argolid in the winter. About 338–37 B.C.E., Argos arbitrated a conflict between Melos and Kimolas over the use of three small islands as pastures (Georgoudi 1974). This pattern continued at least as late as 1938. After the conquests of Mycenae and Tiryns in 460 B.C.E., Argos redistributed their lands to groups from Argos and reconstituted the ancient sanctuary north of Mycenae as the "Argive Heraion." The main temple of the Heraion was rebuilt about 416 by Eupelemos of Argos under improved demographic and economic circumstances (des Courtîls 1992). Earlier the temple had contained a chryselephantine statue of Hera by Polykleitos (Placzek 1982: 413).

The earlist monumental architecture that has survived within Argos is from the early fifth century (Hall 1995). During the fifth century, the small theater rebuilt several times on the southern edge of Larissa was a place of popular assembly (Ginouvés 1972; Marchetti and Kolokotsas 1993). This theater was very close to the sanctuary of Aphrodite as described by Pausanias in the second century C.E.

The ancient streets have become the modern streets in the area of the agora and of the modern central plaza (Marchetti 1994), but the inadequacy of the ancient infrastructure for 10 times the population makes traffic in modern Argos a nightmare! Ancient roads led from the center north to Nemea and Corinth, east to Tiryns and Mycenae, southeast to Nauplia (Reynal 1986), and south-southwest to Sparta. The Corinth-Sparta route has been conserved by the modern highway (Reynal 1986)—both routes constrained by the geography of the area.

The fortress foundations on Larissa—taken over by all the rulers of the Argos area, one after another, with the Turks the last to reuse it—and those of the polygonal rampart both date from the fifth century (Reynal 1986); other stretches of the early wall curtain have been found to the south and to the east of the city (Hesse 1988). Outside the rampart to the south ran a "moat" that may also have served as a drainage ditch (Reynal 1996).

The population of Argos in the fourth century has been estimated at 10,000 (Tomlinson 1972) or 20,000–30,000 (4000 adult male citizens), with the western half of the plain adding another 31,000. In 1995 more than 220,000 lived in Argos and its vicinity (Hall 1995: 13). To grow food for 60,000 people would have required 35 to 52 km^2 of land (at the subsistence rates of Attica). If we say 40 sq km, only two-thirds of the 59 km^2 plain between Argos and the river would have been used. The location of the river probably moved over the centuries, so the figures for the plain's ancient size and for how much was used for subsistance farming are only estimates. The Argives preferred at first to farm the very fertile land northwest of the city. As the population grew, Argos needed more and more of the plain for farming (Touchais and Divari 1998)—at least 200 km^2 of arable land or 585 km^2 including all the marshy grazing areas nearer the sea.

Practical works of the fourth century B.C.E. like fountains and sewers are known, as are recreational buildings like the race track and palestra, and the great theater dates to the end of the fourth or the beginning of the third century B.C.E. (Moretti 1998). It held 10,000 or perhaps even 20,000 people (Marchetti and Kolokotsas 1993). When the Hera games were shifted from the Heraion to Argos at the end of the third century B.C.E., dramatic contests took place in the new theater. One of the earlier sanctuaries spared in the theater's construction was a well-shrine to the River Erasinos, suggesting the possibility of either a long-distance waterline to the city from that southern spring (predecessor to the Hadrianic line 400 years later) or a realization by the ancients that the well tapped the same water table as the river did (Moretti 1998). The defensive rampart and the related streets of the city received attention during the Hellenistic period, especially the relationship between monumental gates at the southwest, southeast, east, and north and the main arterial roads of the area (Pitéros 1998). The southwest gate was found by geoelectric soundings (Reynal 1986). Much of the line of the rampart was recovered by following the ditch outside the wall (Aupert 1981; Hesse 1988). Hesse expected to find a rampart of polygonal masonry like that on Larissa, but did not. Along the wall, which frequently responded to details of the topography, most of the visible remains are Roman, though it is believed that the city was also walled during the earlier Greek period. A silt layer of a bit less than 2 m — just the maximum depth found by Zangger (1993) as having accrued since Roman times — could hide completely many other vestiges of the early streets and ramparts (Hesse 1988).

After Rome annexed the Peloponnesos in the second century B.C.E., the architecture and urban plan of Argos developed according to Roman standards and means of expression. The Hellenistic temple just east of the theater became a Serapion/Asklepeion centered on a large Augustan building (discussed subsequently). A new water supply for the city supplied the Kriterion fountain (Fig. 4.5b) attributed to Hadrian about 120 C.E. by an inscription, verified by the *bipedales* (two foot) bricks used there, typical of his constructions. (See Vollgraff's drawings of the Kriterion fountain — plan, section, view, and detail — 1956.) Its grotto, which may originally have had its own spring (Gaki, personal communication, 1994), was given an Ionic portico and topped by a statue of Hadrian in heroic nudity, flourishing a sword, and is similar to another Hadrianic fountain at Perge in southern Asia Minor. Hadrian had the water system extended to the bath near the theater, to another bath south of the agora, and to a third whose whereabouts are unknown, as well as to a fountain in the agora and to the lower town. Hadrian also rebuilt the theater after an extensive fire (Spawforth and Walker 1986); the theater was repaired several more times during the second-to-fourth centuries (Reynal 1986; Marchetti and Kolokotsas 1993). The tholos fountain in the agora received a new dedication to Adonis in the Roman era, when it was rebuilt. The palestra along the southwest edge of the agora was embellished with porticoes. The race track, however, was displaced from the center of the agora to the north, in a new stadium; the Hera and Nemean games were then held just beyond the Aspis hill (Hall 1995).

At every period of Argos's urban development, from the fifth century B.C.E. to the sixth C.E., the matter of drainage had to be taken very seriously, as the area was unusable without major drainage (Pariente et al. 1998). Aupert's map (1989) of the northern aqueduct and the Xieras (Charadros) River runs the two through Argos on a path equivalent to these huge channels. In Hadrianic times, the drainage of the agora was modernized with three sewers of brick. These huge, arched channels, tall enough for a short adult to walk upright, run through the middle of the agora. With the soil covering removed during excavation, the now-visible tunnels are the most interesting features of the agora (Aupert 1981, 1982; Marchetti and Kolokotsas 1993; Reynal 1986). In size and shape they are similar to the great drain of the Serapeion bath (Ginouvés 1972) (Fig. 4.7) and very similar to long stretches of tunnels found in the Thessaloniki agora, one in the Stadtsmarkt (Gov-

FIGURE 4.7. Photo of the drain tunnel from the Roman bath at Argos, similar to the set of three large drains in the agora. The tunnel is tall enough for a medium-size adult to walk upright in. Drains of this time, made with brick vaults, are assumed to be Roman. (Photo by Crouch)

ernment Center) at Ephesus, and another at the center of Athens, all of Roman age. Architect Reynal (1986) suggested restoring the drains across agora to their original function as part of his plan for a revitalized Argos.

Attention to roads and communication was a prerequisite for the functioning of the Roman Empire. Important data comes from a map of late Imperial Roman roads (4th century C.E.), called the Peutinger Table. Sanders and Whitbread's useful study (1990) of Peloponessean roads comparing the roads shown on the Peutinger Table with the "least paths" method in topology, combines the methods of two disciplines—social geography and geometry—to link the roads and seaports of Achaea (the Roman province of which Argos was a part) into the area-wide communication system. The Peutinger Table and other land maps show singular paths, whereas the sea routes are multiple. That is, there was basically one road from Argos to Corinth, which took 8 hours to walk, but from a port on the bay of Argos one could sail to many ports in the Mediterranean. If both sea and land are taken into consideration, Argos rates as one of the three best-connected cities in the Peloponnesos—more important than any other town on the southeast coast (Sanders and Whitbread, 1990: fig. 2, Roads on the Peloponnesos and in Achaea; fig. 3, Analysis of roads and sea links of Achaea; fig. 4, Analysis of roads and sea links of the Peloponnesos; fig. 5, Connectivity in the Peloponnesos as related to time distances). Argos was nearly as important as Corinth (the capital of the province) and Patras (the western port). In antiquity the port of Argos was at Temenion, located east of the modern river mouth by excavations (Drorvinis 1998; see also the sixteenth-century maps, his figures 5-15 and 9-22).

During the long post-Roman period, the locus of settlement at Argos shifted to accommodate different populations and life patterns. After the invasion of the Visigoths in 395 C.E., the city center moved eastward, and new buildings were constructed. The earthquake of 552 was followed by epidemics and Slavic invasions during the rest of the sixth century, leading to the decline of the city (Banaka-Dimaki et al. 1998; Abaddie-Reynal 1998). The Roman area around the agora was also the site of the medieval Byzantine town, with which is associated the Church of Panagia Katakekkrymeni (the Virgin Mary), which replaced the Temple of Hera on the side of Larissa hill in the seventh or eighth century (Sofianos et al., 1988). The Turkish town of the later Middle Ages lay to the north, just where the route from Corinth met the streets that climbed the Aspis and Larissa (Reynal 1986, Fig. 6, plan of the city). The Turkish government brought in Albanian shepherds to take advantage of the grazing potential of the mountain valleys above Argos. (One outcome of this resettlement is that although Argos is near the sea, it is difficult to find fish in local restaurants—lamb is preferred!). From the government's point of view, it was easier to install the new population in an area adjacent to the old town of Argos than to displace earlier inhabitants. (This pattern is also evident at Agrigento in the Saracen occupation next to the medieval city.) After 1828 when the Greeks gained independence from Turkey, the classical site of Argos was reoccupied. The three modern cemeteries to north, east, and south of the city are those also used in antiquity.

Urban Design Based on Geology

One building's site, in its changing form and uses, exemplifies both the architectural developments at Argos over this long period and the imposition of Roman concepts of urban design. The bath building near the large theater at Argos tell us how the city and its people changed during the Hellenistic and Roman periods, while still fitting into the geological setting in a realistic manner (Aupert 1982, 1985, 1988, 1994). This building was both religiously important and well supplied with water for about a thousand years. At first a Hellenistic Egyptian sanctuary uniting worship and healing, it became a Serapeion/Asklepeion about 100 B.C.E. using water from the slopes of Larissa. In about 125 C.E., the courtyard of the Asklepeion was filled with a bath added under the rule of Hadrian; supplied by his two aqueducts from the springs 5 km to the south and 10 km to the northwest. Overlapping water supplies from sources in two different directions was a basic concept in Greco-Roman municipal water management (see Ephesus case study). A fourth century C.E. inscription suggests that the building was then a gymnasium. In its elevated site, its water supply, its building materials, and its relation with neighboring buildings and the extensive agora, this bath-temple depended heavily on the local geology. Its continuing lack of direct access to the agora preserved an air of mystery that contributed to the effectiveness of the cult.

Because of the early origin and longevity of the settlement, the regular street patterns that appear in other classical, Hellenistic, and Roman city plans are much less evident here. Because Argos, like Athens, dated from very early in the history of Greek city building, its pattern was organic, not a regular grid. After all, the planner Hippodamus did not live until the fifth century, and by then the Argos settlement, though not the formal city, was nearly a thousand years old.

The growth of the city responded to the local geography (Fig. 4.1 and list of stratigraphy): looking up to the almost straight-sided hill of Larissa, utilizing the lower and rounder Aspis hill as a sanctuary equivalent to the acropolis of other Greek cities, and later crowding the Roman buildings of the city together near the water sources. Large as the Roman bath and Greek theater were, they were dwarfed by the Larissa hill looming up steeply behind them. Roman urban design had an "organic" and irregular quality, at least in the capital city of Rome and here at Argos. In the succession of great fora of Rome (first century B.C.E. through fourth century C.E.), large and magnificent buildings jostle one another for prominence, without the punctuation and separation of streets and long vistas that we moderns, since the Renaissance at least, have come to expect as "spacers" in urban design (MacDonald 1986). In Argos, the Roman manner of urban crowding shoved the bath up against the theater and crowded their neighboring buildings so close to them as to give little room for visual contemplation of their individual greatness.

The resource base of Argos, her political role in the Argolid and in the Peloponnesos, and her commercial connections with the rest of Greece, Rome, and Egypt are made explicit in the city's fabric. Blessed with ample resources for settlement—water, fertile land for agriculture, mountain valleys for grazing, the sea for fishing and trade, and the mountains for stone and wood for building—Argos was well situated to become a strong Greek city in many periods. Indeed, Homer refers to all the Greeks as "Argives," suggesting the city's preeminence in his day (ninth century B.C.E.).

Yet settlement here was constrained not only by human needs and human nature, but also by geological processes. As late as the Bronze Age, the sea and the land were adjusting their boundaries, to a depth of 80 m and an extent of 10 km. During the Greco-Roman period, the plain of Argos was very similar in size to the modern plain, but had to be managed intelligently to avoid problems and even catastrophes of flooding and drought. Farmers had to learn what grew well and how to enhance the yield of their crops. Herders had to learn when to leave for the mountains and when to return with their flocks to browse the stubble of the fields or the grasses of the wetlands, as they did until at least 1938. Providers of raw materials had to locate and transport clay for ceramics and terra-cottas, as well as stone and wood for construction, without bogging down in the marshes or slipping off the steep mountain roads. Artisans had to master the subtleties of local materials and the quirks of local and distant consumers. Frequent floods, periods of drought, and possibly earthquakes were vivid reminders that the physical world was dynamic not static and had to be propitiated. Over the many centuries of this period (800 B.C.E.–600 C.E.), within the normal fluctuations of rainfall and gradual alluviation, the Greco-Roman Argives learned to live in balance with the environment.

Their most important achievements were in the field of water resources. Not only torrential rain in winter and the outflow from springs along the bases of Larissa and Aspis, but also the occasional floods of the rivers and the perennial need for irrigation water for the fields to the south made for complicated and demanding water management. No wonder that we find numerous layers of supply pipes, drainage pipes, and channels in the street leading up to the theater. No wonder that during the Greek period (third century B.C.E., according to Reynal 1986; fig. 5) several parallel sets of stone channels directed the flow along the natural slope of the agora; they lie beneath the Roman tunnels of brick in the same areas. The whole arrangement speaks of a determination to control and utilize the potentially dangerous floodwaters (Niemi and Finke 1988). And the Hadrianic aqueducts from distant sources supplemented the earlier lines that drew on the water table within Larissa.

To these water problems and others of seismicity, building materials, and agriculture, the Argives developed solutions that made possible sustained urban settlement. The city's, duration contrasts dramatically with Morgantina (case 2) and Priene (case 10), whose life spans were about 400 and 800 years, respectively.

Corinth

History of the Site

Settlers came very early to the site of ancient Corinth, site, blessed with ample land and water, located at the crossroads of east-west and northeast-southwest roads, with access to the sea to the east and west. Habitation began no later than the Neolithic period, continued through the Bronze and Early Iron Ages, and lasted through the devastating earthquake of 1858 C.E. In the hinterland (the Corinthia), deposits of materials from two Neolithic, eleven early Helladic, seven Middle Helladic, and six Late Helladic sites had already been found before 1920 (Blegen 1920. map). Mycenaean materials from the Demeter and Kore Sanctuary date to 1340–1300 and 1140–25 B.C.E., while the Mycenaean trading post was at Korakou village, the later Lechaion port, just east of the river mouth (Rutter 1979). Beginning around 900 B.C.E., a group of Dorians from Argos settled here and gave their Argive Hera to be chief goddess at the Perachora site on the north shore of the gulf, opposite Corinth, in the most distant area of the city-state, providing further evidence that shrines were located to define and protect boundaries, as at Selinus and Argos.

During the Geometric period, Corinth was an area of villages that gradually coalesced to form a town. Corinth organized as a *polis*, an independent city, by the late eighth century, when the last mythical king was replaced by the first ruler from the Bacchiad family, which was active in the city's colonization efforts: There is some evidence of famine in eighth-century Greece (Camp 1979b)—a problem that land monopoly might have exacerbated, making colonization more attractive. The growth of trade, colonization, and overpopulation were interrelated. Corinth founded colonies to the west of the Gulf of Corinth. One was named Cephalonia, being a major source of the commonest fir tree for ship masts and construction. One was Syracuse, a settlement based on trade in ceramics and on agricultural products.

Since Corinth had a great deal of fine, pale clay ideal for pottery, the ceramic industry developed rapidly, sending traders all over the Mediterranean. The potters traded first with their fellow Dorians in the Argive Plain, and then by 750 with Delphi; by the last half of that century Corinth was the chief trading city of Greece (Roebuck 1972). Many workshops made the city a production center, not just a market. The transfer of forms and functions between terra-cotta and metal vessels stimulated production of decorative metal vessels and objects as well as armor— further contributions to the export trade. After 700, industrial growth stimulated building construction, stone cutting, and carpentry.

By 650, a new family of rulers, the Kypselids, continued the tradition of founding colonies (six more) and enhanced Corinth by their building activity, such as the archaic temple on the site of the later Temple of Apollo (Malkin 1989). The temple was roofed with terra-cotta tiles—one of the first known uses of roof tiles (compare with the Cittadella area at Morgantina). In the seventh century, defensive walls were begun, some still visible at the Potters' Quarter.

Forming a mental picture of how the site of Corinth looked at the beginning of the seventh century B.C.E., before any of the large reshapings of the terrain is very difficult. If only geologists had been as interested in the geology of human settlement as they were in the location of petroleum supplies! However, we do have the Higgins book (1994) and a final report by Scranton (1951) on the Lower Agora from which we can derive a few ideas, and other publications provide numerous fragmentary archaeological clues. Scranton points out that a narrow valley ran near the Propylon, in the southwest part of the agora area, between rock outcrops. A shallower valley stretched from the Propylon site southeast toward the Southeast Building site. All three of the early springs (Sacred Spring, Peirene, and Cyclopean Spring) were supplied from Acrocorinth naturally, and they opened into the hollow at the top of the Lechaion road. To the west and slightly north of this hollow, the ridge of the Temple of Apollo stretched from the north-south valley of the Lechaeon Road toward and past the Glauke Fountain, with a swale cut between the ridge and the lower surface of the southern area where the Roman forum would later be. West of the Sacred Spring, the ground sloped up to the hill of Temple E (a Roman replacement of an earlier temple).

The core of the city continued to change during the Greco-Roman period.

For instance, the South Stoa of the fourth century B.C.E. cut into and partly destroyed a large earlier cistern, which had the sinuous appearance of channels and reservoirs enlarged from karst channels, like those at Agrigento and Syracuse. In the South Stoa area, C. M. Morgan (1939) found house walls of sun-dried clay brick (adobe) dating from before the sixth century B.C.E., as well as Helladic materials from more than a millennium earlier (Broneer 1954). By the end of the Greek era, the last vestige of the valleys south of the Propylaea had disappeared and the ground was leveled to a continuous slope, within which a hollow was provided for the entry of the Lechaion Road up to this level.

During the archaic and classical periods, from the late 700s to 350 B.C.E., Corinth was still a loose set of villages whose residents went to the center for religious, civic, entertainment, and market activities. Several cemeteries of the tenth to fourth centuries have been found within the classical walls, whereas centralized Greek cities enforced burial outside the city limits. The ancient Greeks defined the "city" population as those living within the walls and those in the outlying areas equally. As Duncan-Jones puts it (1963: 85), "the influence of cities was more comprehensive than today, but urbanization [density] was generally looser." Really large towns with thousands of private residences, however, were closer to modern towns in their density (Duncan-Jones 1963, n.11).

Corinth's independence was lost to the Macedonians in 338 B.C.E. for a hundred years, but enough local patriotism remained to spark the struggle against Rome in 146 B.C.E. This ended in disaster, the city being razed to the ground and the fortifications of Acrocorinth dismantled. Local government of the area ceased and exports halted, so that economic self-determination was lacking after 146, although there are indications of farming and some commerce after 111. At that time, the Romans conducted a survey preparatory to the sale of Corinth's farmland to new owners both Roman and Greek. A modest revival began, to serve the incoming surveyors, farmers, Roman agents, and landowners, but this population was without urban structure, since neither a municipal government nor a ceremonial core were functioning (I. B. Romano 1994). After about 100 B.C.E., however, some commercial activity and foreign interaction is evident from archaeological finds.

The site remained sparsely populated until the new Roman town of Colonia Laus Julia Corinthiensis was founded in 44 B.C.E. Julius Caesar and Augustus between them founded over 100 colonies. Their aims were to transfer surplus free populations out of Italy, thereby increasing the peace by reducing the numbers of trained belligerents and at the same time increasing the number of areas growing grain and generating taxes for the central government (Hopkins 1978b: 67). Julius Caesar was the original patron of the new town, though settlement by Roman veterans was completed after his death. Many bore Greek surnames, suggesting that freedmen were returning "home."

Roman Corinth was laid out from a definite plan, based on a detailed and accurate topographical map generated by the surveyors working for the Roman patron, as discovered through the survey work of D. G. Romano's modern team (1993).

The official plan had nine north-south roads and seven east-west, producing a grid plan familiar to the veteran colonists from their army camps (Fig. 4.8). The

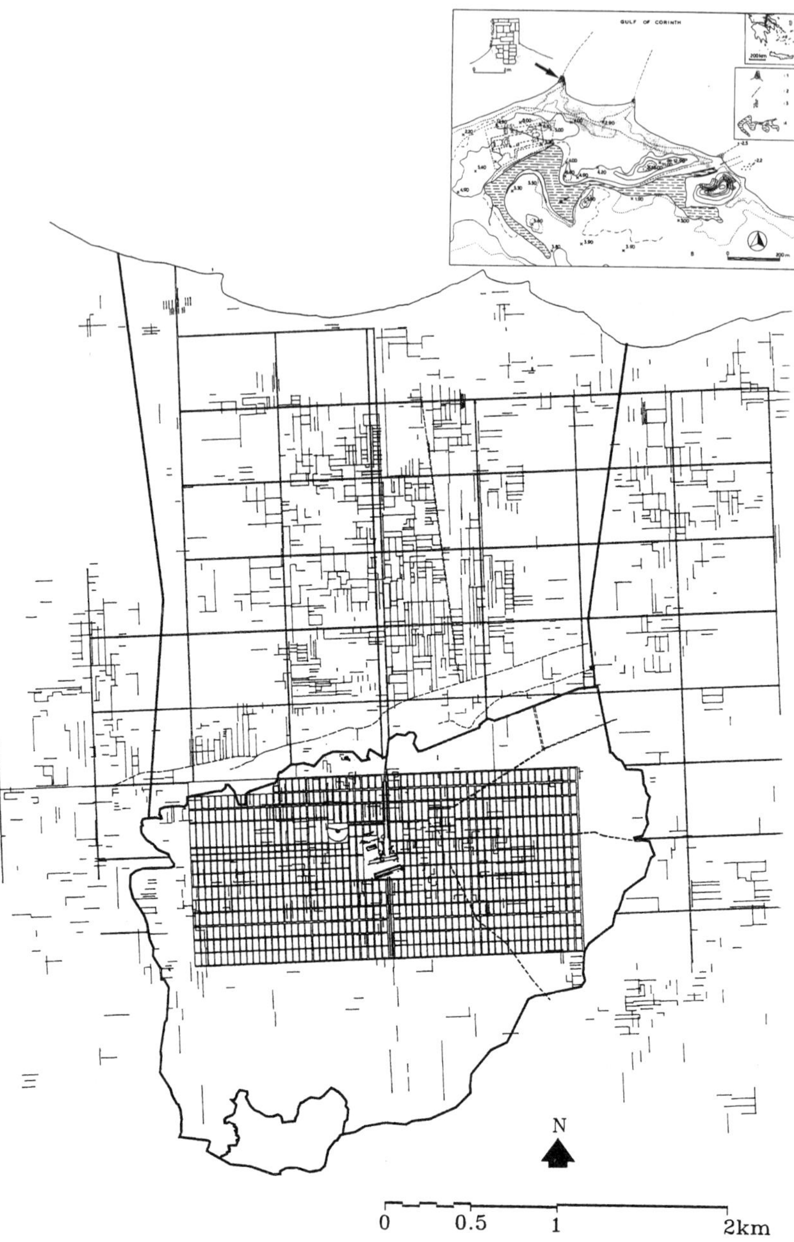

FIGURE 4.8. Surveyed map of ancient Corinth during Roman times. At the northern coast, a map of Lachaeon port has been inserted. The walls of the Greek period, rebuilt by Romans, outline the city. Outside the east and west walls, and within them between the north wall and the coast are ruined house walls, remnants of three overlapping divisions of farmland (centuriation). To the south, between the inner north wall and Acrocorinth was the Roman city, a regular grid centered on the same open space (agora/forum) that the Greeks had used. (City map by D. G. Romano 1998; Lachaeon by Mourtzas and Marinos 1994.)

main street was the north-south Lechaeon (Greek)/Lachaeum (Roman) Road (excavated for 85 m) paved with limestone after the earthquake of 77 C.E. and extending to the port for a total of 3000 m. This road and a major east-west road, running from the coast at the northwest through the Forum to the Krenachi port at the east, divided the new town into the four quadrants of the typical Roman colonial plan. Similarly, the agricultural land to the north was replatted in a typical Roman centuriation (hundred foot) plan; two previous gridded divisions of the farmland, one Greek and one an earlier Roman plan, have left traces on the ground (D. Romano 1993).

The new residents provided themselves with the structures a city required to be a complete provincial capital: temples and shrines, municipal and provincial government buildings, gymnasium, theater, marketplace, houses, shops, workshops, and other production facilities, fountains and aqueducts (Pausanias X. 4.1), and a new defensive wall. One sign of the change of ethnicity is the amphitheater at the eastern edge of the grid. All Greek cities with amphitheaters (others are Knossos, Patrae, and Gortyn) were settled by Roman veterans (Welch 1996); amphitheaters are common in Italy and North Africa, rare in Roman Greece, and almost unknown in the eastern Mediterranean area.

Under the Pax Romana, Corinth was the capital of the province of Achaea. Corinth came to have a second century C.E. population of about 100,000 in the urban and rural areas combined (Engels 1990). A Christian community from the mid-first century C.E. is still commemorated in visits from the faithful. The deaconess Phoebe and other Christians visited by St. Paul lived at Krenchreai about 52–53 C.E. Corinth flourished despite the barbarian raid of 267, the earthquake of 375, and the destruction by Alaric and his Visigoths in 395 C.E. Even then, the site was so important that major repairs were made, such as the agora reorganization of the fifth century (C. M. Morgan 1939), enabling Corinth to carry on until further earthquakes in 522 and 551.

A reduced population remained during the later Byzantine era. Europeans made raids and conquests beginning in 1147 with the Norman sack. Frankish rule in the early thirteenth through the fifteenth centuries was gladly exchanged for Turkish rule in the later fifteenth century (after the Turkish conquest of Constantinople and then Greece). Foreign domination continued until the Greek War for Independence in the early nineteenth century.

As recently as 1858, ancient Corinth—then the center of the Greek silk industry—was destroyed by an earthquake. The residents moved to the coast and built New Corinth, which in turn suffered from the earthquake of 1928. The new village of Old Corinth grew up later around the ruins to serve the excavations and the tourists.

Geology and Hydrogeology

How do we account for 7000 years of settlement here although the soil is not rich and earthquakes are common (Higgins and Higgins 1996)? The longest view and most complete geological description of the Gulf of Corinth and its coasts is that of Armijo et al. (1996), noting high seismicity in all periods. For five to ten million years activity along a series of parallel faults has extended the Aegean arc from eastern Anatolia to the western Peloponnesos, producing many grabens. The Gulf of Corinth is one of these: a symmetrical half-graben or asymmetrical graben (Keraudren and Sorel 1987) running NE, with an uplifted south footwall where the Corinth terrace is located and a down-flexed north hanging wall. Its extension to the east lies in the Saronic Gulf. The Argolid is a related graben, long ago attached to the Gulf of Corinth.

The geology of the northern and eastern Peloponnese is especially affected by extension of the earth's crust in the Aegean region. The isthmus of Corinth shows marked rotation because of the stress of extension (Mourtzas and Marinos 1994). Similar rotation on the eastern side of the Aegean, at Miletus, Priene, and Ephesus, is likely. In the Aegean area, the recent tectonic deformation appears to be related primarily to vertical movements of small tectonic blocks. The region, however, has few volcanoes along its eastern edges and few hot springs heated by faults related to ancient rather than currently active volcanoes.

The Gulf of Corinth increases in width, depth, and sediment fill from west to east. The complicated interplay between sea-level changes and tectonic uplift is measured near Corinth at 3 mm/yr—not the normal 2 mm/yr of the region. Keraudren and Sorel (1987) recognized 20 terraces representing 19 marine transgressions since 450,000 B.P. The two terraces at New and Old Corinth are, fortunately for our understanding of the site, the best defined of the area. The terrace of Old Corinth is 235,000 years old, dating from just after the main interglacial period, whereas the lower one at New Corinth is only 124,000 years old.

The geology of Corinth and Acrocorinth is shown schematically in Figure 4.9 (Higgins and Higgins 1996). The stratigraphy of Corinth is as follows, beginning with the topmost modern layers:

Jurassic limestone that forms Acrocorinth

Jurassic shale; red and green marl containing iron, lower on Acrocorinth

Beds of fine clay and Pliocene marls

Terrace layers

    Conglomerate and pink limestone breccia

    An upper layer of soft reddish sandstone incorporating decomposed limestone

    Conglomerate forming the ledges above the springs along terrace edges

    Layers of very hard clay and impermeable, fine, hard sandstone or sandy Pleistocene limestone

Alluvium and beach deposits

Dufaure and Zainanis (1980) described the Temple of Apollo terrace as a simple structure from a single terrace-formation cycle, made of basal conglomerate with beach sands and aeolian sands probably in the form of sand dunes. These sands were transformed into *poros* (sandy limestone of the Pleistocene era), which was used for construction of the Apollo Temple and other structures (Higgins and

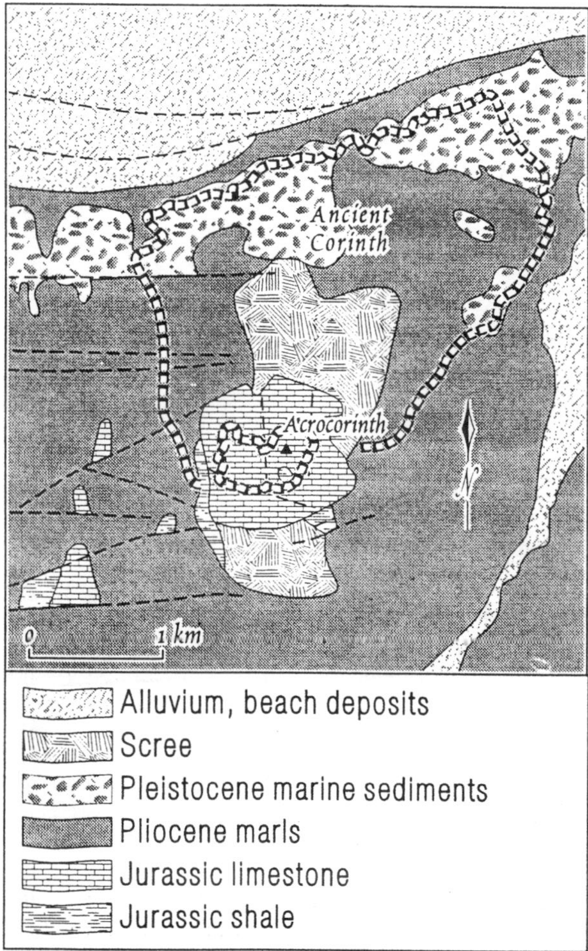

FIGURE 4.9. Schematic map of the geology of Corinth. The chain symbol represents the defensive walls of the city; dashed lines stand for faults (Higgins and Higgins 1996).

Higgins 1996). Other terraces, for example, the large Old Corinth terrace, have a more complex geological structure involving two sequences of conglomerate followed by marls and sands (Fig. 4.10).[5]

Quarries   Beginning during the archaic period when the Corinthians started to build monumental architecture, they explored their area for suitable stones. They were concerned with the ease of extraction, the durability of the stone, the ease of carving, and the aesthetic qualities of the material. "Marble and poros can be sawn and hence lead inevitably to an ashlar style," noted Carpenter and Parsons (1936), whereas "the heavier limestone must be split and hence tends to arbitrary and irregular shapes." During their earliest efforts, builders of the Temple of Apollo extracted small limestone blocks from the quarry adjacent to the temple. The rather weak stone without joints was quarried as large blocks that needed to

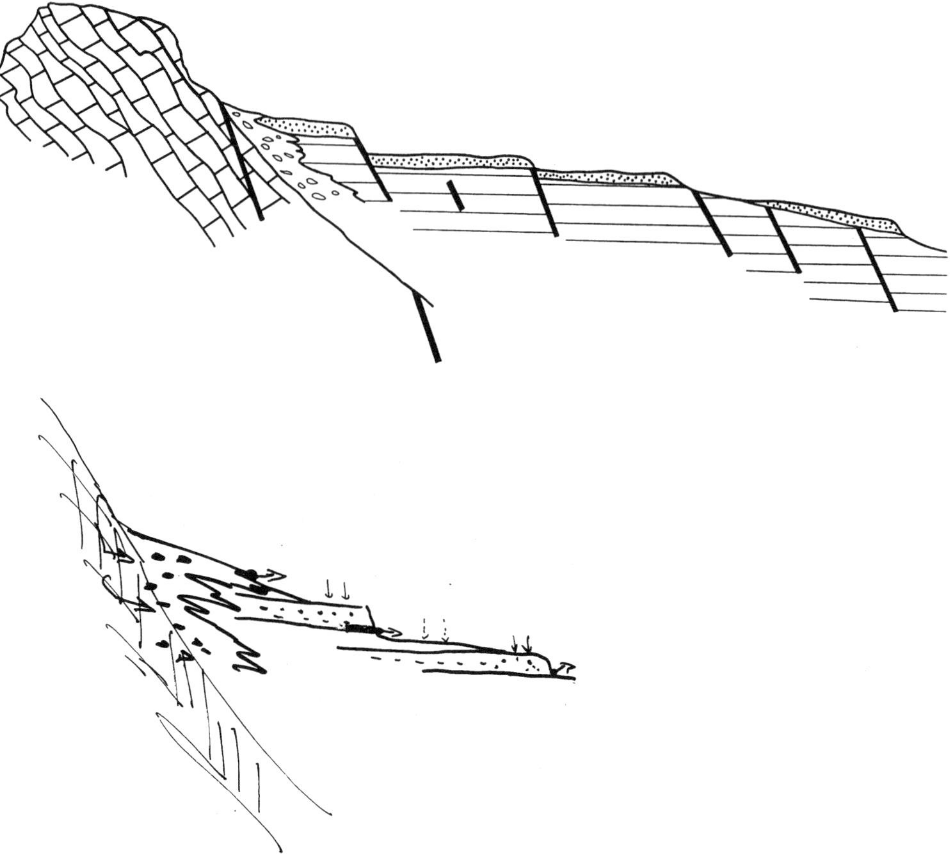

FIGURE 4.10. Section from Acrocorinth (at top) to the Gulf of Corinth (at bottom). The limestones in Acrocorinth are represented as masonry-like blocks. Detail of section. B Dots represent skree and conglomerate. Heavy arrows are underground water and resurgences. Smaller vertical arrows indicate rain. Large and small dots represent skree, partially cemented into horizontal conglomerate. (Both sketches by P. G. Marinos 1995)

be faced with stucco to preserve the surface (Higgins and Higgins 1996). The quarry seems to have been enlarged during the Roman period, when new trenches were cut around Glauke Fountain (Williams 1987), but I have not found specific reference to Roman buildings made from this stone. Another two quarries dating to the sixth century B.C.E. lie 140–180 m due west of the northwest corner of the Temple E site; these were sources of the column drums of the Temple of Apollo (Robinson 1976). In addition, sixth-century quarries are located at the northwest corner of Temple E and at the northeast end of Temple Hill; other rock outcroppings within the city were also quarried. A vast ancient quarry lies 1 mile east of Hexamilia village, according to C. M. Morgan (1939), and still other quarries were

4 km east of the central core. A detailed study of quarries lying just outside the archaeological area was made by C. L. Hayward (1995). Like other Romans, the Corinthians also used imported stones such as white or gray marble and cipollino from Euboea, but pink limestone breccia is found on Acrocorinth (Higgins and Higgins 1996). Breccia forms when carbonate-rich water cements limestone fragments derived from the collapse of caves that are a standard feature of karst geology.

Acrocorinth and Karst Acrocorinth was once an island composed of mid-Jurassic limestone. The ancient sea left deposits of marl and shale along the flanks of Acrocorinth, and these deposits have become the bottom layers of each series (Fig. 4.10). Marl contains much iron, as revealed in the red and green colors seen in road cuts along the hill. The very hard clay is also seen at the bottom of wells and in subterranean galleries of the water system, and it forms the floor of the Sacred Spring house.

In the karst of the Corinth site, the rainwater falls on the mountain and flows northward in the skree and more deeply within karst channels inside the mountain, until its passage is blocked by impermeable Pleistocene marls. Then the water rises along faults toward the surface, spreads out in sheets oozing along contacts between the marl and conglomerate, or follows larger channels that the water has carved in the rock. Water flows outward at the Peirene Fountain and other resurgences.

Utilizing the natural flow of water in underground aqueducts was first worked out by the Armenians as early as the eighth century B.C.E. They amplified natural resurgences by enlarging channels. This technology was passed on to the Persians and thus to the Greeks (Crouch 1993: chap. 10). In the case of Peirene, the natural spring emerges from the seam between very hard clay and impermeable, fine, hard sandstone or sandy limestone, with conglomerate above.

That the level of groundwater supplying Peirene varied is suggested by side niches or pipes to carry water at several levels within the galleries. The west tunnel to Peirene was cut in clay below the sandstone, in an area without conglomerate. Each channel in this tunnel has several short branches to collect water (Roebuck 1972). The east tunnel is similar in concept, though not in details. Its source of supply may have been the same that presently gives water to Hadji Mustafa fountain—lovely ice-cold water once used as refrigeration in the shops of the South Stoa (Broneer 1954). The early Greek facade of the fountain was destroyed by the late Greek and Roman arrangements still visible at this spring (Hill 1964). This very cold water of Peirene reappears at Lerna, where it drains over the cliff to the north (Roebuck 1951) and supports a water-loving fig tree. All in all, there is ample water for the center of the city at the edges of terraces and along ravines.

A particularly interesting spring is the Upper Peirene, located on Acrocorinth in a small saddle southeast of the summit, 5 m below the present ground level. The spring is fed by rain which seeps through the rock until it is blocked by shale; the water runs along a north-south fault bisecting the hill. Erosion of rocks along both sides of the little fault has produced a hollow in which the ancient springhouse and a modern cistern have been built. The ancients understood that there was a connection between Upper Peirene and the Peirene in the agora, but they thought of the connection as a simple tube; today we speak of an aquifer with water oozing slowly through rock.

The ancient inhabitants had an keen knowledge of local hydrology (Kardulias and Gregory 1991), as evidenced in their use of natural and built features in the channels. On the surface they intervened to prevent erosion, regulate sedimentation, and provide abundant water. Within the site of Corinth, Kardulias and Gregory did not find terracing but rather the preservation and enhancement of slope integrity. They postulate minimum cultivation along the sides or bottoms of ravines in ancient times, although some ravine bottoms with rich soil are currently farmed. The work of Kraft (1998) and Brückner (1997 a & b) about landscape changes suggests that the floods of 2000 years and concomitant alluviation have probably wiped out ancient traces of temporary weirs and dams in these ravines, burying remnants deep in alluvium. Still, ancient check dams to retard erosion are visible at the bases of some ravines (Kardulias and Gregory 1991).

Additional water was brought to Corinth from the karst system of the Tripoli plateau, for example the long-distance waterlines at Argos. Under Emperor Hadrian about 120 C.E., an aqueduct about 85 km long brought water from the Stymfalian Kefalari to Corinth to supply the great new bath of the same period, lying east of the Lechaeum Road on the second terrace, and to supplement the water supply of the first terrace between the South Stoa and the foot of Acrocorinth. Lolos has analyzed this aqueduct (1997: his fig. 5); details of lithography along the aqueduct indicate that construction varied with soil conditions. There are built tunnels, rock-cut tunnels, support walls in limestone areas, and deep subterranean trenches in marl and alluvium. Hybrid solutions were necessary in areas made up partly of alluvium. Lolos's maps show that the aqueduct curves around the northwest flank of Acrocorinth to enter the city.

Coastal Change at Two Ports   Corinth had two ports with access to the Mediterranean world (2 and 3 on Fig. 4.11). One harbor faced east toward the Aegean and Ionia, the other west toward the Gulf of Corinth, Magna Grecia, and Sicily. Exports were bronze work, architectural terra-cottas, textiles, cosmetics, and stone (e.g., for the Temple of Apollo at Delphi) (Roebuck 1972). Since the sea route around the southern end of the Peloponnese was extremely windy and slow, it was often more cost-effective to transship materials and even whole ships across the isthmus by the paved transit way (*Diolkos*) than to go around. The Diolkos was 6–7 km long—longer than the modern Isthmus Canal—but only 3.5–5 m wide, except at Poseidonia on the west where the paved width was 20 m for loading purposes (Raepsaet and Tolley 1993). Beach rock is visible below and above the paved road, which itself is made of slabs of beach rock (Mourtzas and Marinos 1994). The three road layers encapsulate the history of deposit and uplift since Greek times. Since beach rock forms just under the surface of the water (Fig. 4.12), the lowest layer was once under water, then rose enough to be a dry-land base for the middle layer of paving, which again sank enough to become the basis for the present top layer (Marinos, personal communication). These changes in the heights of water and land affected the ports also, as we will see.

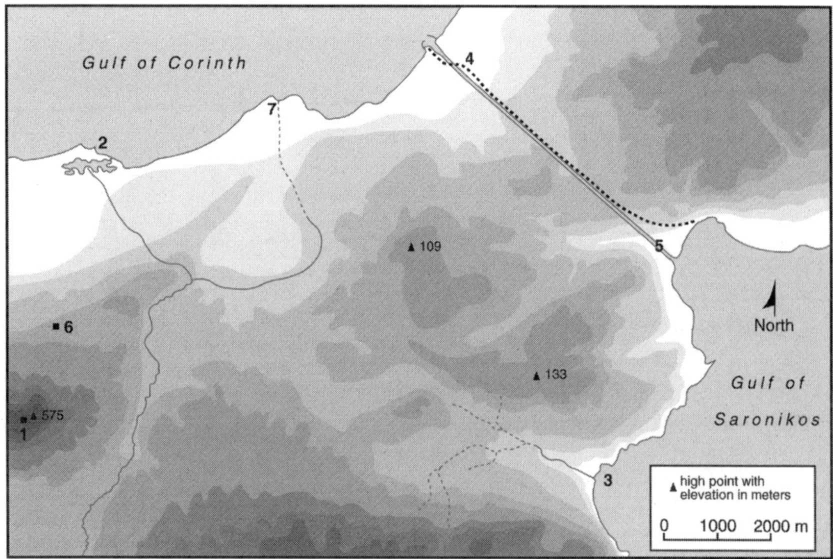

FIGURE 4.11. Map of Corinth's hinterland and the Isthmus. The dashed line (4) along the modern canal (5) at the upper right is the Diolkos or carriage way. Modern Corinth (7) is shown at the shore of the Gulf; ancient Corinth (6) lies on an upper terrace, north and slightly east of Acrocorinth (1). The thin line east of Acrocorinth is the river that runs to the port of Lechaeon (2) to the north. At right, near the scale, lies a second port, Kenchreai (3). Isthmia, a religious center for Corinth, was at the right end of the canal and south of it. (Raepsaet and Tolley 1993, redrawn by Langford; by permission of the French School at Athens)

Construction of the Diolkos began in the sixth century B.C.E. It was rebuilt in the fifth and fourth centuries and again probably in Roman times. The engineering implies sophisticated technology. As early as the sixth century, the Corinthians had experience in moving heavy loads when they quarried stone for the Apollo Temple at Delphi, bringing blocks and drums down a slope of 3.25% for a distance of 20 km from the quarry at Kirrha. At provincial Selinus, where blocks of up to 70 tons were moved from Cave di Cusa and Cave di Barone, the process of heavy transport was already taking place in the seventh century. The technology thus existed to move ships of 10–30 tons over the Diolkos. A carriage 37 m long and 6 m wide was appropriate for the maximum ship size and could be accommodated by the 5-m-wide road by having the carriage body extend beyond the width of the wheels. To pull this load, 21–40 pairs of oxen would have been needed, but this number was impossible to manage on the narrow path of the Diolkos. However, 117 people would have been the maximum number needed, using cables and rollers, plus fat or pitch as lubricant, and they could be accommodated by the available width of passage. The typical manpower of such a ship was quite sufficient to pull the ship, loaded on a carriage, over the hill from one

FIGURE 4.12. Beach rock on the southern edge of the Gulf of Corinth, near new Corinth. Three levels are visible: (1) amid the stones on the beach; (2) rising several inches above the level of the beach; and (3) just covered by the surface of the water. Beach rock is cemented immediately below the water's surface. The three rock levels here, which may form in as short a time as ten years, thus indicate three sea levels. (Photo by Crouch)

port to the other (Raepsaet and Tolley 1993). Passage took about 3 hours at 2 km/hr. Extending the paved road into the water meant that the carriage could be inserted below a ship and the two gradually pulled together out of the water. Since a modern sailboat of 40 feet in length is rated to hold 8 tons, the combined weight of a Greek ship and its freight would come within 30 tons, the inferred maximum load for the Diolkos. Thus we surmise that ancient ships of this length and load could be handled by the system (sailor L. Polk, personal communication).

Corinth's two ports, Kenchreai and Lachaeon, were also affected by the variation in sea and land levels. At Kenchreai, Greek, Roman, and fourth-century Christian buildings are presently submerged in up to 2.3 m of water (Fig. 4.13). A classical temple now at this depth some 150 m offshore had been elevated on the south point of the harbor mole, which also supported warehouses. Moles were built or rebuilt during the Hellenistic era, and a tower (possible lighthouse) was added at the inland end of the north mole. New moles and buildings were added or replaced during the Roman era, so that the harbor had porticoes all around, as shown on coins and in an *opus sectile* (mosaic) panel (Papahatzis 1984). At the time of the earthquake of 77 C.E., sea level at this port was 2.3 m lower than at present. It is thought that the earthquake lowered the land another 0.7 m. Buildings of the early Christian period were sited farther inward along the mole, at the

FIGURE 4.13. The submerged part of the southern jetty of Kenchreai, the eastern port of Corinth. Walls of the Isis Temple that stood at the end of the jetty are now mostly covered with water, but the apse of the Byzantine church that replaced the temple is more visible. (Photo by Mourtzas and Marinos 1994)

new ground level. Earthquakes in 365 and 375 C.E. resulted in sinking the moles by another 0.8 m, so that the early Christian structures are in shallow water now. An earthquake during the sixth century submerged the land another 0.8 m, causing the final abandonment of the port (Sakellariou and Faraklas 1972).

From Kenchreai, the road goes uphill and westward to Corinth, leading to the central core just south of the forum. The main road to the north from the forum led directly to the Lachaeon/Lachaeum port 2.5 km north of ancient Corinth (Figs. 4.8 and 4.14), much closer than Kenchreai. This port was originally the mouth of a river, like the ports at Selinus. By the sixth century B.C.E., the old system of pulling boats up on the beach was no longer satisfactory for the volume of trade, so that formalization of the port began. During the fifth century, Lechaeon was the principal harbor of Corinth. During the last third of the fourth century and beginning of the third century B.C.E. it was used as the western base of operations for the Macedonian fleet.

Imperial Roman changes at Lechaeum included repairs to the jetty and dredging to widen and deepen basins, as well as Latinization of the name. Since this harbor was the outfall of the river system that drained the area around the city (Raepsaet and Tolley 1993) (Fig. 4.11), it was similarly subject to alluviation. The study of the "harbor question" at Paestum, by geologists de Wit et al. (1988), suggests the type of information that could be gained if a sedimentologist studied this port. At Paestum, travertine deposits closed off natural drainage and even

FIGURE 4.14. Basin and approach channels of the inner harbor of Corinth's northern port, Lachaeon. The arrow points to the remains of a building on a small island in the harbor, probably a Greco-Roman temple rebuilt as a Christian basilica. (Photo by Mourtzas and Marinos 1994)

induced peri-marine marshy sediments (clays, peats, and calcareous muds) independent of the sea level. From 10 B.C.E. to 10 C.E., the newly stagnant water brought malaria to Paestum because drainage was blocked by the travertine and silt deposits. If a similar process were under way at Lechaeum, the new Roman rulers would have had to dredge out the harbor to keep it in service. Indeed, dredged sand was been piled up 10 m tall, with the piles serving as port markers. In Roman times there were two rectangular harbors outside the jetty, open to the northeast, with two piers (one to the west and the other between the basins). A canal led from the western harbor to the inner port basin. An eastward entrance led directly from the sea into the outer basin, between walls extending into the sea in a northeast-southwest direction. The inner port had two basins with a small island in the western basin (perhaps for shrines—a Temple of Poseidon with his bronze statue and a shrine of Aphrodite are known from the literature). A stone quay 7000 m long encircled the whole harbor, and the total surface of the port was 46 hectares (ha) (Sakellariou and Faraklas 1972). Although almost no excavation has been done here, surface finds around the port indicate settlement during the entire Bronze Age and then again in the early Christian period.

Like Kenchreai, the Lechaion port has endured both uplift and sea encroachment. Recent tectonic uplift makes visible parts that had long been hidden under the sand, whereas earlier subsidence lowered parts of the area. Some presently submerged parts were once 0.7 m higher, as can be seen from holes caused by stone-eating organisms that live above the sea surface, whereas emerged beachrock now cemented together with blocks from the construction of wharfs and piers indicates that sea level was once 1.1 m higher (Mourtzas and Marinos 1994). Thus the history of both ports results directly from geological change.

Construction, Destruction, and Urban Design

Changes at Corinth are partly the result of human actions such as fires, the destruction caused by war, and urban renewal. In addition, natural events such as earthquakes and uplift of land, floods and alluviation, and sea intrusion altered the area. Changes brought about by the interaction of humans and their geological environment are particularly interesting. The following are major events that changed the appearance of Corinth:

1. Formalization of the Peirene spring as a fountain no later than the end of the eighth century. Earliest wells also date from this period, as do the unimproved harbors.
2. An eighth-century drought may have induced a temporary decrease in population, especially of rural areas, when the first colonies were founded about 734–720. (Camp 1979b).
3. Prosperity from ceramic production and other industry in the mid to third quarter of the seventh century, and new governmental forms after the downfall of the tyrants seem to have been the stimuli for building the first monumental religious architecture, the Temple of Apollo on the ridge that became the core of the city.
4. Sixth-century destruction by fire of the Temple of Apollo and its replacement. Also construction of the rampart circuit, more temples, and the earliest racecourse in the central area.
5. In the fifth century, amplified water supply and drainage system necessary for increased population. In the fourth century, beginning of construction of large secular buildings: gymnasia and baths, racecourse, and theater. The South Stoa became the largest classical secular building in Greece (165 m long).
6. Defeat by the Romans in 146 B.C.E. meant complete destruction of the city, which was abandoned for a century.
7. Grid-planned Roman city refoundation in 44 B.C.E. (Fig. 4.8). Repairs, remodeling, and insertion of Roman buildings into the area within ramparts and adjacent to them, notably the amphitheater. Gridded residential areas and a larger grid of farm lands. Important earthquake in 77 C.E. Gradual reorganization of forum area, first for Imperial Roman purposes and then for early Christian purposes.
8. Barbarian raids of 267 and 395–400 C.E. bracket the period when many Corinthians were becoming Christian. Earthquakes in 365 and 375, as well as during the mid-sixth century and in modern times.

Against this background of notable events, quieter changes took place when people altered their understanding and use of the environment. People do not consistently notice features of their surroundings, nor do they use those features in the same way century after century.

Urban Design

In every phase of Corinth's history, the looming bulk of Acrocorinth and the definite edges of the terraces were the major physical features that determined how the Corinthians perceived their world and how they occupied it. Sometimes, the edges of terraces were boundaries between villages or urban quarters or between the city and the countryside; again the edges were important as the locus of springs and shrines. Locations of buildings were determined by the valleys that penetrated the space from the north and by the major springs (Martin 1983; his figs. 1 and 3).

The spatial pattern of Corinth in the Roman period has been explored and mapped by D. Romano (1993) (Fig. 4.9).[6] Roman Corinth is a grid parallel and perpendicular to the main axis. At the center, the plan of about 44 B.C.E. called for leaving empty a block of six by four insulae for the new civic center (forum) approximately in the area of the old Greek agora. These insulae were measured in the *actus* of 120 Roman feet, set out in modules of 1 × 1, 1 × 2, 1 × 3, and 1 × 4 parallel to the long axis of the road to the Lechaeum Port. Yet the regular Roman plan accepted the necessity of variations from the grid to respect the topography and the surviving or rebuilt monuments. The grid was planned to fit within the Greek long walls, half on the lower terrace leading north to the edge where the Asklepeion stood and half on the next higher terrace leading south to the foot of Acrocorinth.

The longest occupation of any site in Corinthia is on Temple Hill at the center of Corinth. Excavators found Neolithic pots in natural cavities of the rock surface, near foundation cuttings for the Apollo Temple and the seventh-century road (Robinson 1976). The springs of the agora area—especially the Sacred Spring, which was in use in the seventh century—were the early foci of residences and workshops, but they also acquired shrines. The pottery found at the Sacred Spring relates it to worship of a nymph who may have been Kotyto, daughter of the pre-Dorian Timandrus, which would date the use of the spring to the tenth century B.C.E. (Steiner 1987). Findings include funerary vessels, provision for purification through lustration, and evidence of animal sacrifice; therefore, the ritual may have been one of expiation for collective guilt, possibly the death of the nymph and possibly the dewatering of the spring. A fifth-century inscription warned of a fine against any unqualified person who entered the springhouse (Cole 1988).

For several centuries, the focus of the important festive and ritual open area at the city center was as a racecourse. As the core area was gradually regularized by stoas and colonnades, the houses and workshops were displaced towards the edges of the settlement. The Potter's Quarter west of the central area was already enclosed in a rampart in the seventh century. Potters settled on this promontory between two ravines to utilize for ceramics the fine, low-temperature-firing clay found here. A second kind of clay suitable for transport amphorae came from different areas, often associated with lignite (3 km away, to the SW and SE at Nikoleto or Solonos); a third kind was made by washing the terra rossa soil from weathering limestone, which gives a pink clay (Higgins and Higgins 1996). Pottery production depended not only on the resources of clay and wood for fuel, but also on the city's crossroads location, on the fully developed shipping industry, and on the built-in market of Corinth's colonies. Until 575 B.C.E. (for 150 years after the foundation of Syracuse), Corinth was the chief exporter of pottery in the Greek world. Then in one generation, she lost dominance to Athens, which took over by producing more and better quality wares (Roebuck 1972). Were the Corinthian beds of clay or forests for fuel exhausted by then?

Between the Potters' Quarter and the core of Corinth stretched a tunnel and reservoir system possibly fed from the springs of Acrocorinth. It was over 600 m long with two tunnels turned into reservoirs totaling 200 m long. We have seen in Syracuse that the locations of potteries and the reuse—often for reservoirs—of empty spaces caused by the excavation of clay are part of the Greek tradition of resource management. Corinthians who settled Syracuse in the later eighth century carried with them traditions about using clay. Stillwell (1948) reported that in Corinth the artisans used the channel network for water to mix with clay for pottery making during 700–650 B.C.E. and later; he surmised this from the fact that the major channel was set into crushed seventh-century pottery. Yet another section was reused as a Roman cistern holding 100 m³, or 26,400 gallons (Robinson 1969). Roof and ground runoff was saved in a cistern outside the city wall here in the sixth century. Quarries (clay beds?) were found in the east ravine and a large one 0.8 km (0.5 miles) to the west of the east ravine. The edges of the east ravine were weakened by mining for clay especially during the fourth century B.C.E., so much so that the industry and the residential area had to be abandoned (Carpenter and Parsons 1936).

Another quarter of Corinth that shows a strong relationship to the geology is Acrocorinth, 575 m south of the city core, which takes the form of two rocky peaks with a grassy saddle between them, viewed in Figure 4.15 rising behind the Peirene fountain house near the Agora. Sometimes Acrocorinth was most important as a fortress, sometimes as the goal of religious pilgrimage, nowadays as mountain pastures for flocks. First defenses of this acropolis date possibly to the seventh but more probably to the sixth century B.C.E. The top of Acrocorinth, encompassing 240,00 m² or 60 acres, was enclosed by strong walls, hundreds of meters around—a fine citadel. The upper Peirene spring contributed to the defensibility of the acropolis, as did various quarry hollows that were used as reservoirs after they had yielded stone for the rampart.

Two large valleys descend from the citadel to the north, and the walls of Lower Town lead up to the fort entrance at the northwest corner gate. The defensive walls that surrounded Corinth from no later than the post–Persian War period in the early fifth century were rebuilt piecemeal in the fourth century. The length of the rampart around the city proper was 10 km/6 miles/55 stades. When the acropolis is included, the walls extended 72 stades. In addition, long walls to Lechaeon added another 4 km (2 km on each side) or 96 stades total. Corinth thus had the largest city area in mainland Greece—2.5 times the size of classical Athens. The west wall still retains a tower and two bastions from the Greek era, all made of hard limestone. Some of the stones weighed up to 3 tons. The north

FIGURE 4.15. Peirene fountain house with Acrocorinth behind. In the foreground is a spirally grooved Roman column at the edge of what was once an ornamental pool; one standing column and several bases on the far side of the pool suggest the rich decoration of the pool during Imperial times. The four arches of the Roman facade of the spring house are visible at the center; behind the facade, Ionic columns of the Hellenistic rebuilding are not visible; behind them ran long tunnels that collected and dispersed the water from Acrocorinth from at least the sixth century B.C.E. (Photo by Crouch)

wall along the terrace edge north of the Asklepeion, probably made of reused blocks of poros like the west wall, was destroyed by the Romans in 146 B.C.E.; they were particularly determined to raze crucial sections such as road heads and gates.

Because of changed political and military circumstances, the late Roman rulers built new ramparts in the fourth and fifth centuries C.E. East of Acrocorinth, they used a mixture of conglomerate, rubble that included older building and statue fragments, and mortar that included large amounts of ground terra-cotta. Coins found under the later destruction layer are from 396, suggesting that the destruction was caused by Alaric and not by the earthquakes of 365 or 375. According to the historian Procopius, the Emperor Justinian found Corinth's wall ruined again by an earthquake, possibly that of 551 (Carpenter and Parsons 1936; Gregory 1979).

The placement of monumental buildings on the land during the pre-Roman period was in keeping with Greek urban design principles as analyzed by Doxiadis (1972). Along the road from the northern port, the Temple of Asklepios towers above its terrace edge in full three-dimensional glory. From this terrace the Temple of Apollo on its ridge and farther on the sheer mass of Acrocorinth are visible. Even in Roman times, this dramatic sequence was retained, though the lesser streets were more geometrically organized into parallels and perpendiculars, whereas the Greek plan had utilized several major diagonal streets from the core to the gates, from which roads led to neighboring cities. One such street ran east-northeast from the agora along a terrace edge, past the later Roman amphitheater. Another ran from east of the Asklepeion down the notch in the terrace edge and directly north toward the coast just west of the Lechaeon Harbor. Current thinking is that the long trapezoidal area between this Greek thoroughfare and the Roman version of the Lechaeum Road was set aside for commerce—a kind of ancient strip mall and business park (D. Romano, private communication).

Another combination of buildings and geology resulted in the utilization of the hollows west of the Asklepeion Temple: the first hollow as the Lerna complex with a series of fountains, banquet rooms, and sleeping rooms and the second hollow as a bath or fountain. Cole (1988: 163) distinguished three different kinds of water use at the Asklepeion: purification by sprinkling outside the temple of Apollo/Asklepios; bathing in the lustral chamber before incubation (sleeping in a temple in hopes of experiencing a vision and a cure); and healing baths in the Lerna complex. Visually the westernmost hollow appears to be a collapsed doline, and certainly the excavators found evidence of more than one instance of collapse of conglomerate ceiling slabs (Fig. 4.16). This complex was called the Fountain of the Lamps by Wiseman (1970), because of thousands of votive lamps—1600 whole ones and fragments of several thousand more—found in it. The "fountain" was actually a Greek bath of the fifth to fourth century B.C.E., rebuilt by Romans late in the first century C.E. when a swimming pool was added. Both Greek and Roman versions were supplied by karst water from under the conglomerate shelf. This ledge, however, collapsed in the late fourth century C.E., possibly from the earthquake of 375. The complex then became a votive shrine with coins dating from as late as the sixth century. The long duration of the use of this complex is attributed to the steady supply of water and its location 200 m west of Lerna along

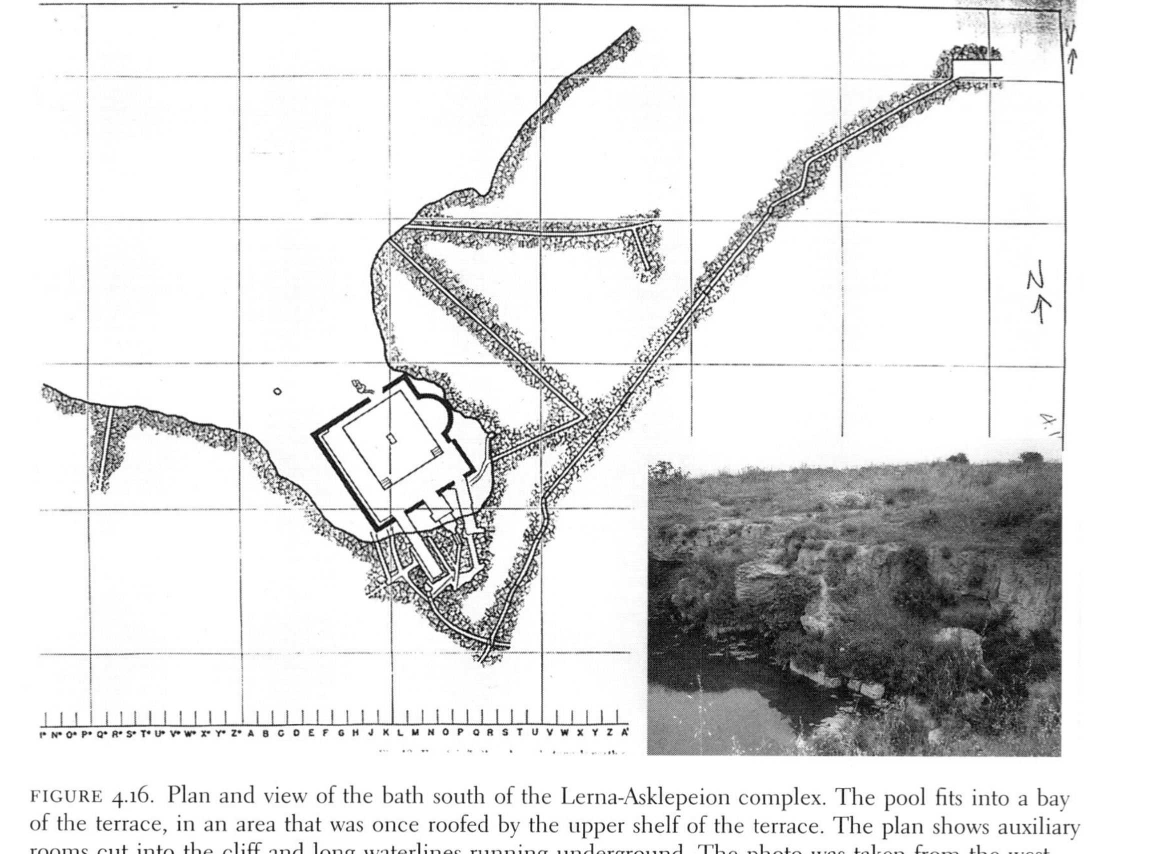

FIGURE 4.16. Plan and view of the bath south of the Lerna-Asklepeion complex. The pool fits into a bay of the terrace, in an area that was once roofed by the upper shelf of the terrace. The plan shows auxiliary rooms cut into the cliff and long waterlines running underground. The photo was taken from the west. (Plan by permission of the Corinth Excavations, American School of Classical Studies, Athens; photo by Crouch)

a major public road from the coast. The bath was a public facility, richly decorated in the Roman period with high-quality sculptures, some of them found by the excavators under half a meter of fill, where they had been dumped into the swimming pool (Wiseman 1970, 1972, 1978, 1979).

So richly endowed with public buildings was Corinth that modern excavators have concentrated on them with little effort to recover the residential areas. Even studies of single houses are scant. Only four houses had been excavated at Corinth up to 1990, and these were elite houses near the civic center, the most convenient location for daily life and for interactions of the powerful (Engels 1990). A few villas sited in the agricultural land outside the city but near the ramparts are also known. Only in the Potters' Quarter has much evidence of workshop activity come to light. Therefore, it is premature—after more than 100 years of excavation—to pronounce on these aspects of Corinth's arrangements for everyday life.

Similarities with Syracuse

A brief comparison between the mother city Corinth and the colony Syracuse, concentrating on the relevant geological factors of similarity, concludes this case study.

Stones and Stratigraphy   The lithography is certainly not the same at Corinth and Syracuse. Most evident is the lack of conglomerate in Syracuse, such as that used for the later Roman wall on the east of Acrocorinth and that forming the ledges above the springs along many of the terrace edges at Corinth, for example, at Peirene and the Fountain of the Lamps. At Syracuse, rather, we find two layers of limestone, the upper one more intensely karstified. Instead of a similarity of stone types, the two cities exhibit a strong similarity of process: They have analogous water behavior and accessibility, which could be managed with the same kind of water system elements. The conglomerate and sandstone at Corinth, like the two kinds of limestone at Syracuse, are bedded on clay. Water emerges along the contact surface between the stone and the clay. As water management skills developed, new ideas were probably exchanged between the two cities.

Karst Features   Although the karst water supply system of Syracuse drew on an extensive hinterland of limestone mountains, that of Corinth was for many centuries more narrowly supplied from the massif of Acrocorinth. Only in the second century C.E. was Corinth connected by an aqueduct to springs at Stymfalia at the north end of the Tripoli plateau; the springs provided a rich supply of karstic water also tapped for Argos. A similar, long-distance waterline was part of the supply system for Syracuse from the fifth century B.C.E. Both systems relied also on large reservoirs within the city. The spring system in Corinth provided ample water at many points, especially along the edges of terraces. In 1920, Blegen cited 12 springs; Landon (1994) mapped 24. At Syracuse, the modern analysis of the difference between springs and aqueduct outfalls is still incomplete, but from early times the residents were imitating and manipulating natural channels and springs for the benefit of neighborhoods as well as for clusters of monumental buildings (for Corinth, Broneer 1954; Hill 1964; Richardson 1902, 1910; Robinson 1969; Roebuck 1951, 1972; Williams 1972, 1974, 1976, 1977, 1980, 1987; William & F. F. Thompson, 1987; Williams & Fisher 1971, 1973, 1975; Williams & P. Russell 1981; Williams & Zorvos 1982, 1984, 1991; Wiseman 1978; for Syracuse, Collin-Bouffier 1987; Schubring 1865).

Use of Topography and Resources   Uses of topography and water resources at the two cities are similar: at Corinth, the cluster of fountains at the transition from the lower to the upper terrace of the core of the city, and at Syracuse, the arrangements made for water at the grottoes above the theater. When colonists left Corinth for Syracuse in the late eighth century, some resurgences of water were known at the crucial center of Corinthian life. The Sacred Spring may date from the tenth century B.C.E., and the ancestral version of the Peirene Spring seems to go back before the archaic period. Travelers from the northern port to the Tripoli plateau or to foothill towns like Nemea could pause at Peirene for water and to exchange trade goods, as could travelers from Patras or Sikyon on their way to the eastern port. The combination of gentle terrain and available water made an attractive stopping point and node of settlement.

At Syracuse, the grottoes above the theater were for a long time outside the true city; this is evident from the many grave niches on the facades of the grottoes. Activity near the inland grottoes increased during the seventh century, when the theater area was being quarried for stone for monumental buildings, and during the fifth century when large numbers of people from Gela were moved to Syracuse. The new population, settled on the mainland between the grottoes and the shore, required more water. In the fifth century B.C.E., the Galermi aqueduct brought water to supplement the springs of the grottoes; water was carried to the lower city, which it still supplies. (See the Ephesus chapter on technical advantages of pouring the water into a reservoir and then drawing it off again.) The connection between quarrying and water supply is nicely symbolized in these grotto fountains and in the Glauke Fountain at Corinth, which seems to be a very old remnant of the quarry from which the sixth century B.C.E. Temple of Apollo was built. In the Roman period, if not before, the fountain was supplied by a pipeline. Water continues to flow in the Syracuse grottoes just as it does in the Peirene Fountain at Corinth, validating the location decisions of the early settlers. Thus Corinth and its eighth-century B.C.E. colony Syracuse are ideal subjects for comparing the impact of karst on settlement development:

> Clay and limestone, and to a lesser extent conglomerate, are found at both cities. The details of many similarities and differences between the two are to be found in the kinds and clusters of the stone and clay and in the patterns of hydrogeological potential.
>
> Water technology at both cities displays geologically related cultural development.
>
> Urban history gives clues for geological dating; each city's urban design is explainable in geological terms.

In addition to physical details about the geology of both cities, we must leap to abstraction for a richer understanding. The karst or karstlike behavior of water in stone has abstract similarities at Corinth and Syracuse, but it is different in operational details. Further, a chronological anomaly at Corinth can be used to correlate geological and architectural change. That is, most unusually for the cities of our study, Greek and Roman Corinth do not form an uninterrupted continuity. The Greek settlement here was destroyed and left deserted for a century. Then the Roman city was built anew and populated by immigrants from Italy.

Conclusions

Our new synthesis of the human history and geology of Corinth is intended to complement the data from older studies and from the other nine cities investigated herein. We have evoked local resources to explain Corinth's success: beds of fine clay for the pottery that dominated the seventh century B.C.E.; stones that were so readily available for construction of monumental architecture and that could be exported by ship to neighboring cities; water that refreshed the pilgrims to the Temple of Aphrodite at the top of Acrocorinth and the merchants and their customers at the agora, served by so many natural springs and artificial fountains; ports and the Diolkos that moved trade goods to and from the city in spite of earthquakes and sea intrusion; surrounding farm lands that produced food for the city and income for the farmers to spend in the city (Engels 1990); and timber brought from the mountains, the Tripoli plateau, and colonies such as Cephalonia for construction of buildings and ships.

We can no longer look at Roman Corinth as a place were some decadent Roman squatters occupied the ruins of what had been a glorious Greek city. Rather, the resources and locational advantages of the site drew new Romans during the second century B.C.E. — as they had Greeks in the seventh century B.C.E. From the most casual to the most monumental aspect of ancient life, the ecological constraints determined much of the uniqueness of this city. The geography and geology of Corinth are not simply a neutral background for human cultural history, but rather the interactive setting for 1400 years of Greco-Roman life.

Delphi

Introduction and Description

Few sites of the ancient world have Delphi's aesthetic effect on modern visitors — so profound as to amount to a spiritual experience. After climbing the steep slope past tier after tier of buildings set along a zigzag path to the top of the ancient ensemble, the visitor has the experience of sensing the Phaidriades Mountains towering in the background while looking out and down past the remaining monuments to the extensive forests of the mountains and the invisible Pleistos River in its narrow valley, then far below to the myriads of olive trees in the plains, and finally to the waters of the Gulf of Corinth, flashing like gilded silver. All of these forms bathe in luminosity that changes with the hour and the weather but never relaxes to banality. The numinous light and the awesome scenic beauty suggest that the divine—beneficent or malevolent—is here closely available to human contact.[7] The steep slope of the sanctuary hill made it a test of devotion to climb all the way up to the stadium. The slope also provided numerous building sites that clearly displayed ensembles of individual statues, small treasuries and fountains, large porticoes, and huge structures such as the Apollo Temple, the theater, the Roman bath, and the stadium.

Delphi was a place where the danger from earthquakes and rockfalls was and is part of the attraction. The particular aesthetic quality of Delphi contributed to the healings attributed to Apollo, known as a healing god but also one who could punish with plague and other ills (Dietrich 1992). Since he was deity of the sun and of light—once visiting the sanctuary as a flashing meteor (Morgan 1990 citing Homeric Hymn 440–445)—it is appropriate that his sacred precinct be marked with unusual light.

The site is particularly well documented, with literary and archaeological evidence as well as modern analysis by engineering geologists and others. In this account we concentrate on the tie between the geology and major events; a detailed social and monumental history of the site is available in the *Guide de Delphe* (Bommelaer 1991).

Delphi is bracketed between mountains on the north and a river valley on the south (Fig. 4.17). The towering Pheidriades north of Delphi are the southern edge of the Parnassos horst, which was rapidly (in geological terms) uplifted north of the Gulf of Corinth graben; Higgins and Higgins (1996) estimates the combined movement of horst upward and graben floor downward at 3 km in 10 million years, with earthquakes having especially large effects. The stratigraphy at Delphi is as follows, beginning with the topmost modern layers:

Parnassos horst made of shallow-water limestones (neritic) over flysch, alternating with brown-red siltstones, mudstones, and sandstones with lots of calcareous materials (dolomite); karstified.

Schist and flysch above the stadium covered by limestone fragments and alluvium.

North of sanctuary: Dark, Late Jurassic limestone of the Phedriades Mountains over mid-Cretaceous pale limestone.

Younger dark or black bitumen-bearing limestone.

Thin-bedded pale limestone with chert layers (Cretaceous to Eocene).

Base of active fault zone covered by four generations of scree and talus cones consisting of carbonate stones and mud stones, cemented by carbonate-rich waters to form ancient conglomerate and breccia.

Red shales (of unknown age) and some Cretaceous limestone and recent screes, below the road.

The southern edge of Delphi is the Pleistos River valley, where erosion of fragmented and pulverized rocks occurred along a fault that is nearly parallel to

FIGURE 4.17. Schematic map of the geology of the Delphi area. The limestone mountains that arc behind Delphi are shown as a brick pattern in two sizes to signify stone from two different periods: the most recent (late Cretaceous–Eocene) farther uphill and the earlier (Jurassic–early Cretaceous) closer to the sanctuary. The name "Delphousa" actually belongs to a spring at the left end of the thin rectangle that symbolizes the stadium, whereas the spring shown with a tadpole shape is actually Kerna. The Castalian (Kastalian spring) is at the east side of the sanctuary, just above the modern road. The sanctuary of Apollo is denoted by the large rectangle, and the other set of shrines and athletic buildings below the modern road by the rounded white shape labeled "Marmaria." The Pleistos River, though hidden from viewers on the site, is shown here in its alluvial bed. The line of springs represented by a fringe actually runs more diagonally from the stadium spring through the sanctuary and under the Apollo Temple, and then to the Kastalian springs (Higgins and Higgins 1996).

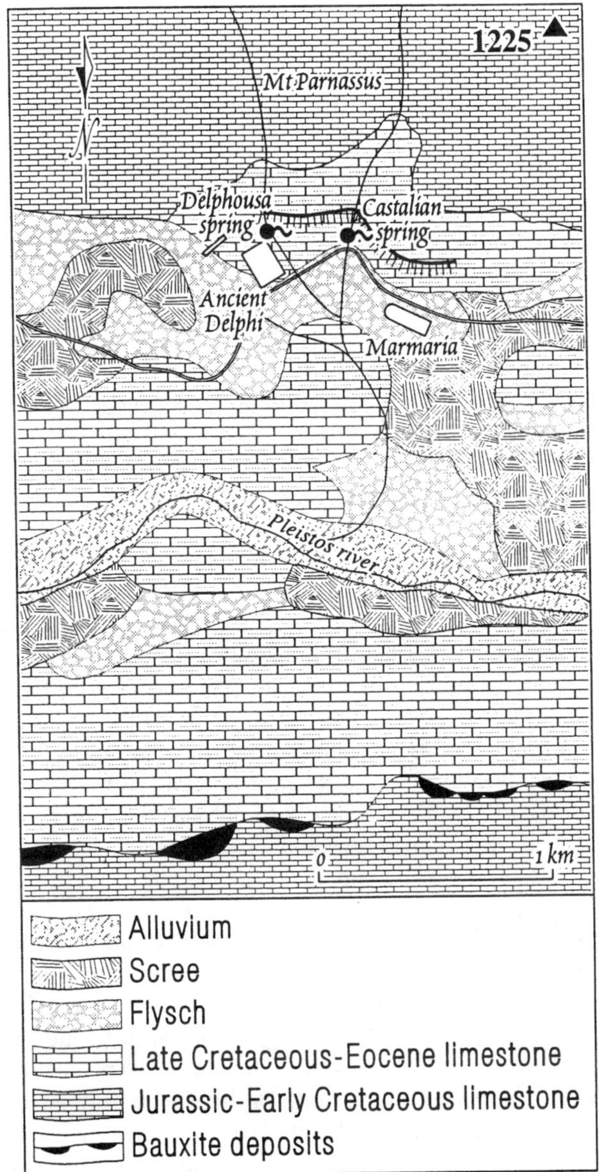

1225
Mt Parnassus
N
Delphousa spring
Castalian spring
Ancient Delphi
Marmaria
Pleistos river
0   1 km
Alluvium
Scree
Flysch
Late Cretaceous-Eocene limestone
Jurassic-Early Cretaceous limestone
Bauxite deposits

the gulf graben. The river lies just below the sanctuary, in an EW syncline corridor of flysch, the location of recent major tectonic activity (Jacquemin and Jacquemin 1990). From the west end of this ravine, another fault runs NS from the gulf, along which the trade route into the mountains runs close to the site. Delphi thus had access to the wide Greek world via main roads and via her ports to the sea lanes of the Gulf of Corinth, as is shown by imports found at Delphi, which are more extensive than those found at the contemporary sanctuaries of Olympia and Dodona (Müller 1992). These routes remained important for communication for about 4000 years.

The Parnassos Mountain horst above Delphi consists of limestones over flysch alternating with brown-red siltstones, mudstones and sandstone (Kolaiti et al. 1995). The edge of the Phedriades Mountains runs NNE-SSW and slopes SSE and WNW at 40–60° angles. Visible just above the modern road, a horizontal thrust fault cuts off the limestones (Higgins and Higgins 1996). Older limestones upthrust over younger flysches, particularly near the Kerna spring in the area just below the stadium (Fig. 4.18), which was destroyed in a rockfall of January 1980 (Pentazos 1988).

Delphi's geometry of inverted faults affects the joints of both natural and hewn blocks, as evident in the construction and reconstruction east of the theater and at the foot of the polygonal wall. Secondary thrusts and reverse faults are visible in the tectonic graben of the Pleistos River and in the active normal fault zone of Arachova-Delphi, which separates the horst of Parnassos from the graben of Itea at the coast. Known quakes occurred in the following years: 730 B.C.E. (Neeft 1981), 600, 480 (Herodotus VI or VII, 37–39), 373 [estimated as 7 on the Richter scale; hereafter R], 348 [R 6.7], 279 B.C.E. [R 6.8], 551 C.E. [7.2], 996 [R 6.8], 1402 [R 7], 1580 [R 6.7], 1870 [R 6.8], 1887 [R 6.3], 1965 [R 6.3], 1965 [R 6.3], and 1970 [R 6.2]. (For quakes not otherwise attributed, see Kolaiti et al. 1995.)

Rockfalls are a major geotechnical problem at Delphi (Figs. 4.18, 4.19). During the last million years, schist and flysch have been covered by limestone fragments and alluvium, for example above the stadium (H. Jacquemin 1990), and these in turn have been covered by relatively thin clastic screes (fallen rocks) and talus cones. This layer ranges from 2 thick in the uphill areas to 15m downhill and is continually renewed by stone peeling off from the steep slopes. Intense faulting and systems of discontinuities induce rockfalls and slides. The base of the major active fault zone to the north of the site, mainly in the Phedriades, is covered by four generations of scree and talus cones from successive toppling events and surface weathering. Schists along a NS axis east of the sanctuary are affected strongly by earthquakes because they are relatively slippery and weak. Unfortunately, the Apollo Temple site is in the midst of a large area of fallen rocks from the surrounding mountains and as a result is susceptible to shaking and to being bombarded by new rockfalls. Large blocks of breccia surround the sanctuary of Ge, especially at the rocky crevasse of the Sybil (Bousquet et al. 1989). Because of all the loose scree fallen into the "bowl" formed by the half-circle of surrounding mountains, any movement of the earth has more drastic results than would be the case if the underlying rock were solid granite. An earthquake shakes the scree like a bowlful of hard candy.

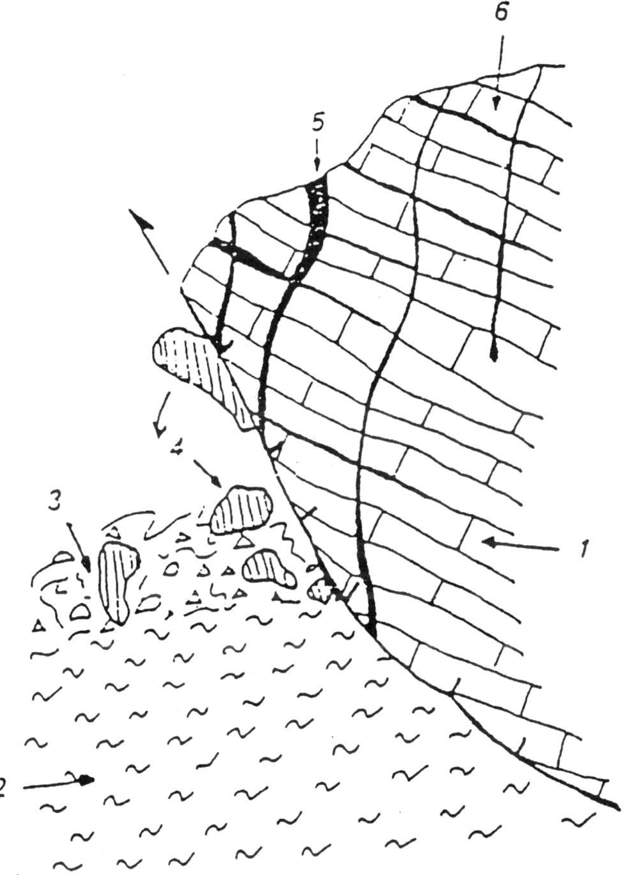

FIGURE 4.18. Thrust diagram. Stability at Delphi is a contest between upthrusts and rockfalls. (1) limestone; (2) schist; (3) scree, consisting of (4) fallen boulders and smaller rock fragments; (5) fractures, sometimes filled with water, sometimes with breccia; (6) bedding planes along which water flows (Papageorgakis and Kolati 1992).

A striking example of a rockfall is the large, fallen limestone boulder outside the south wall of the sanctuary of Apollo; this boulder fell in the seventh century B.C.E. or later and may have been part of the original Kassotis Fountain, visible at the lower left in Figure 4.19. A similar boulder (from a rock fall in 1905 C.E.?) is visible in the foundation of the tholos at the east end of the site. The nearly vertical cliff face above the Kastalia Fountain is the result of another series of rockfalls; (Koroniotis, Collios, and Basdekis 1988: their fig. 1, section; Kolaiti et al. 1995: their fig. 12).

Other tectonic problems are foundation subsidence, ground creep and defor-

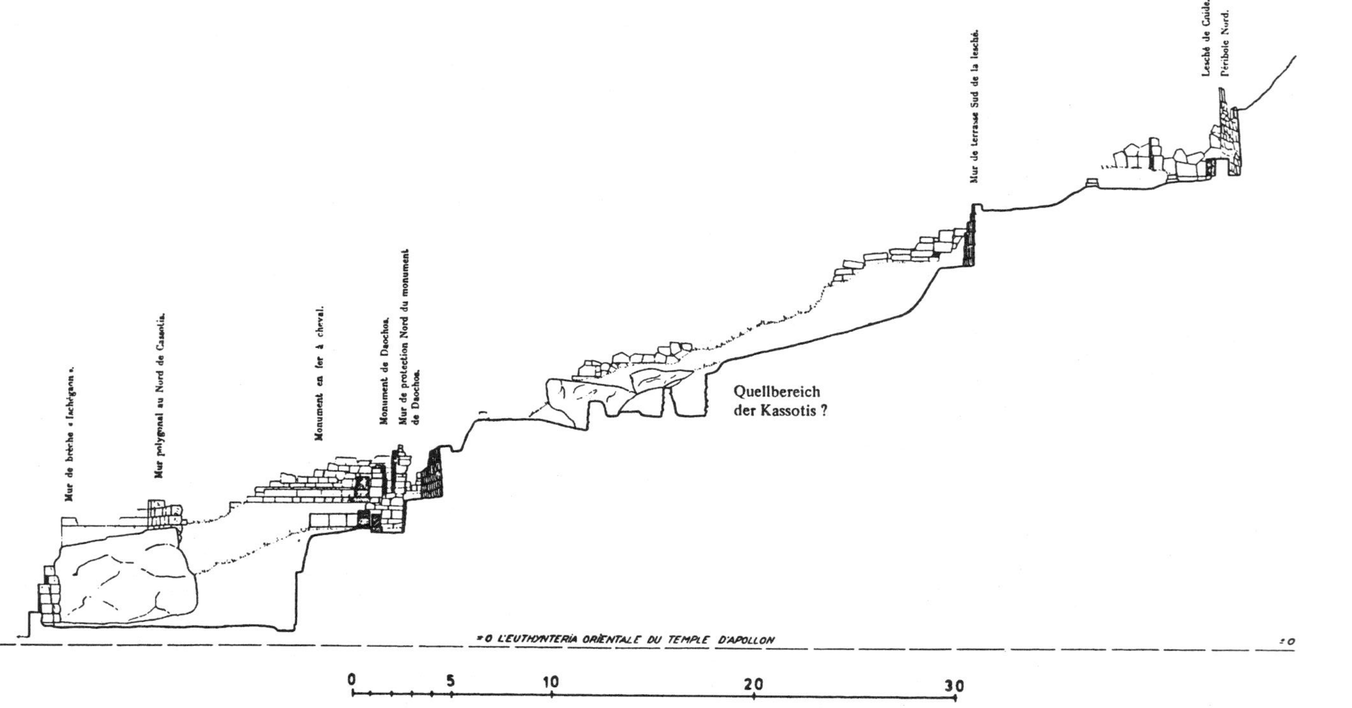

FIGURE 4.19. Section through Apollo sanctuary area, with many retaining walls and fallen boulders of various sizes. The Apollo platform is at the bottom left. (From J. Pouilloux, FDelphes II (1960, plan 22), by permission of the French Archaeology School, Athens.)

mation, and landslides. Creep can be caused by the combination of steep slopes with loose deposits (screes, talus) over flysch. The action of the flowing ground-water table at the contact between flysch and the overlying limestone (discussed subsequently under karst) also contributes to creep, which has noticeably affected the retaining walls of the Apollo sanctuary, the gymnasium area and whole Marmaria area, the stadium seats, and especially the wildly slanted tree trunks just below the stadium. Similar to creep is foundation subsidence and deformation (Fig. 4.20), both very common here especially with foundations that span two or more soils, such as the north side of the foundation of the Apollo sanctuary and the base of the Treasury of Corfu.

These clues point to ongoing movements at the site—movements that are not as abrupt and drastic as earthquakes. Earth movements are frequent, with average return periods determined by Kolaiti et al. (1995) as follows: Those with a magnitude of 7 have a return period of 70 years; magnitude of 6, a period of 8 years; magnitude of 5, a period of 1 year. Most destructive were the earthquakes of 600 B.C.E., which completely destroyed the sanctuary; 373 B.C.E., which did so much damage to the Apollo Temple by rockfalls from the Phedriades Mountains that the building had to be replaced; and 1870, in which the major Arachova-Delphi fault zone was reactivated, the movement being felt in most of Greece and in Aegean Turkey (studied by Ambraseys and Pantelopoulos 1989). This quake, rated by Kolaiti et al. at 6.8 magnitude, caused significant damage to the monuments at Delphi and in nearby towns. It was one of a series of quakes extending from

FIGURE 4.20. Deformation of a wall, produced by earth movements, in Apollo sanctuary area. (Photo by Crouch)

July 31 to November 30, 1870, in the Delphi area, where some 300 earthquakes ranged in magnitude from 5.3 to 6.7. We mention the details of this particular earthquake series because it is impossible to obtain precise data like this for the earthquakes of 373 B.C.E. (R 7), 348 B.C.E. (R 6.7), and 551 C.E. (R 7.2), which, however, seem to have been at least as strong.

On August 3, 1870, the main (second) shock was one of the largest in the nineteenth century. After 15 seconds of violent shaking, with great explosions and landslides, the road to Delphi from the east was buried and the Kastri village in the ancient Delphi site was completely destroyed, with all 205 houses and two churches fallen, 28 people killed, and 80 injured. Rockfalls from cliffs above and scree falls below wiped out olive groves and vineyards. The newly begun archaeological excavations caved in. New springs appeared but the Kassotis Fountain as it had been rebuilt in 1802 was ruined, and the spring near the Kastalia Fountain stopped flowing for a while. A third shock, 11 hours later, was as strong as the second, but for only 10 seconds during which it was widely felt in Greece and Turkey. Continuing aftershocks demoralized the inhabitants, who lost heart for reconstruction. Violent shocks occurred on October 25 and November 30. Then came three months of quiet, followed by another series of earthquakes between March 17, 1871, and August 1, 1873, with more occurring during the extended

FIGURE 4.21. Fault running EW through the site to the western end of the stadium, at the upper right. Ruins of the Temple of Apollo lie just below the center of the view, with the fault running diagonally through them. Steep cliffs occupy the lower right corner and the tourist road wanders-along the left side. (Photo © French Archaeology School, Athens)

series until the last shock in 1971. Ambraseys and Pantelopoulos (1989), using the Mercalli intensity scale, which can be derived from descriptions of human reactions and buildings destroyed, think that the highest intensity was IX on the Mercalli scale (see Table 1.1 in Chapter 1); the main shock was comparable to a Richter scale rating of 6.7 ($\pm$.3), and there was another of 6.3. There was one quake of VIII (5.7) and two of VII (5.3, 5.4). All in all, this series was similar to the February 24, 1981, sequence in the Gulf of Corinth.

Examination of the terrain and of photographs taken over many years by the French School of Athens reveals a fault running diagonally from the southern corner of the east side to the northern corner of the west side of the Temple of Apollo (Fig. 4.21) and beyond in both directions. This is likely to be one of a set of parallel faults, with a second set to the northeast, the locus of the springs of the site, and possibly a third set at the base of the ring of mountains, along which they were uplifted. All of the Delphi springs "appear to fall along a straight line, which suggests that they are associated with a steep fault that runs about northwest/south-east" (Higgins and Higgins 1996: 79–81). Springs are often found along the face of the up thrown side of a fault. These springs are fed by water (from underground reservoirs) that travels along direct conduits near the surface, where increased fracture density has weakened the bedrock (Kastning 1977: 199).

History of Settlement

Mycenaean Era   Mycenaeans lived at Delphi (twelfth–eleventh century B.C.E.) toward the east end of the Marmaria area and northeast of the present Apollo sanctuary.[8] The Mycenaean settlement was destroyed when a landslide, possibly that of about 1000 B.C.E., wiped out the houses of the lower village—a type of drastic earth movement repeated for nearly every settlement here. Afterward, there was a break in occupation. Excavations of 1992, directly above Mycenaean evidence, in the area east of the Apollo site at the of Pillar of Rhodians, date the next occupation to the tenth century (Pariente 1993).

Delphi's trade expanded in the early second millennium B.C.E., with the pressure on resources to make elite goods like armor and dedicatory tripods (Morgan 1990). Increased trade both alleviated land shortages at home and fostered colonization abroad. Then, in the eighth century, the change from pastoral to agricultural economy exacerbated the shortage of land for farming in Greece, putting stress on city institutions and increasing the motivation for colonization. Delphi capitalized on these changes through her famous oracle.

In the mid-Geometric era (ninth century B.C.E.) the settlement was known as Rocky Pytho. By then the village may have named the two sacred stones as the Omphalos (world's navel) and Stone of Kronos (one of the Titans who ruled the world and this sanctuary before Apollo took it over), although the first mention of the Stone of Kronos is in Hesiod (eighth century B.C.E.). It may be that the old devotion to the Omphalos and the Stone of Kronos was a religious response to the threat of loose stones wreaking havoc. Devotion to the Earth Mother, Ge/Ga/Gaia, seems equally early. Her shrine was near the buried Mycenaean houses, where a spring or seep is still visible at the Rock of Sybil on the terrace of the later Apollo sanctuary (Higgins and Higgins 1996). In the eighth century B.C.E., a settlement and sanctuary coexisted in the core area (Morgan 1990), but an earthquake in 730 (Neeft 1981) destroyed the main part of the settlement, which was abandoned and the materials used to build a new sanctuary. Houses were built elsewhere on the site.

Oracles   From the early first millennium B.C.E. onward, the peculiar geology of the site contributed to its sanctity. The first evidence of the cult of Pythian Apollo is from the eighth century, when the new god was said to have killed the female dragon/snake who had guarded the prophetic spring of Kassotis near the center of the site. Divination at Delphi likely occurred by 750 or 725 (Morgan 1990), with Apollo appropriating the oracle to his own use. According to Pausanias (X, 5, 9), the first shrine of Apollo was made of laurel boughs.

During the archaic period, Delphi was the common sanctuary of a small group of city-states—Corinth, Chalkis, Thessaly, and Sparta—but not of all the Greek city-states. The Pythian oracle that drew them to Delphi is now remembered mainly for its role in the process of ancient Greek colonization. About 10% of founders of new cities consulted Delphi in the eighth and seventh centuries. Delphian Apollo had competition from Apollo of Didyma (near Miletus) and other oracles. Many Greek cities—among them Megara, Eretria, and Miletus, who together accounted for a large number of colonies—went to other oracles or sought no oracular guidance. Cicero, speaking 700 or 600 years later, claimed that Delphi sanctioned colonies for the whole Greek world (Londey 1990). The archaeological data, however, does not support Cicero's assertion.

Oracles to guide the colonization process began in the eighth century and continued until the fifth.[9] One hundred thirty-nine colonies were founded during this era, of which only fourteen or fifteen have known oracles from Delphi (Londey 1990, Table 1); a few others date from the fifth to fourth centuries when there was a second wave of colonization. The earliest colonies related to Delphi are those from Corinth and Chalcis. They settled at Croton, Zancle, Rhegium, and Syracuse, beginning west of the Gulf of Corinth, quite convenient to Delphi, and dating to the eighth century. From the eighth century, Corinth was Delphi's major partner in the wider Greek world. When, for instance, Corinth was colonizing Syracuse in 750–680, its founder Archias was guided by Delphi (Pausanias V.7.3). Enna, a colony of Syracuse, also had connections with Delphi in 664 or 552 or both. In the seventh century, Gela, Thasos, and Cyrene, as well as Megara's colony of Byzantium (founded in 668 or 659), consulted Delphi even though this oracle was not as conveniently located for them.

The god manifested his will through a Pythia (priestess), who sat at the edge of a chasm and perhaps inhaled a gas that caused a trance. Mystery religions and inspiration oracles (Dietrich 1992) are features of the archaic era, but not of prehistory, whereas Pythian frenzy, which took place in the fifth century is not recorded until the fourth century by Plato who said that the Pythia "raved" as she pronounced oracles from Apollo's tripod. Maurizio (1995) studied the Pythia's role in terms of the worldwide phenomenon of spirit possession and concludes that it was common in archaic Greek society to be in touch with the gods via divination.

Dietrich (1992: 41) claims that "the traditional account of intoxicating fumes swirling around the Pythia from a chasm in the ground arose from a confusion of the Stoic 'pneuma' doctrine and an ageless belief in subterranean caves as places of communication with divine chthonic powers." Geologist B. Bousquet, who has long studied the site, thinks the presence of trance-inducing gas is impossible (according to Mulliez, personal communication). Yet a possible geological explanation was given by Minissale et al. (1989, as summarized in Higgins and Higgins 1996): "Discharges of carbon dioxide are not uncommon in areas of limestone and such gas could perhaps induce a trance. Earthquakes and construction activities can change the natural plumbing system, so the tradition of the vapors cannot be excluded on geological ground." However, no vapors are found here today. The useful though seemingly ephemeral appearance of intoxicating vapors, just at the time when oracles were the socially preferred method of communicating with deities, boosted Delphi into the first rank of religious pilgrimage sites. A study in 2001 indicated that toxicant gases from a rock fissure affected the Pythia. Investigators John Hale (archeologist) and Zelle Zeiling de Boef (geologist) found physical evidence in the travertine stone under the temple and in the waters of a nearby spring, where methane and ethane were present (*Washington Post*, Feb. 7, 2002).

Construction and Destruction   Early in the seventh century, musical competitions were organized every eight years by the town of Delphi in honor of Apollo's victory over the python. Athletic contests at Delphi, as at Olympia, attracted participants from several Greek regions (Morgan 1990). Major games open to outsiders are known from 586 on. The earliest stadium was probably located on the coastal plain, far below the sanctuary (Mulliez, personal communication). The terrace at the top of the site seems to have been cleared somewhat later for an informal stadium where the spectators sat on the hill and drank from the conveniently located spring at the northwest corner (Fig. 4.22). Also archaic were the steps made of poros blocks that led from the Apollo temenos up to the Kassotis spring at the north (Roux 1976: 137). Likewise, the custom of storing at Delphi a portion of the moveable wealth of the league's cities began in the seventh century with the erection of the earliest treasury building, that of the Corinthians, at the beginning of the Sacred Way (Hellmann 1990).

Before the building of the Apollo Temple, the earth goddess Ge is thought to have had an oracle and a spring in the prehistoric village where the Apollo Temple is now. The oldest fragments of Ge's shrine have been identified as belonging to the seventh- or sixth-century predecessor of the Athena Temple and its altar (Bommelaer 1991). Its site was a seemingly appropriate location on a solid breccia ledge. A separate shrine stood on what is now called the Marmaria Terrace (10–12 in Fig. 4.22), at the east entrance to the sanctuary area. Proximity to the eastern edge of the Phedriades arc and the cleft from which the Kastalia flowed meant that the Marmaria terrace collected flows of mud and was often threatened and battered by dislodged blocks from the slopes above (H. Jacquemin 1990).

First efforts to regularize the upper hillside for building sites seem to date from the sixth century, after problems of monumental construction without formal terracing had become apparent. In the decades after the 600 B.C.E. earthquake, one of the strongest ever felt at Delphi, much new construction took place. Confusion and misery following this earthquake may have triggered the First Sacred

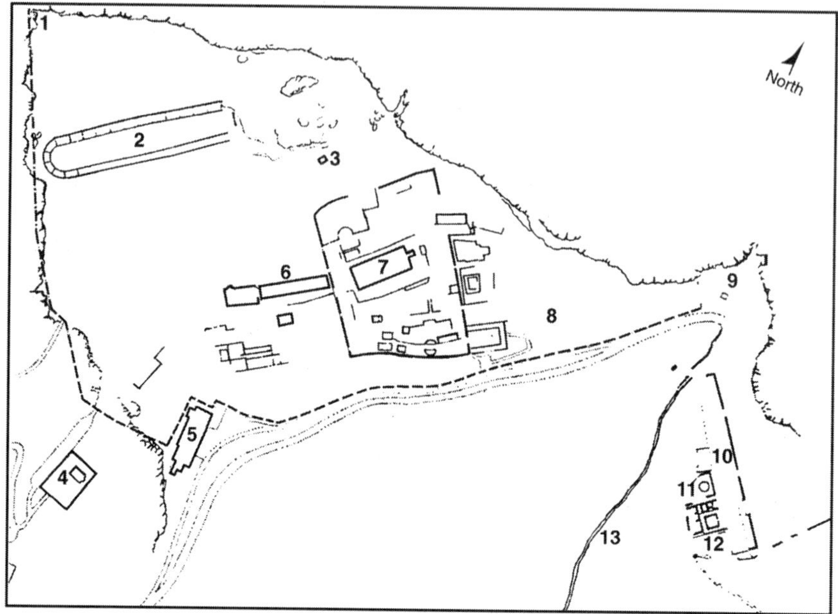

FIGURE 4.22. General plan of Delphi area: (1) quarries outside the rampart at the upper left; (2) stadium (its fountain is not shown); (3) Kerna spring; (4 and 5) modern museum and church, west of the archaeological site and close to the road; (6 and 7) Apollo sanctuary with the west portico to its left; (between 8 and 5) ancient city; (9) Kastalia fountains; (10) portico (Xysto); (11) palestra; (12) gymnasium on Marmaria terrace; (13) river from Kastalian spring. (By permission from the French Archaeology School, Athens)

War of 596–86 B.C.E., after which Delphi was administered by a group of states aligned in the Delphi Amphictyony (League), which was the governing body for centuries (Morgan 1990). This change of governance signified Delphi's evolution into a pan-Greek sanctuary.

The first formal building in the area of the later stadium was a fountain known as Delphousa from the cleft in the limestone out of which it flows. The fountain utilized a water source from deep in the rock crevice above stair 13 at the west end of the (later) stadium. A drain for this water was reported by Foucart (1864). Aupert (1977) dates the fountain at 580–25 or 550–535 B.C.E. from its marble and poros Doric elements; it is similar to a fountain at Pergamon of the same era (fig 20.1 in Crouch 1993). Marble was used because of its greater resistance (impermeability) to running water, whereas poros is soft and permeable but usable for upper walls and other less visible parts of the structure; therefore, each material was used where appropriate for its function. Some sintered blocks of the archaic fountain were described in Aupert's 1979 study, which showed that channel, fountain, and drain are older than the stone stadium. At that time, the area seems to have been occupied by artisans (Bommelaer 1991). This old fountain is the highest one at Delphi; several others run in series from the stadium past Kerna and Kassotis down to the gymnasium, likely following a fault parallel to the one running through the Apollo Temple.

By about 675–650 B.C.E., the wooden shrine to Apollo was replaced by the first stone temple, of which the remaining physical evidence is roof tiles and in situ blocks in the southwest corner (Morgan 1990). The written sources insist on the oracle being located in or at the temple, but the architectural form of the surviving (fourth-century) podium gives no clue to the precise placement of the oracle in the temple, whether underground, sunk into the podium, or in the opisthodimos (western treasury room). The omphalos stone was placed somewhere at the temple for safekeeping, and a duplicate—possibly a rockfall—was later erected outside on the front (east side) of the temple, in a more convenient spot for veneration (Pausanias X.5.9; Amandry 1993). During modern times, another omphalos stone was found under the temple terrace.

After an arson fire destroyed the oldest stone temple in 548 (Childs 1993), the second limestone temple was built in the 530s. An exiled Athenian family, the Alkmanionidai, won favor by not only paying for the new temple but also exceeding specifications in substituting more expensive Parian marble on the east facade—an extra expense partially compensated by reusing blocks and column drums from the old temple in the walls and podium of the base of the new temple. Old and new are also represented in the pediment figures of the temple. The figures of the west pediment, in stuccoed poros, depicted a traditional combat called a Giantomachy; the east pediment in marble showed the arrival of Apollo in Attica, done by the sculptor Anténor. "The different materials of the two pediments make a comparison of them difficult," writes Childs (1993), who thinks that the west pediment is before and the east pediment after 530, whereas the sculptures may be as late as 513 or 510 (or 505 according to Bommelaer et al. 1990) after the body of the temple was complete. The west and north walls around the entire Apollo sanctuary also date from the sixth-century rebuilding, as does the 85-m-long polygonal wall below the temple. Polygonal walls, according to Higgins and Higgins (1996), are more stable in earthquakes than ashlar masonry; this one still survives.

In the last third of the sixth century, the Kassotis Spring was located in the oldest and most uphill of its three locations. Pouilloux and Roux (1963: 86–89) think that a channel was built from the spring into the *adyton* (inner shrine) of the Apollo Temple. By the end of the century, many treasuries had been added along the main street.

Divine intervention was experienced again at Delphi when a raid by the Persian army in 480 was aborted because of thunder and a huge avalanche, accompanied by an earthquake of magnitude 6.3 (Kolaiti et al. 1995). Destruction from this disturbance necessitared the rebuilding of the south (downhill) wall of the Apollo precinct. Because southern walls were farther from the underlying rigid stone and more precariously balanced on fallen rocks, they have always been more susceptible than the north terrace walls to earth movements (Fig. 4.19). The south retaining wall toppled under the pressure of a torrent of mud. Once it was repaired, the stoa of the Athenians was erected between that wall and the pilgrim's route, some time early in the fifth century. This stoa was a resting place for those coming to Delphi for divination. Tombs of the classical period were found at the site of the present Museum, indicating that this area was then outside the sanctuary.

The late sixth-century temple was destroyed by an earthquake of magnitude 7 in 373 B.C.E., which caused heavy damage at Delphi. An aftershock or independent earthquake of magnitude 6.7 or a landslide in 348 completed the destruction of Apollo's temple. Repairs begun in 373, on the upper retaining wall of the precinct using blocks from the sixth-century temple, were suspended during the Third Sacred War (356–346) and then resumed.

Replacement of the temple (Fig. 4.23) has been studied by Bommelaer (1983) and its dating by Child. The highest costs in rebuilding were for the transport of stone brought from Corinth in the 340s, according to the copious records about this construction project that have survived. In the fall of 338, the supervisors paid for painting the epistyle above the omphalos shrine with encaustic—a stucco with pigmented wax that is burned in to fix it in place—to improve and protect the surface. A new baldachin—a sort of miniature temple—in the portico, probably in the opisthodomos (Amandry 1993), covered the omphalos. In the 330s, the builders bought 571 pieces of wood for the roof. Each log weighed nearly a ton. The logs were transported by land to the shore, loaded on ships, brought to Delphi's port Cirrha, then dragged 6 km to Delphi. This last step of the passage required at least two yoke of oxen per log and cost 500 drachmas (Bousquet 1977). The sculptures for both new pediments were paid for in 327: at the east, Apollo seated on the omphalos, with his twin sister Artemis, his mother Leto, and his attendant Muses; at the west, Dionysos and the Thyades (local Maenads). Many pieces of these figures were found in the museum store rooms (Bommelaer 1990; Dietrich 1992). In 337 they bought ivory for the god's statue inside the temple. Pausanias mentions a gold statue of Apollo—probably gilded bronze (Bommelaer et al. 1990) and probably from the classical era—that had survived the destruction of the pre–fourth-century temple.

Outside the temple building, the altar located east of the temple and originally built in the sixth century by the people of Chios also had to be rebuilt. The podium of the fourth-century temple preserved a curious arrangement of channels under the great stone base—channels large enough for a small adult to walk in. Several modern investigations have failed to find any water beneath the temple base, yet F. Courby's early excavation in 1913 (1925: 176–81) reported some water under the Temple of Apollo, which he thought was supplied by a nappe of underground water. Reports about water under the temple could result from the variable height of the groundwater, which in turn depends on the precipitation history for the preceding decade. (See subsequent discussion on effects of karst water on the results of earthquake.) The site of the oracle has left no visible sign in the extant temple except some sagging of the floor—possibly the result of laying the new cella's paving on fourth-century fill that was not well compacted. Higgins and Higgins (1996) suggests that the sagging is the result of settling in the original chasm, some of the original fill having been washed out later by karst waters.

The Marmaria Terrace was badly damaged by flying stones in 373; immediately afterward a new circular structure of pentellic marble was built, and it is

FIGURE 4.23. View of re-erected columns of the Apollo Temple. The column drums are chipped along both bottom and top edges, evidence of the twisting and shaking they underwent from the earthquakes that overthrew them. The columns stand on a tall podium, which adapts the steep hillside to the support of the building. Three steps (stylobate) elevate the columns above the platform.

dated 380–370 according to the style of the details, for example, the extremely early use of the Corinthian order (Bommelaer et al. 1990).

Building activity in the second half of the fourth century may relate to another earthquake of 348 B.C.E. (magnitude R 6.7), which caused great destruction and therefore required extensive reconstruction (Fig. 4.24). During this century, the spring or fountain called Kassotis, where the Pythia drank immediately before the oracle ceremony and which had earlier been located uphill at the area west of the Sanctuary of Neoploteme and the Stone of Kronos, is known to have moved down hill, closer to the Apollo Temple.

In the central area of the site, the tradition of constructing treasury buildings ended about 300 B.C.E. with the completion of the Treasury of the Cyrenians (Hellmann 1990). New construction during this period included the West Portico of 334–279, a gymnasium on the Marmaria Terrace (Jacquemin, Rolly, Queyrel, and Biélis 1990), and a palestra near the gymnasium.

The gods utilized earth movements[10] again to repulse marauding Gauls in 278–279 B.C.E., with thunder, an avalanche, and a magnitude 6.8 earthquake, necessitating more renovation. The formal stadium was moved from the plain below the city to its present location above the sanctuary area in 275 (Aupert 1977). In the last half of the third century B.C.E., Attalos I of Pergamon built the West Portico, on a wide terrace west of the Apollo Temple. It is the only one of the six one-story porticoes that has an interior colonnade (Bommelaer 1990). The portico on the East Terrace, also built by Attalos, has the typical Pergamene detail of alternating tall and short courses of masonry. The king rebuilt the altar of the Apollo Temple too. His successors as kings of Pergamon completed the 500-seat theater on the hillside between the Apollo Temple and the stadium.

After two severe earthquakes during the fourth century, a magnitude 6.8 quake of 279 B.C.E. continued to affect water supply as well as building stability. The flow of the Kerna spring up by the stadium changed, at first supplying the Kassotis fountain by overflow. An otherwise unidentified earth movement was responsible for the Kassotis fountain being waterless during the period 210–100 B.C.E., according to an inscription. A similar geological change may be associated with the conversion of the Attalos Portico into a cistern for the storage of rainwater, during or after the fourth century C.E. (Mulliez, personal communication).

From the second century B.C.E. to the second century C.E., the archaic fountain set below ground level near the modern road at the east of the site was supplemented and then replaced by the Kastalian nymphaeum at the uphill site across the little stream (Fig. 4.25).

Later Hellenistic times were quiet seismically. After 279 B.C.E., there were no major earthquakes until the mid-sixth century C.E. Delphi exhibited "seismic silence" only during this quiet period (*Homage a Pierre Birot* 1984: ch. IX.).

From the late first century B.C.E., Delphi was still a free city but more a museum than a functioning oracle (Jacquemin, Vaatin, Sauron, and Déroche 1990). Tourists came to see the forest of honorific statues, read the many inscriptions, admire the view, and attended the games. In the first century C.E., Emperor Claudius tried but failed to repopulate Delphi with Greeks from many cities. Emperor Nero looted 500 bronze statues in the third quarter of the first century

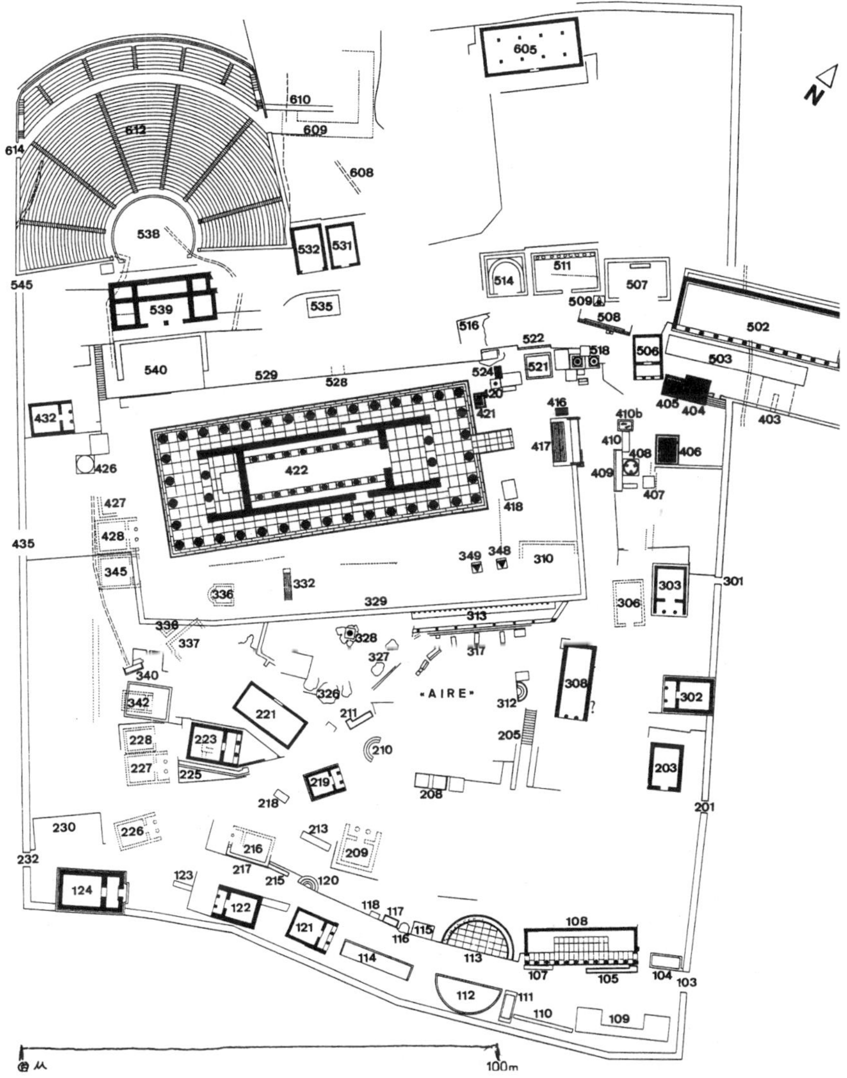

FIGURE 4.24. Delphi sanctuary plan with numbered buildings (courtesy French Archaeological School, Athens). Those of importance to our study are: 108, Unnamed portico; 209, 216, 219, 223, 226, 302, 306, 308, 342, 345, 432, 506, Treasuries; 221, Possible bouleuterion; 313, Portico of the Athenians; 326, Large fallen rock, possibly of the Sybil; 329, Lower terrace of the Apollo Temple; 336, Oikos or chapel of Ga (earth goddess), with seep of water; 340, Fountain of the Asklepeion; (309, off map at right, is the Roman East Bath); 403–503, Terrace of Attalos I, with portico (502); 417, Altar of Apollo; 422, Temple of Apollo; the West Portico is off the map, immediately to the west; 538, Theater; 608, Pipe from spring farther uphill; 610, Possible Kassotis fountain.

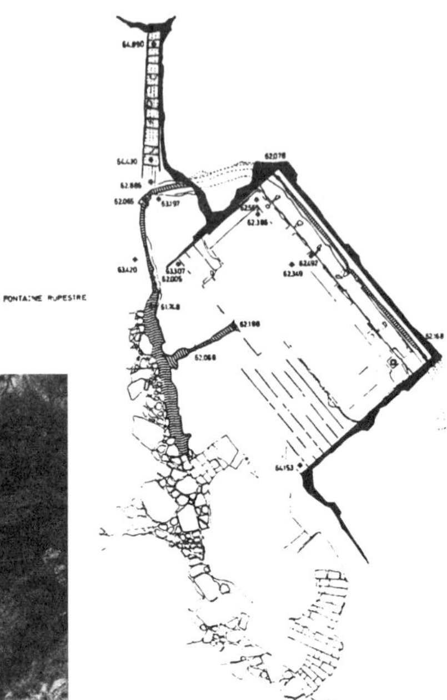

FONTAINE RUPESTRE

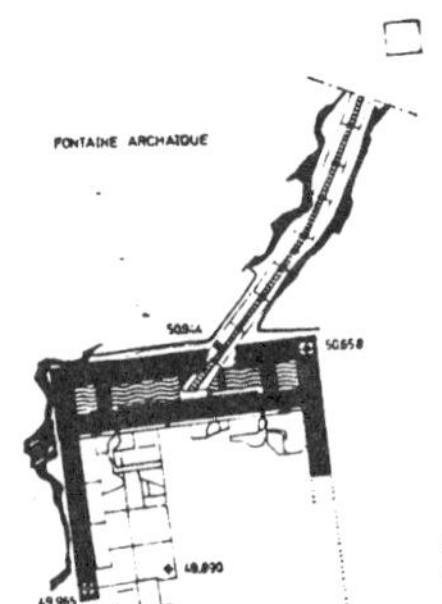

FONTAINE ARCHAIQUE

1 ___________ 10m

C.E., which set the precedent for confiscations by later emperors. Repopulation of the site began after Domitian had restored some of the buildings. In 83 B.C.E. invaders from Thrace burned the Temple of Apollo, which was restored by Domitian in 84 C.E. Emperor Trajan, however, brought prosperity again to Delphi. Then about 125 C.E. under Hadrian, the Temple of Apollo was restored. This seems to be the time when water from a source to the west of the theater was led to the fountain at the theater. However, the heavy travertine deposits (30 cm) visible today on the south side of the fountain (Fig. 4.26) have accrued only since excavation in the first decade of the twentieth century, when a cone of travertine that filled the entire little plaza of this fountain was removed (Higgins and Higgins 1996: 80). Only small amounts of travertine are found at the spring itself, which Higgins describes as located amid red limestone breccia over shale. The spring still supplies part of the water for the modern Delphi village.

The Kastalia ground basin at the east end of the site is the most studied resurgence at the site. Probably during Roman times, the Greek ground basin was replaced by the Roman nymphaeum on the opposite side of the stream and a bit farther uphill. This change is likely to relate to the uphill movement of the water, pouring out of the east wall of the Papadia ravine just above the modern road (7 in Fig. 4.22). At some later point it was necessary to provide additional water through pipelines from the east to the pool in front of the Roman fountain. The Romans also rebuilt the gymnasium on the terrace below the fountain, adding hot baths to the Greek palestra and replacing the old columns with new marble ones, possibly Ionic. The round tholos farther east on Marmaria, however, was left in ruins.

In the 170s C.E. the stadium was rebuilt with tiers of stone seats by Herodes Atticus. His workers took apart the stadium fountain's Hellenistic facade of marble blocks and gave the fountain outer walls and an arched facade of brick. One meter inside the facade is a 0.9-m-tall parapet of hewn sinter (*calcaire plaque*, Aupert 1977), so that the existing basin is partly natural and partly artificial. Even during the Roman period, there is no trace of a water distribution system in the stadium, but there were drains from at least Hellenistic times.

In the Roman era, most of the site, including the sanctuary, was covered with

←————————————————

FIGURE 4.25. Kastalia fountain, view and plan. The plans are placed in proper position and distance from each other according to the scale of 10 meters shown at the bottom of the plan. The Greek basin is sunken into the ground at the bottom of the plan. It was replaced or supplemented by the nymphaeum-type fountain (at upper right) with a series of wall spouts filling a waist-high pool reached by climbing a set of stairs. Later this fountain was supplied by a lower channel from the east. The photo shows the upper fountain and pool in 1985 before stabilization efforts were undertaken in the 1990s. Both versions of Kastalia were fed by the creek in the ravine, which produced a tall waterfall as recently as the eighteenth century. (Photo by Crouch; plan by permission of the French Archaeology School, Athens)

FIGURE 4.26. Sinter at the fountain above the theater. This travertine has been deposited in modern times, after excavations began in the early twentieth century. (Photo by Crouch)

civic buildings. There were many public and private baths, indicating the abundance of water. Pausanias (10.9.1) noted that the residential area of the Roman site lay just below the sanctuary, on the same hillside where today domestic courtyards, bathrooms, wells, cisterns, and tombs are visible above the modern road.

Following Nero's example, the emperors Constantine and Theodosius plundered the site in the fourth century, to "clothe [Constantinople] in the nakedness of all the cities of the world."[11] The edict of Theodosius in 381–385 against paganism and his decision in 394 C.E. to shut down the Pythian Games were the final blows against the worship of Apollo.

Like many another Greco-Roman towns, however, Delphi continued as a place of residence and a local center for trade and commerce through the fifth and sixth centuries, because of the natural resources and strategic location of the site, as well as the availability of so much preshaped building stone. The road through the sanctuary area was repaved with old inscriptions (usually face down) and ashlars, and the great houses of the Imperial Roman period were rebuilt and redecorated. The latest new installation was a mosaic of the late sixth century C.E.—perhaps necessitated as a repair after the earthquake of 551 (R 7.2). There are three Christian basilicas at Delphi: one of the fifth century on an unknown site, one outside the city at a cemetery, and one at the palestra of the gymnasium; cemetery and gymnasium sites were common for early Christian churches (see Ephesus case study). Devastation by the plague of 542 in which one-third to one-half of the population from Wales to Central Asia died caused the final elimination of classical culture (Thirgood 1981).

The Temple of Apollo was fired for the last time during these years and roughly rebuilt (Jacquemin, Vaatin, Sauron, and Déroche 1990), only to be destroyed in the Slavic invasion of about 600 C.E. Extracting metal clamps from the old buildings during the Middle Ages hastened their decline, because the clamps had increased the earthquake stability of the ashlar walls (Higgins and Higgins 1996; Bommelaer et al. 1990).

In the Kastri village at Delphi, because of poor construction with unreinforced adobe on steep slopes, and because of landslides and rockfalls, the cumulative destruction of the 1870 quake series was almost total. Houses collapsed because of upthrusts and ripple motion, leaving the people with no choice but to camp outside for months. The villagers may have been glad to resign their property to the new French excavations when they moved west of the sanctuary onto a slope where their houses and shops stood on firm stone. The new houses were only lightly damaged by the 1905 avalanche, which affected the site from Marmaria to the Temple of Apollo. The damaged ruins of the so-called Temple of Athena were again bombarded, knocking down the 12 columns of the temple, which have since been sketchily restored. The Apollo Temple, less badly ruined, was restored in 1939–41 (Figs. 4.23, 4.24). This same earth movement finally stopped the flow of the stadium fountain.

Karst   Given the lithography and the tectonic activity of the Delphi area, the effects of karst processes are unusually important. Delphi's area of 1774 km² has 56 closed karstic basins (with internal drainage into the mountains) and 32 open drainage basins (along the surface of the mountains) that support 57 towns or villages. The escarpment at Delphi has recent surface gullying, but the local karst is much older. Great volumes of water from the snows and rains of winter are stored and transmitted in, through, and over the mountains. The investigations of Burdon and Papakis (before 1961) found six aquifers in the 4000-m-thick limestones and dolomites. Because they are 13% more porous than limestone, dolomites are important reservoir rocks. Springs occur, as in the sanctuary, along the NNW-SSE faults, where the subterranean waters flow until dammed by either impermeable rock or alluvium within the karst, or by the seawater barrier at the coast. Other waters collect in perched nappes in the limestone, wherever it lies above flysch.

Delphi itself sits between a karst aquifer and a flysch aquiclude. The area has numerous downhill talus cones made of limestone and dolomite fragments, cemented if older but unconsolidated if younger, that behave as good aquifers. These cones feed lower karst systems lying over limestone, and springs situated over flysch. Springs above flysch usually have only small reservoirs, but are numerous (Burdon and Dounas, before 1961).

Annual infiltration, though only half of all precipitation, greatly exceeds the output of the springs. The groundwater balance of the area, based on precipitation of 583 mm/year, relies on an intake of 51% of the rain and snow. Burdon and Papakis took water samples from springs in the summer months, deducing that annual yield is probably 50% of input or 415 M m³/yr or 47,400 m³/hr, including 3,370 m³ from perched nappes in the limestone at places such as Delphi. The karst springs of the area show maximum discharge right after maximum precipitation from winter storms but minimum discharge in July–September, when there is maximum need for irrigation water and for human uses to alleviate the heat of summer. Without the reservoir capacity of the karst system, human habitation would be impossible in this area. The advantage of the underground reservoirs of a karst system is twofold: no evaporation and no silting up.

A parallel study by Burdon and Dounas (also before 1961) analyzed the water at Delphi along with other springs in this area and found it to be a typical calcium-bicarbonate water from Aquifer A in the limestone-dolomite. Their figure 3 shows that spring 13 at Delphi is clearly within the "predominately calcite aquifer," having a very low magnesium content because it derives from limestone not dolomite.

Springs and their associated fountains at Delphi from the top of the sanctuary downward have been as follows: The stadium fountain, formalized probably between 550 and 535 B.C.E., flowed until the early twentieth century when an avalanche coupled with drilling for the modern Delphi village water system changed the underground water network (Aupert 1977) and dried up the fountain. Just below the stadium is Kerna, renovated in the 1990s. Overflow from Kerna fed Kassotis fountain. Kassotis spring moved downhill from its sixth-century B.C.E. site near Kerna to near the Painting Gallery east of the upper tier of the theater during the fourth century, and finally to the level of the temple of Apollo. The theater fountain (Fig. 4.26) was fed by pipe from a source to the west. The seventh (?) century B.C.E. Kastalia (7) in the ground east of the site was replaced after the third (?) century B.C.E. by the more elaborate fountain in the rock wall. An overturn or overthrust of the karstic Mesozoic carbonate sequence above the impermeable flysch (Norton 1965: 19) exposes limestones of the Senonian era, and the transition to the flysch is a pattern of gray to red thin bedded argillaceous (clayey) limestones followed by red marls. The spring itself comes out of a thrust fault, shown by the V-notch in the rocks, which dips 30° to the north, where limestone lies above shale. This formation forces the karstic water to the surface—a phenomenon we have already seen in the upper Peirene spring at Corinth. In the past, as illustrations from early modern travelers prove, the stream's flow was much more ample, forming a major waterfall and an important tributary to the Pleistos River (Pouilloux and Roux 1963). Amandry (1977) reported that archaic Kastalia was restored in 1958–59 and already completely hidden under 150 cm of sinter by 1977; the sinter had been removed again by the mid-1980s when I first saw it. After the new fountain was carved into the cliff face, the old one was used as reservoir.

Fountains not known to have changed location include one at the southwest angle of the polygonal wall at the center of the Apollo sanctuary, called the fountain of Asklepios. Another small fountain was excavated in 1893–94, on the terrace south of the Temple of Apollo. This fountain, which was related to channels that penetrated the foundation of the temple, may be the Fountain of the Muses described by Plutarch. Pausanias reported incorrectly that this stream entered the soil and communicated to the inner shrine where Pythia gave her oracles. Such inaccuracies in the ancient texts were based on ignorance of precisely what happened to the water. The excavation of Pouilloux and Roux (1963) showed that Kassotis spring's water flows southeastward, not through the Temple of Apollo but around it.

Stability   The combination of karst activity in the mountains above Delphi and the scree underlying the buildings of the sanctuary induced frequent stability problems, according to engineering geologists Kolaiti et al. (1992, 1995). Groundwater movement caused displacements, rotations, toppling, collapse of masonry, cracking of stones, and both deformation and subsidence of foundations and retaining walls. Intense rainfall increased destabilization by both lubricating movements between rocks and dissolving supporting clay and stone. The main evidence of instability is the rockfalls, which were already occurring before the sanctuary was built and which continue today (Figs. 4.18, 4.19). Fracturing by faults along the joint system and tectonic upthrusts displaced bedding planes and weakened the unity of the strata, which was already endangered in its lower bedding planes by clay-filled joints in the Eocene limestone (Constantinidis et al. 1988). The solution and erosion of rock was exacerbated by surface weathering, which temperature fluctuations at the site increased.

Climatic Factors

Hydrogeological and climatic conditions, especially the flow of groundwater through relatively permeable limestone to the impermeable flysch below, increased pore pressures. Diffusion of water under pressure in seismic karst zones dissolves much carbonate during the long periods between earthquakes. In the loosening that accompanies the earthquake, the carbonate is then deposited in karst "ribbons" that facilitate dispersal of scree. Talus beds thus register reformative tectonic events and processes. Recent fault activity at Delphi indicates free networks in the rock, but Péchoux (1977) insists that climatic factors are necessary to explain the visible changes. When a tectonic crisis occurs at the exact time that the cracks within the rock are fully saturated with water, there is less damping of seismic waves than when the stone is dry. During an earthquake, a competition takes place between exterior pressure and churning inside the rocks that have karst passages. If a humid phase of climate has led to much water stored in the rock, an earthquake can cause the deposition of lots of debris. In historical times, Delphi has been relatively dry and therefore relatively stable (Péchoux 1977).

Materials and Quarries

In Mycenaean times, Delphi was already associated with the *polje* of Livadi, to the north in the Parnassos area. That site was abandoned at the end of the Helladic period because of catastrophic flooding, when mud flows stopped up the dolines. Later, in archaic and classical times, much stone for monumental buildings at Delphi came from the several quarries at Livadi.

Although most buildings of Delphi used local stone, in the eighth, seventh, and fourth centuries some Corinthian stone was used in the Temple of Apollo. This was probably because quarrying near the sanctuary and moving stone from uphill quarries or along mountain contours was considered less cost-effective than trading for Corinthian stone, which could be shipped across the Gulf of Corinth relatively inexpensively and then carried up the existing road from the port. The cost of reinforcing this road (see the Selinus case study) would have been less than the cost of cutting new roads from local quarries to the site, even though the transport cost was 420 drachmas per block for the 7 miles from the harbor uphill to the sanctuary in the third quarter of the fourth century (Burford 1960).

There were, however, small quarries near the site and a large quarry at Prophet Elias 5 km west and a little south of Delphi at 210 m of altitude (Delphi is at about 600 m). In Hellenistic and possibly Roman times, these quarries were mined for Early Cretaceous limestone with some chert and calcite (Papageorgakis and Kolaiti 1992).

To differentiate marbles that are visibly alike, Maniatis, Mandi, and Nikolaou (1989) studied marble samples found at Delphi by electron spin resonance (ESR), which gives data free of the influence of weathering or pollution and is oblivious to color and grain size. Stone from ten ancient and modern quarries was studied. Creating a data base of quarry materials, they studied 25 samples found at Delphi with different grain sizes and colors. They found that nine examples were from Livadi, all being gray and fine, the source identified by a particular analytical pattern unlike others such as Hymettus (near Athens) and Thasos (in the north Aegean). Another seven pieces could be assigned to known quarries: four from Aliki on Thasos and three from Penteli near Athens. Unfortunately, this publication does not ascribe these samples to particular buildings at Delphi.

Other archaeologists describe the materials used in particular buildings here. Bommelaer (1990) details the following uses for stone in porticoes:

1. The Portico of the Athenians (109 in Fig. 4.24) was very like the Treasuries of Marseilles and Athens, which were both of limestone, built in the early to mid-fifth century B.C.E., having steps of limestone, Ionic columns of marble (atypical for the fifth century), and an entablature of wood.

2. Portico 108 in Figure 4.24, about 400 B.C.E., whose back (north) and side (west) retaining walls (*analemma*) are of conglomerate; the foundation for the colonnade (stylobate) was raised on a vertical socle of conglomerate and surfaced with limestone, whereas poros was used for the colonnade, antae, pillars, and entablature; walls and socles were of conglomerate. With so many materials, this building lacked a unified design conception.

3. The North Portico (10 in Fig. 4.22) of the gymnasium had a wall of poros, from Delphi not Corinth. The early poros colonnade was later

(probably in Roman times) replaced with marble. Where the building walls served as retaining walls, they were of limestone, and the stylobate was of conglomerate with a limestone surface.

4. The Palestra Portico (south and a little west of the xysto of the gymnasium) had Ionic columns of limestone.

5. The West Portico (just outside the Apollo sanctuary, along the way leading from the west exit, 6 in Fig. 4.22) had retaining walls of conglomerate like the gymnasium and the Attalos portico, whereas its stylobate was limestone and steps were conglomerate.

6. The Attalos Portico (502 in Fig. 4.24) used conglomerate for parts that were not visible but limestone for the facings. The bases of the columns were stuccoed, and the upper moldings were of marble. This portico illustrated a rare use of tuff. The long rectangular construction in the court before this portico (which may be an altar) is also of tuff.

Problems of nomenclature are evident in the claim that the 530s Apollo Temple was constructed of Corinthian tuff, whereas that stone, a carbonate rock, is today usually called tufa (see Glossary in Appendix B). When the temple was rebuilt in the fourth century, limestone from the quarry at St. Elias 2 hours away was used to pave the platform and for steps of the colonnade, but the orthostates and the paving of the cella were gray limestone. Higgins and Higgins (1996) names the local stones "limestone, sandy limestone, and sandstone"; the latter may be what the French excavators call *calcaire* — probably calcarenite, a limestone made of cemented sandlike grains. Higgins claims that the columns were of poros limestone, which had to be covered with stucco because of its softness. The east front of the temple, however, was of Parian marble (Childs 1993; Bommelaer et al. 1990).

The roof and inner ceiling of the fourth-century round tholos in Marmaria were made of marble, which was appropriate for the complex three-way curves of the roof panels because of its strength. The building used dark and light marble for the walls, with a limestone pavement (A. Jacquemin 1990).

The materials used for the various treasury buildings have been noted by M.-C. Hellmann (1990). The series utilized tuff (probably limestone), marble, and calcarinite. The Siphnos Treasury, flamboyantly Ionic, was made of gray calcarinite from Parnassos for the subbasement and terrace (still in place), Parian marble of fine grain for the sculptures, Naxian marble of large grain for the architrave, antas, and perhaps gutters, and Siphnian marble — unfortunately of poor quality — for the walls. All the white marble was vividly painted in red, blue, green, and yellow-orange.

Conclusions

Delphi's history involves geological, archaeological, and documentary factors. The depth and extent of scholarly investigation at Delphi began with the ancient Greek inscriptions and the recorded visits of Pausanias and other ancient writers in Roman times. The physical activity of the site, although unfortunate for local residents and their buildings, is useful to correlate geological change with human history. At this site, many historical dates correlate with geological events, although processes are difficult to date. But alas even dates inferred by historians and archaeologists are not unquestionable. Archaeologist Amandry (1988) points out that the chronology at Delphi is based on the association of objects recovered via archaeology, and on their stylistic evolution as related to historical events, but because of subjectivity such historical correlation may be illusory. He thinks that we can be certain of nothing at Delphi before the middle of the sixth century B.C.E. and should always allow +/− 20 years for ancient dates. Indeed, archaeologist Childs (1993), after a manful effort to attempt to bring "different classes of information into accord" in a field where there are "so few facts," was led to question "how the past generations came to conclusions we have generally accepted for decades" and to sigh that "perhaps the monuments and the texts do refer *on occasion* [emphasis added] to the same past events and enhance our understanding of them."

For well over a hundred years, archaeology has added details of the built environment and its changes over time, and now engineering geologists and other scientists are contributing their insights. For Delphi we can begin a true integration of geological history with construction history and social history. The strong impact of the physical setting is inescapable, but equally strong has been the perseverance of human activity here. The balance within the rocks was broken at strategic moments, either to protect by avalanches and lesser rockfalls or to wreak destruction. Less noteable but more essential for life at Delphi was the abundant supply of water. Abundance of stone for construction made possible the visual appearance of the sanctuary and its functionality.

What are we to make of human persistence in rebuilding on this site? I suggest that the sanctuary was a focus of Greek anxiety about the natural and supernatural worlds. Temples, shrines, statues, tombs, pilgrimages, dramatic festivals, and athletic games combined as religious management of the threats. The Temple of Apollo, then, was located here because of the danger, not in spite of it.

Eastern Greco-Roman Cities

Miletus

Introduction

Only since the last two decades of the twentieth century have professional geologists been working specifically at Miletus, investigating the shiny white limestone on top at Kalabak Hill; the limestone of Theater Hill and Humei Hill; and the limestone of Zeytin Hill (at the west of the hill called Degirmen, west of Kalabak), with sandstone and tuff to the east. The stratigraphy at Miletus is as follows, beginning with the topmost modern layer:

Iron deposits in mountains east of Miletus and both iron and brown coal at Mt. Mykale (today Samsun Dag) north of the Meander River valley.

Volcanic tuff.

Soluble limestone: 200 m of Yatagan or Balat strata of marble with tufa, sand, and gravel from which springs emerge.

Early Pliocene limestone, shiny white limestone cap 60–100 m thick, karstified but with few on-site springs, forming ridges, hills, and a thick layer of scree.

Miocene marls, sandstones, conglomerates, and clays with springs.

Pink-yellow sandstone, sometimes with tuff, around sides of hills.

Older clayey limestone deposited in a lake environment.

Former large bay, now a swampy river plain with rich alluvial soil.

Prof. B. Schröder and his team have done geological research in the area (1990–94) and published their findings swiftly, which I acknowledge with gratitude. Suspected faults run north and south of Kalabak Hill (recognized in C. Schneider 1997). The town of Akköy to the south, where the German archaeological house is located well above the mosquitoes of the archaeological site in the swampy plain, sits on limestone that forms a peninsular ridge, surrounded on three sides by sandstone with tuff.

The west coast of Asia Minor is subject to strong relief, steep gradients, and high precipitation; the high amount of energy available allows rapid change in topography (Gage 1978: 621) (Fig. 5.1). In Ionia, on average, the sea has risen or the land has sunk 1.75 m since antiquity (Bintliff 1977: 24; 1992).

The Greek cities of Asia Minor were without exception built on or next to karst terrain (V. Klemes, 1988, personal communication when he was president of the International Association for Karst Hydrology). Local water supplies at Miletus were enough for early residential quarters arranged around Kalabak Hill, but after the Persian War the reconstructed city (early fifth century B.C.E.) required water brought from distant springs.

The Miletus Chronological Table in Appendix A summarizes the geological events and processes that affected historical events and construction of buildings at this site. Water management issues are briefly described here as examples of the relationship between geology and urban development.

History and Geology

Prehistory   The prehistory of the Milesian peninsula begins with the regional tectonic activity of Eurasian and African-Arabian plates. These large plates are separated by two small plates, the Anatolian, which includes most of Turkey, Cyprus, and the sea between them, and the Aegean plate, which includes western Turkey, the Aegean Sea, Crete, and the Peloponnese of Greece. (Fig. 2.2; see also fig. 4 in Baird 1971). For millions of years, the Anatolian plate has been moving westward, whereas the Aegean plate moves to the southwest and the African-Arabian plate moves northward under the Anatolian plate, forcing it upward (G. Evans 1971). The westward movement was demonstrated by the catastrophic earthquake in the Izmit area (east of Istanbul) in August 1999 and its aftershocks. Strong motions, noted occasionally at Miletus, interact with the fluctuations of sea level and variations in rainfall to produce a very active geology (Schröder 1990).

Horst and graben movements from plate collision produce a "mosaic-like block structure between fault lines running in different directions" (Kayan 1996; Vetters 1998). The early research on Mediterranean plates indicated that "in the Mediterranean, a large strike-slip component of motion between the two blocks may produce rotation of small intervening blocks" (Morgan 1968; Vine and Hess 1970), such as the small islands that grew together to form Miletus. Mourtzas (1988, fig. 15, chart and map) reviewed the literature and studied the physical evidence for rotation in Samos, which lies offshore from the ridge of mountains between Ephesus and Priene. Eight other small, uplifted lithosphere blocks identified by Pirazzoli et al. (1982) have had, like the north part of the west side of Samos, tectonic behavior with complicated physical development since 6000 BP, including clockwise rotation around a horizontal axis in an EW direction, so that the

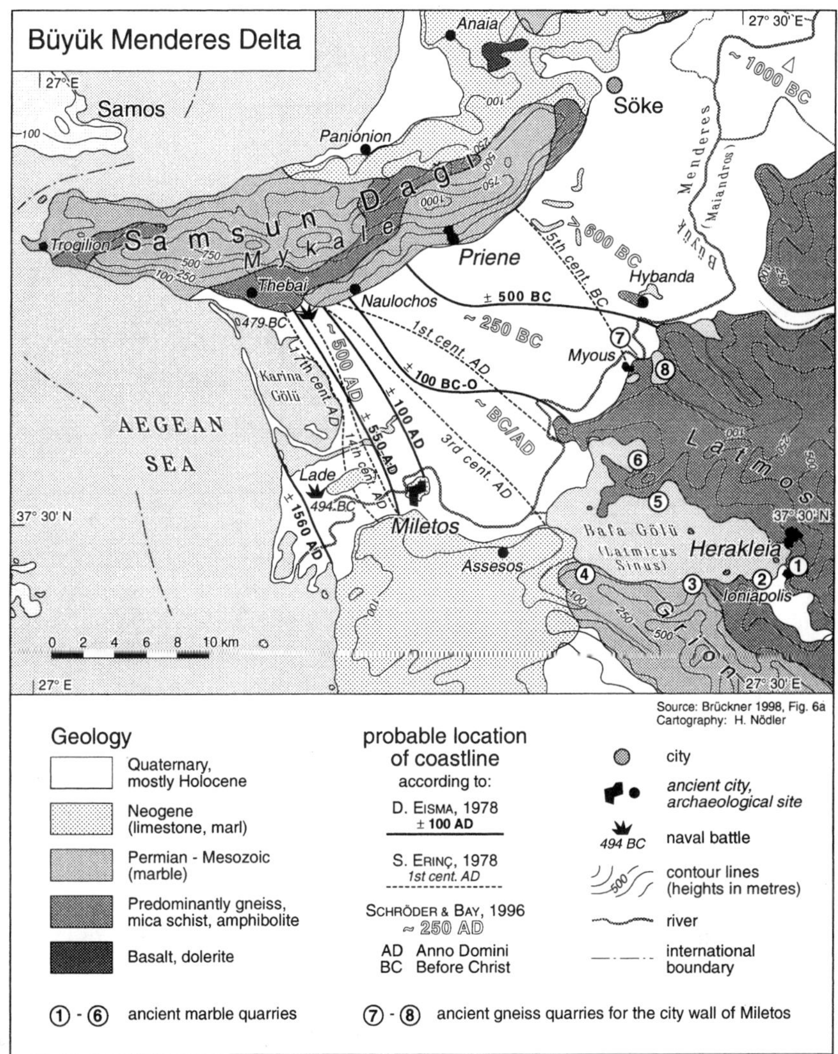

FIGURE 5.1. Map of sedimentation process of the Great Meander River (Büyük Menderes) between Miletus and Priene. Locations of ancient quarries have been added from Schröder and Yalçin (1995, fig. 5) by permission. (Brückner 1998, based on the literature, supplemented by field studies)

north side was uplifted and the south submerged.[1] The south part of the Samos block is another separate tectonic element that also moves up and down while rotating; the eastern section of this block generally has subsided. (Compare these movements with block rotations reported along coastal Peru by Ortloff et al. 1983). Given the proximity of Samos to Miletus (Fig. 5.1), it is reasonable to expect similar rotation at Miletus.

The mountain blocks along the west coast of Turkey form long fingers connecting the sea with the central plateau, rather than with the abrupt mountain walls of the north and south shores. The valleys between these mountain fingers have been particularly affected by changes in sea level and by erosion.

Some of the western rivers of Anatolia run level with or higher than valley floors, increasing the danger of floods (Karatekin 1953: 79–82; statistics in his tables 1, 2, 3, 5 and chart 1). The Great Meander Valley where Miletus stood was most deeply under water before the last Ice Age (Fig. 5.1). Higher sea levels than today are suggested by the higher terraces upstream, where brackish water fauna are found (G. Evans 1971); habitation remains from this period might exist high on the slopes. After the sea receded, the sea level minimum at 18,000 B.P. was 120–130 meters lower than now, so that islands such as Samos and Lade were connected with the mainland, whereas today the Mediterranean Sea is 100 m deep near Samos, 30 km from the modern coastline (Schröder 1990). By 6000–5500 B.P., however, the sea was at a height close to its present level, because the water locked up in glaciers had melted. The sea flowed into the valley, past the peninsula of Miletus all the way to Aydin (ancient Tralles), which is now some 65 km inland. From 5500 B.P. on, the area of later Miletus had become several islands (Brückner 1997 a&b; 1998); then about 1500 B.C.E., the islands became a peninsula (Voigtländer 1985; Brückner 1998) (Fig. 5.2). Sea transgressions led to horizontal changes along the coast, depending on the previous inclination of the land, the rate of sea-level rise, and the volume of alluvium brought down by erosion

Erosion has been actively clearing debris from western Anatolia since at least 1000 B.C.E. Accumulated sediment has drastically affected the setting of the horsts, incorporating some of them (e.g., Lade) into the mainland. We will discuss this further under the Hellenistic period, when erosion began to be a human problem.

The climate, like the location of the sea coast, has seen variations. Anatolia lies between 36°N and 42° N, and the climate of the Aegean and Mediterranean coasts is hot, with most rain falling during the winter and spring. Typically, the Mediterranean climate is semiarid with only 10–15% of rain falling during the summer. Temperature and precipitation reach their maxima at opposite times of the year. The area is hottest in July and August 28° C), when almost no rain falls but a cooling breeze from the western sea mitigates the temperature while increasing evaporation (Tuttahs 1998: fig. 16). Then in December and January, precipitation is at its maximum but temperature is only 8–10° C because of the cold winds from the northern mountains. At Miletus, rainfall is usually between 630 mm/yr and 730 mm/yr, with a maximum of 940 mm and minimum of 340 mm.

Pollen analysis (Wille 1995) of Brückner's corings from the Lion Harbor noted an increase in deciduous oaks from 4000 to 500 B.C.E., indicating very cold winters during that period. The forests allowed for good hunting for several centuries from 800 B.C.E., but remains from domestic animals dominate the record (Peters and von den Driesch 1992: 117 ff). The coldness eased considerably between 500 B.C.E. and 300 C.E., when the forests were used to feed pigs, to graze sheep that were the basis of wool wealth, and to cut wood, which unfortunately contributed to erosion. Some areas still show evidence of ancient mistreatment, although as we

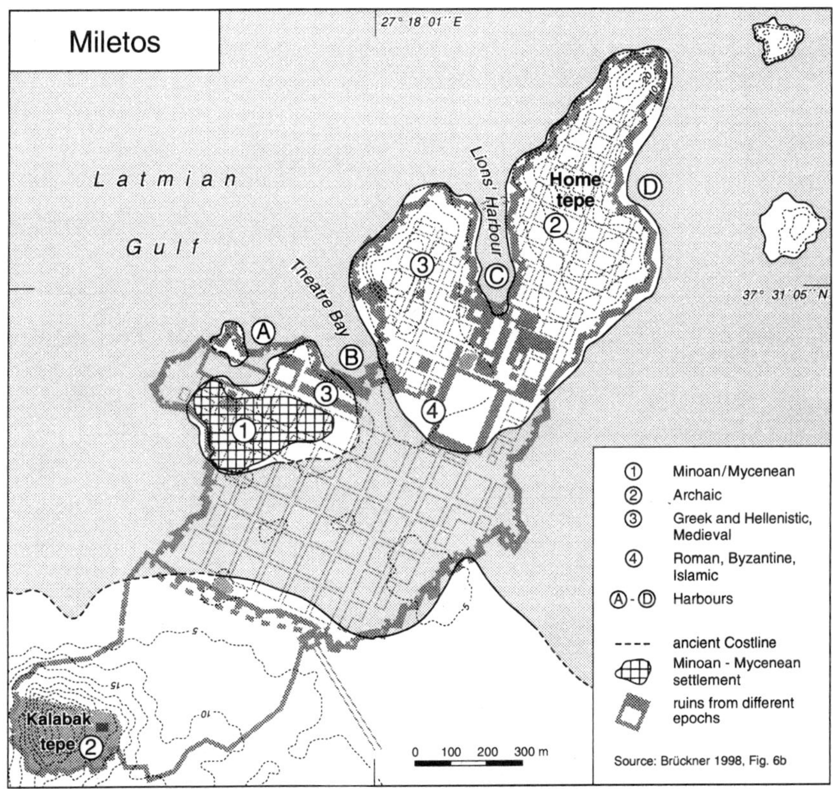

FIGURE 5.2. Geography of the Milesian Peninsula. During the peak of the Holocene transgression, islands existed in the Latmian Gulf; later, the islands coalescing as the historical peninsula. Water once flowed between the northern island (Home Tepe) and the central island and between the latter and the southern hill (Kalabak Tepe). Settlement dates are as follows: (1) second millenium B.C.E. (2) seventh–sixth centuries B.C.E. (3) fifth–second centuries B.C.E. and fourth–fifteenth centuries C.E. (4) first century B.C.E. through eighth century C.E., with some occupation into the fifteenth century. Harbors: A, Athena temple area; B, Theater area; C, Lion Harbor; D, Northeast harbor. Monuments and peninsular seawall from Greco-Roman period in dark gray. The fifth-century B.C.E. grid plan is shown as double dotted lines. Note the ancient coastline, a solid but thin black line (dashed where inferred) enclosing land areas shown in white. (Reprinted by permission from Brückner 1996, Fig. 6b).

have seen at Argos, degradation of the forest is likely to have occurred during times of political chaos.

Like all the large rivers of this coast, the Great Meander River—also known as the Büyük Menderes (Turkish) and the Maeander (Greek)—runs in a trough or graben formed by faults (Evans 1971) and therefore Miletus is in perpetual but intermittent tectonic danger. The Great Meander's plain is comparable to the

Little Meander River plain at Ephesus, with even more obvious meanders (Fig. 5.3) but extends more than twice the distance into the hinterland. The river provided the easiest route into the hinterland, as we have seen with the rivers at Selinus and Syracuse.

Given the rich alluvial soil and the climate patterns, the valley is well suited for agriculture, as evidenced by the long duration of human settlement. A relatively

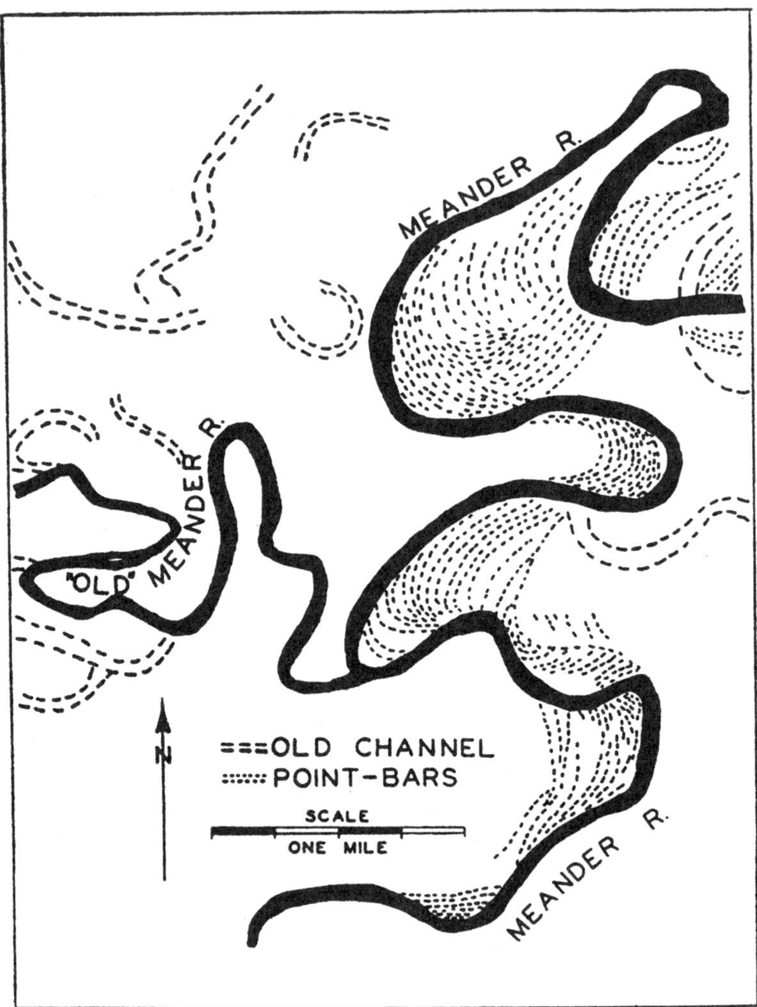

FIGURE 5.3. Meanders of the Great Meander River south of Söke. The river's name became synonymous with "winding." Heavy black lines indicate the course of the river, double dashed lines abandoned river beds, and parallel groups of dashed lines sedimentary deposits along the river's course (Russell 1954).

small area of arable land, the valley bottom and the lower slopes of the mountains, has increased over the centuries as the delta has grown westward (Emlyn-Jones 1980), supporting increased population. Tectonic change widened the mouth of the valley at least once, when a strip of land north of the Karina lagoon at the present coast was lowered by faulting in the Balat earthquake of 1955.

The site is isolated from the interior by mountains that also protect it from inland invasion. The mountains along the two sides of the valley, being of carbonate rocks, are rich sources of water, which is accessible via karst systems (Evans 1971). Mature karst hills at Miletus step down toward the Meander, each capped with hard white limestone over more soluble layers of softer limestone. This kind of capped karst conveys water via vertical shafts through the impervious rock to the lateral flow through the softer limestone below, often taking the form of long integrated caves (W. B. White 1977); the water reappears on the surface as springs. The open shafts and tunnels of karst transport water rapidly, compared to its slow movement through permeable stone. (See W. B. White's tables, "Hydrologic Classification of Carbonate Aquifers," and "Effect of Hydrogeologic Setting on Carbonate Aquifers" 1977: 177, 179.)

At Miletus and Priene, the different forms of karst resulted in quite different water potential, since Priene's water system continued to rely on the water stored in the karstified marble mountain and readily available from springs, but the older karst with fewer on-site springs at Miletus required development of long-distance water supply lines as early as the sixth century B.C.E. Fortunately, by then Miletus had both the population to warrant such a water system and the wealth to pay for it.

Greeks and Others   Humans moved into the Miletus peninsula during the Stone Age. Neolithic settlements at Milet, Balat, Didyma, and Alatinkum were all located at springs (Schröder and Ü. Yalçin 1992). The earliest evidence of settlement on the peninsula dates from the second half of the fourth millennium B.C.E. (Brückner 1996), with earlier evidence destroyed by sea rise and alluvium deposit. Surveys of the 50 km² area that includes Miletus have found more than 250 sites dating from the Chalcolithic (late fourth millennium B.C.E.) to the late Byzantine era (fourteenth century C.E.) (Lohmann 1997).

In the second millennium, easy access by sea, fertile ground, and a river route to inland customers drew traders who bartered for surplus products from agriculture, herding, and forests. As early as 1900 B.C.E, indigenous peoples met Minoan traders at the Miletus site, thus beginning its extended role as a gateway city (Burghardt 1971; Neimeier, oral communication).

Settlement and trade may have been further stimulated by the valuable iron deposits from the mountains east of Miletus and by both iron and brown coal at Mt. Mykale north of the Meander River valley (Philippson 1910).[2] Probably iron weapons and tools were available at the site of later Miletus for trade in the second half of the second millennium (Fig. 5.4).

Cretan (Minoan) remains (Strabo XIV. 1.6; Pausanias VII. 2.5; excavated by Neimeier since 1995) were supplanted by thirteenth-century Hittite (Emlyn-Jones 1980: 14) and Mycenaean evidence from the theater bay (Parzinger 1989; Tuttahs

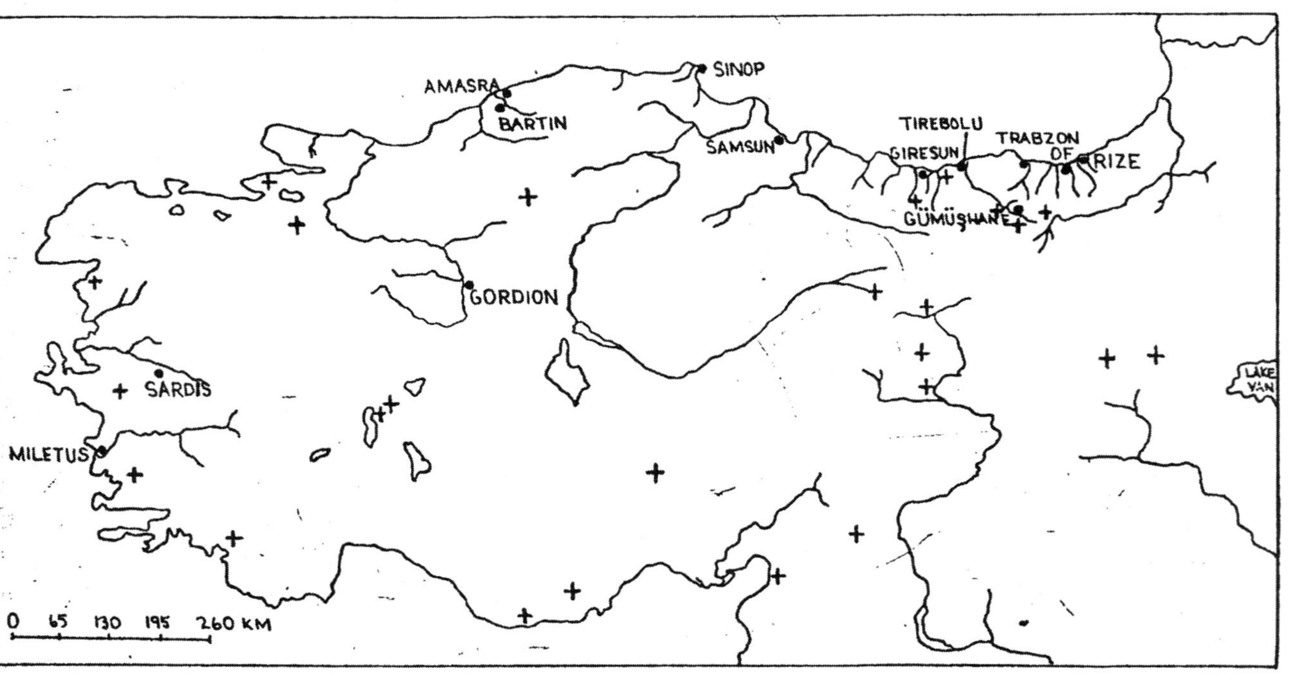

FIGURE 5.4. Map of iron deposits (indicated by crosses) in Asia Minor (Przeworski 1939). Note the deposits north and south of Miletus, and along the northeast coast between Sinop and Trabzon.

1998; Brückner 1996; and Voigtländer 1985: his fig. 10), including rock-cut tombs (Wiegand 1899), which were also found on the south end of Asayoluk Hill. House walls, some with "fragments of elegant wall paintings" (fig. 17 in Gates 1995) reminiscent of those from Thera from a mixed Minoan-Mycenaean-native settlement dating from about 1900 B.C.E., were found in excavations under the Roman streets near the theater port.

Early Bronze Age settlement extended through the twelfth century (Voigtländer 1985). These settlements at Miletus were destroyed in the twelfth century but settlement evidence exists from about 1000 and from the eighth century (Voigtländer 1985; Parzinger 1989). During this time, Miletus was a way station for the route to the hinterland. In 1000 B.C.E., seawater still extended east of Söke; indeed, during the period of Ionian settlement, the sea reached beyond Magnesia-on-the-Meander (Schröder and Yalçin 1992) to ancient Tralles (Brückner 1998). Commercial boring near Söke (30 km from today's coast) found marine and lagoonal sediments, at -7 m to -70 m under a 10-m layer of alluvium from the river. Brückner (1993, 1997a) distinguishes among marine, lagoonal, swampy, and terrestrial deposits.

Literary evidence agrees with scientific evidence. The earliest surviving account of the migrations from Greece to Ionia was by historian Pherekydes of Athens (about 500 B.C.E.; cited by Strabo XIV, 1,3), who wrote that Miletus was founded twice: First, the area was settled by the Cretan/Carians of the early and mid-second millennium, discussed previously; Milatos on Crete gave its name to the site. The colonists married the women of the Carians they ejected from the area (Herodotus 1,146). Second, in the eleventh century, Ionian colonies were established in 10–35 places along this coast. These later developed into a league of twelve cities plus some smaller places. As the settlements matured, they took in non-Ionians, then turned to trade and colonization (Emlyn-Jones 1980; Cobet 1997).

During the eighth and seventh centuries, Miletus traded and colonized in the Black Sea area, possibly because the iron deposits (shown in Fig. 5.4) and/or the fuel on which their superior weapons depended were exhausted in the hinterland (Drews 1976).[3] At Delphi, also, in the eighth century, pressure on resources for elite armor and weapons led to expansion of trade routes (C. Morgan 1990). The Black Sea and long-distance transport of iron as ingots were known to the potential Greek colonists at least by the early eighth century B.C.E. (Homer *Odyssey* 1.1884).

Traders from Miletus sailed specifically to the northeast coast of Anatolia, and after them came persons adept at recognizing profitable deposits of iron and silver, as well as miners and other colonists to find and exploit the needed metals.[4] Therefore, the development of Miletus was as strongly affected by the geologies of the northern Aegean and Black Sea and by the ore-bearing and food-producing areas of her distant colonies and trading partners as by her own Ionian geology. Food-growing colonies were established after those for production of metal, since both fish and food crops were easier to access closer to home (Drews 1976). Eventually, Miletus had 75–90 colonies, some on the coast of Thrace but more in the Black Sea area (Drews 1976; Brückner 1996; Emlyn-Jones 1980). In *Natural History* (V, 112), Pliny reckoned 90 Milesian colonies.

With the success of this colonizing, smelting, and trading effort came the preeminence of Miletus in the metallurgy of copper, bronze, lead, and especially iron. Many remains of iron-working have been found at Miletus in the twentieth century. Most of the analyzed samples date from the archaic era. The great variability in iron samples suggests a variety of sources (Yalçin 1993).[5]

Overseas trade brought wealth to Miletus during the eighth and seventh centuries, and this wealth was used to enlarge the city and improve its facilities for shipping. In the earliest times, boats were beached for loading, so that elaborate harbor facilities were not necessary. Later, the volume of trade required deliberate engineering of the harbors, providing work stations, berths, warehouses (Tuttahs 1998), and loading docks. To the little Athena harbor were added eventually the larger theater harbor, the huge Lion Harbor, and the small harbor on the upper northeast side of the peninsula (Tuttahs 1998; Brückner 1998; K. Weber, personal communication). Ancient sailors preferred ports with east and west harbors, so that they could choose a berth depending on the wind, as was the case at Syracuse. The northeast harbor would have been a useful supplement from earliest times, because it was most convenient to the upriver trade.

Trade expanded during the first half of the seventh century into the Levant and Egypt, with more colonies along the southern route to Egypt, then into southern Italy and Sicily in the seventh and sixth centuries. Miletus traded not only in fish and wheat from the north and papyrus and linen from Egypt, but especially in tin and bronze from the far west (Spain, England), iron and copper (Black Sea region, Syria, Cypress), and gold and silver (Lydia in southern Anatolia, Black Sea region) (Emlyn-Jones 1980). The metal trade brought in riches comparable to the trade in wool, wool products, and oil (Yalçin 1993). Miletus also exported timber to Athens and was a source of cypress for Delos. The rich physical base was important to the success of the city. The Milesians benefited from their bay, which was large enough for a large fleet, as well as from the smaller harbors related to that bay.

By 600 B.C.E., the city's wealth had greatly increased. Expansion from landed wealth to wealth based on trade caused strife among the citizens regarding the best form of government and provoked the envy of neighbors, which led to the invasion by the Lydian king Alyattes about 600 B.C.E. (Herodotus 1, 17, Pecirka 1973). In the seventh–sixth centuries, Miletus led the Mediterranean world intellectually. Astronomy was a valuable tool for Milesian sailors and explorers, and the geographic knowledge of Hekataios (born 560–550) was essential for far-sailing traders (Tarrant 1990; cf. Edgerton 1974). Milesian cartographers produced the first geometrically organized maps of the inhabited world (Vernant 1983).

Water Management   Within the city ramparts, more sophisticated water management permitted increased population and wealth. Miletus is fortunate to have the most thorough modern description of its water system (Tuttahs 1998), summarized here insofar as it relates to the geology. Wells in sands and gravels tapped the aquifer, northeast of the foot of Kalabak Hill and possibly farther north. A few ancient springs have also been located. The spring grotto of the theater, for instance, is at the bottom of a water-bearing limestone layer in Kale Tepe. The same limestone limited the development of the embayments regularized as harbors of the city (see fig. 4 in Schröder 1990).

Eroding water was also the active principle in changing the surface of the land. Wiegand and Krause (1929) reported that weathered karst traces were everywhere. "A relatively small amount of human intervention may have great effects on the ecological balance," because as soon as the vegetation is destroyed by deforestation, farming, or burning, debris accumulation and heavy rains set off a new cycle of massive erosion (Hughes 1994, especially chapter V with citations to ancient authors; see also summary in Brice 1978, chapter 9). Although most or all of the erosion of ancient Greek lands commonly has been attributed to the overgrazing of sheep and goats, "the significance of pastoralism is not that it destroys high forests but that it makes permanent what destruction went before," whether mining or commercial lumbering. "Erosion," continues Hughes (1994: 78) is "a complicated and highly localized process." As a result of its carbonate lithography, which is relatively resistant to erosion, and the karst process, which carries what little soil there is away into underground passages by vertical drainage, Miletus was not subject to the same kind of erosion that occurs in areas with deeper soils and noncarbonate rocks.

The Great Meander River is 350 km long. Its headwaters arise in an alluvial mountain-valley region with small lakes, often karstic, lying at about 1300 m altitude, encircled by mountains rising to 3000 m (Roberts 1990; Karatekin 1953). These springs are supplied from a basin of 23,600 to 24,250 km^2 (Karatekin 1953, Snell 1963) in the Afyon/Dynar region.[6]

The Great Meander River affected Miletus in all its incarnations. The first stretch of the Great Meander descending from the sources is rather steep, but then at Aydin the meanders begin, with typical fluvial plain features of channels, levees, flood basins, point bars, and dunes. Low mound shoals form on convex (inside) curves; these are made of coarse eroded material from an outside bend opposite or upstream (Fig. 5.3). The large tributaries and smaller wet-season creeks join the Meander mostly from the northern flank, carrying large amounts of water and rocky sediments and pushing the river southward for 65 km west of Sarayköy. The large Ak and Çino tributaries from the south and smaller wet-season creeks from the north draw on the large catchment basins in the hinterland for water and for easily erodible materials, loosened by tectonic activity. Streams deliver this material to the valley floor, forming alluvial cones where the streams enter the valley. In its central reach, the Great Meander becomes a braided river for several km (fig. 3 in Kissel et al. 1987). As it meandered toward the coast, the river deposited hundreds of cubic meters of sediment, which silted up ports, raised water tables, continuously shifted the shoreline, and, by natural damming at confluent points, changed the valley into a chain of marshes (Karatekin 1953; Brückner 1997a). During the Hellenistic period, the delta of the river crept closer and closer to the Miletus peninsula. Between the fifth and first centuries B.C.E., the site of neighboring Myus was completely engulfed by sediment.[7] A little later, just before or after the turn of the era, the Bafa Sea was cut off from the Latmian Gulf by alluvial deposits and became a saltwater lake; through the process of flooding and overflowing, the water gradually became fresher. The same process of sand bars or alluvial cones blocking the entrance to a bay created ancient lakes and lagoons in several places along the northern edges of the valley (Evans 1971).

During the sixth century, many people resided southwest of the Minoan/

Mycenaean settlement, on and near the Kalabak Hill where house ruins and an industrial area for the production of pottery have been discovered (Yalçin 1993; Gates 1994, 1995, 1996; Tuttahs 1998). The water supply of Kalabak Hill drew on the spring network south of Kalabak in the deep Yatagan strata (Schröder 1990; Schröder and Yalçin 1991). Clay found 300 m south of the city in the Balat Formation was the source for Miletus ceramics.[8]

In this challenging setting, Miletus expanded to an estimated population of 60,000–70,000 in the sixth century B.C.E. (von Graeve, personal communication). Miletus reached its peak prosperity just before the Persian War, during the first five years of the fifth century B.C.E. (Herodotus VI, 28; VI, 8). The center of Mileus was probably near Lion Harbor, contemporary with the harbor at Lade Island directly west of Miletus, dated by Herodotus (VI, 7) from about 650. Public monuments associated with Lion Harbor were early versions of the Apollo Delphinios sanctuary and the North Market. The city contributed ships to the Ionian revolt— 80 of the 353 Greek ships in the battle at Lade Island in 494 B.C.E.[9] Alas, Miletus was captured in the summer of 494 after the Lade battle; the inhabitants were enslaved and sent to Ampe on the Persian Gulf. Excavations at the site "show a burnt layer everywhere" (Emlyn-Jones 1980).

The residential quarter of Humei Hill, which has not yet been investigated intensively by archaeologists, may have developed in the early to middle-fifth century. Not until a generation after this war did the Persian overlords permit the reconstruction of the city to begin. The fifth-century rampart, eight km long and five m tall with inner and outer faces, around the peninsula required about 25,000 m³ of stone (Schröder et al. 1994) and may have been constructed for strategic reasons to help the Persians dominate the restive Greeks of the area (see also Figs. 1 and 2 of Cobet's report, 1997). The town may have grown again to 20,000 by the end of the fifth century (Tuttahs 1998). Post-revolt Miletus stood on the present archaeological zone, according to T. Wiegand (1899). During the middle of the century, however, von Gerkan looked for early settlements outside the peninsula, hoping to find Old Miletus at Akköy or on Kalabak Hill (Parzinger 1989). His work was the precursor of late twentieth-century survey work, but it was done before scholars realized how many meters of alluvium cover the ancient valley, separating much of the Greco-Roman surface from the present ground level.

In my urban history analysis of Miletus (Crouch 1993, chapters 5 and 23), accepting uncritically the archaic date for the grid plan, I named the Miletus-Priene type as one of the five basic Greek urban plans. At the center of the site were large complexes (developed or redeveloped in Roman times) that adapted irregularly to the topography but regularly to the grid. Kleiner (1968) thought that the pattern of the archaic houses west of the Bouleuterion was the basis of orientation of the later grid. In 1968 it was thought that in the northern sector lay small blocks of large houses and to the south lay large blocks of small houses, both set in regular grids of streets. The even distribution of water system elements contributed to the regularity of the plan.

Until the early 1990s, there had been little physical evidence for the famous grid plan. The excavation by Held (1993, his fig. 3) of the terrace where the Athena Temple sits has now produced real evidence for the street pattern in that sector: a regular grid made from streets of different widths (4.5–7.7 m with blocks of 32 × 44 m divided into quarter blocks of 16 × 22 m for house lots). Building elements and stratigraphy date this from the fifth to the third century B.C.E., although the first building period of the Athena Temple was pre-Persian War, as verified by recent excavation (Held 1998). The destruction of the temple during that war led to construction of an expanded temple precinct over a large group of insulae, incorporating flanking streets, on a grid that Held dates as postwar. Through Roman times, flexibility in adapting the street plan to the temple precinct was characteristic of the Milesian accommodation of monumental architectural complexes.

The planner Hippodamus, who flourished in the third quarter of the fifth century, could not have designed Miletus's plan of the 470s, but he absorbed it (with his other ideas of basic reality) from growing up in the city (Artistotle *Politics*, 1267, 22–1268a14; Burns 1976; Diodorus XII.10; Konstantinopoulos 1968; McCredie 1971; Pedersen 1988; Strabo XIV.6.54 or XIV.2.9; Vernant 1983; Voigtländer 1985). He went on to design towns all over the Greek world.

The grid plan of Miletus deviates 20° from cardinal orientation to encourage winds and to block the sun (Tuttahs 1998). The plan was followed over the many centuries of rebuilding, flourishing under the Romans, during decline, and even into its new life as the Turkish town of Old Balat. Once urban land has been platted and assigned to particular owners, it is difficult to change the order.

In the early 1990s, geophysicists Stümpel and Kiel (personal communication) using georadar, began to detect first the waterlines in the Stephania hills and later the town grid. They have confirmed the size of insulae in the southern part of the city to be as von Gerkan had described, but on Humei Hill they found insulae double the length of those he reported for that area (von Graewe, personal communication). This is a fine example of noninvasive scientific work supplementing archaeology.

The ports—so important strategically—were refurbished in time to play important parts in the Peloponnesian and Ionian wars of 412–405 (Tuttahs 1998). During the fifth and fourth centuries, the city exported inlaid furniture, wool, and purple dye (Altenhöfer 1974).

At the end of the fourth century B.C.E. (a period of political reorganization), the ramparts of Miletus were completed. Where the Jeralex aqueduct enters Miletus by crossing part of the south wall, two water towers were added to Holy Gate (Tuttahs 1998, his figs. 80 and 81).

During the Hellenistic period, the population grew from thirty thousand to perhaps forty thousand (Tuttahs 1998.) This led to expansion of the settlement, occupying an area 1700 m north-south by 1400 m east-west at the widest point near the Athena Temple (Altenhöfer 1974). A Hellenistic quay has been found some 2.3 m below modern surface of the Lion Harbor (Eder 1994). In about 100 B.C.E., a new wall was inserted between that old south wall and the Athena Temple area. This new wall cut the old residential area in half. Further excavations are needed to explain this change. Did a massive flood contribute to the need to tighten the boundaries of the city, or was this a political or military decision?

Because of all this construction, there was a continuing need for stone and hence for developing quarries, particularly those accessible by water (Fig. 5.1). There were six quarries for white and grey marble around the Bafa Sea (Peschlow-Bindokat and German 1981) and two earlier gneiss quarries at Myus, 20 km away by sea. Schröder and Yalçin (1991, 1992) estimate that the 50,000–60,000 m^3 of gneiss known to have been quarried from the Myus quarries were more than enough for the 8-km long rampart, 5 m tall with inner and outer faces. In all, the wall required about 25,000 m^3 of gneiss. Marbles were used for the public buildings of the town.

Milesians also paid attention to water management during these centuries. The wealth of Miletus made possible extensive water and drainage system elements. Evidence of sophisticated water technology exists in the stone elbow joints used for pressure lines, common on the site. Miletus had constructed long-distance water supply lines as early as the sixth century B.C.E. For details, see Tuttahs (1998).

Roman Miletus

By 133 B.C.E., Miletus came under Roman rule but still saw itself as having its own history and importance: A first-century C.E. inscription claimed for it the title of metropolis of the Greek cities in the Pontus. The urbanized area of Miletus (Fig. 5.5) was some 100 ha (Eder 1994). As the first century B.C.E. work on the walls suggests, the city fabric incarnated new political and physical organization in Roman times, with large buildings such as sports facilities and baths (Voigtlän-

———————————————————→

FIGURE 5.5. Plan of the monumental core of Miletus with the Great Maeander (Mäander) River along the east edge of the site and the harbors labeled clockwise from the lower left: Athenabucht (Athena Harbor; note that this harbor is placed south of the small peninsula where the Athena Temple is located, not north of it, as, for example, in Fig. 5.2), Theaterbucht (Theater Harbor), Löwenbucht (Lion Harbor), and Nordostbucht (Northeast Harbor). At the bottom right, the Nymphäum aqueduct from Stephania and the Jeralex "Druckleitung (pressure line)" aqueduct enter the city. The two main sewers are Ost-West (EW) Kanal and Sud-Nord (SN) Kanal. Map labels, bottom to top: Kalabak and Kazar hills, the latter with a cemetery (nekropole). The western aqueduct on the left, from the Jeralux Heights, becomes a pressure line ("Druckleitung"), enters the city through the Holy Gate in the south wall of the later Hellenistic era, passes the museum, and supplies the Faustina bath. The eastern aqueduct of Roman times, which replaced an earlier Greek waterline at right, comes from the Stefania plateau and brings water to the great nymphaeum fountain, from whence it is delivered to the water towers (wassersteigeturm) of the South Market (Sudmarkt). The E-W sewer (Ost-West-Kanal) runs south of the West Market area, into the sea north of the Athena bay (now dry land). The N-S sewer runs west of the South Market and the North Market, emptying into the Lion Harbor. Local supplies of water from the limestone of Humeitepe and Kaletepe (hills) was enough to supplement the cisterns of the houses and public buildings. The Büyük Meander flows close to the right (NE) edge of the peninsula, where additional harbors may have been near "Islam. Brücke" (Islamic viaduct or arch), at "Nordost (NE) Brücht" and at the Mycenaean harbor (abutting "Heroon II") opposite the theater. (Courtesy V. von Graewe and G. Tuttahs)

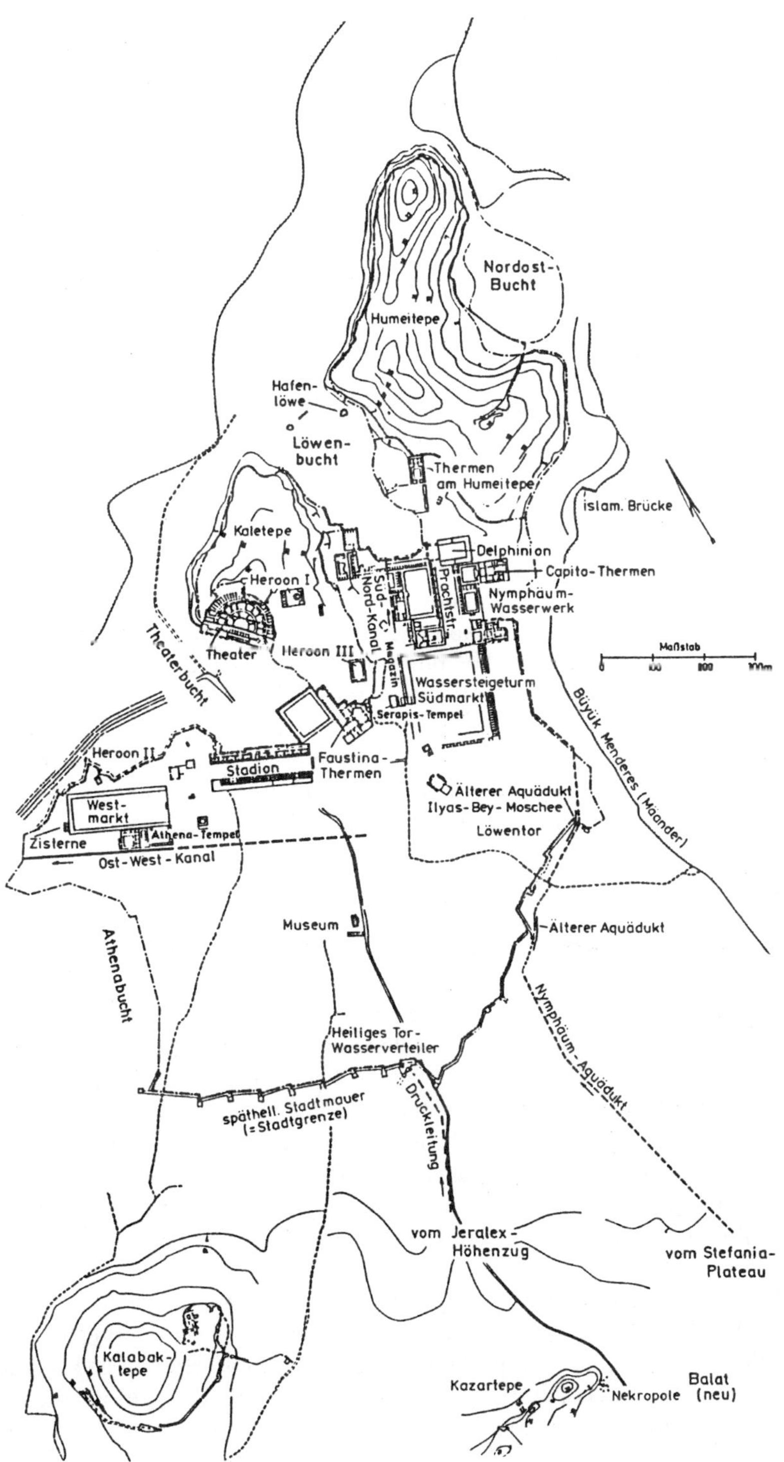

Nordost-Bucht
Humeitepe
Hafen-löwe
Löwen-bucht
Thermen am Humeitepe
islam. Brücke
Kaletepe
Heroon I
Delphinion
Capito-Thermen
Süd-C
Nord-Kanal
Prachtstr.
Nymphäum-Wasserwerk
Theater
Heroon III
Magazin
Theaterbucht
Wassersteigeturm
Südmarkt
Serapis-Tempel
Heroon II
Stadion
Faustina-Thermen
West-markt
Älterer Aquädukt
Ilyas-Bey-Moschee
Löwentor
Athena-Tempel
Zisterne
Büyük Menderes (Maander)
Ost-West-Kanal
Museum
Älterer Aquädukt
Athenabucht
Heiliges Tor-Wasserverteiler
Nymphäum-Aquädukt
späthell. Stadtmauer (=Stadtgrenze)
Druckleitung
vom Jeralex-Höhenzug
vom Stefania-Plateau
Kalabak-tepe
Kazartepe
Balat (neu)
Nekropole
Maßstab
100   200   300m

der 1985), as well as refurbished harbors (Brückner 1998). The old grid was adapted to the new monumental structures and complexes (Voigtländer 1985), with new streets, such as those near the Athena Temple, and with waterlines and sewers (Tuttahs 1998). It is likely that the early version of the Holy Gate in the south rampart and that gate's paving of the Sacred Way leading south to Didyma, some 18 km to the south, also date from the first century C.E., although processions to this very old Carian cult center bearing a Greek name (Emlyn-Jones 1980) may have begun at the end of the second millennium B.C.E.

The Roman population may have been 80,000, although von Gerkan (in 1935) thought the maximum population to be only 40,000–50,000, based on his understanding of the water supply (Yalçin 1993). Roman Milesians may have lived in apartment houses; if the density were like Pompeii's, Miletus could have accommodated some 500 or 600 people per hectare, for a total of 40,000 in the central 90.3 ha (Tuttahs 1998). Pergamon, by utilizing both long-distance waterlines and cisterns, may have supported a greater density of 710 people per hectare (Brinker 1990; von Graewe 1997/98).

Ample stone resources for Roman building consisted not only of gneiss, but also of marble and three quite different kinds of limestone. The Yatagan or Balat strata (Schröder 1990), a 200-m-thick stratum of marble, tufa, sand, and gravel from which springs emerge to supply the high waterline for the Greco-Roman era yielded several kinds of stone. The newer limestone layer above, visible at Akköy, was laid under water and is about 100 m thick. Humei Hill north of the urban area is topped with this limestone, as is Kalabak Hill, where it is 60 m thick. In the larger Meander massif to the south of the river are eye gneiss, schist, and metagranite on a base of metamorphic rock such as mica schist or metaquartz, with upper layers of calcite and dolomite marbles (Tuttahs 1998).

Stones were both brought from the Bafa Sea quarries and cut from the limestone strata within and adjacent to the city. A small quarry was sited, for instance, north of the theater, but Higgins and Higgins (1996) say that this limestone was too soft to be satisfactory for construction, also noting that a gray-green schist with veins of quartz, found with marble, was used in some buildings. From Mt. Mykale above Priene to the north came glimmer-quartzite for plaster. Schröder (1990) suggests that stone may have been exported as well as imported during the Roman period, but we lack a detailed study such as that by Vetters (1998) of the quarries at Ephesus. Miletus was probably the transshipping point for a whitish marble with streaks and patches of red and purple that originated at Docimium in the Phyrigian Mountains and was lavishly used in North Africa and Italy (Ballance and Brogan 1971). Further study of the provenance of stone may determine whether any of the marbles used at Miletus were from Ephesus.

Harbors The renovated harbors (Fig. 5.5) were very important to the communication of Miletus with the wider Roman world and to its wealth during the Roman era. The coastal geography of the Miletus-Ephesus area is being studied by J. Kraft, I. Kayan, and H. Brückner (many publications on their findings during the 1990s). Brückner's coring work in Miletus (1996, 1998), which concentrated mainly on the three western harbors, yielded the following discoveries:

1. The West Market/Athena Temple area was a group of small islands at the last maximum sea transgression. By Minoan times (first half of second millennium B.C.E.), they may have been attached to the mainland by an isthmus (tombolo).

2. Theater harbor soundings indicated a variety of materials from all periods. Marine sediments were found at 12 m below the present surface (Brückner 1998). During Greco-Roman times, the harbor was surrounded by harbor walls and quays that facilitated the delivery of materials from ships. Late in the period, it became a stinking swamp (Brückner 1998). Did sewage also contaminate the aquifer (Tuttahs 1998)?

3. In 1993 Brückner and others concentrated on the Lion Harbor, most ruined of the three harbors, in an attempt to model the ancient environment. This harbor measured 300 m long by 100–120 m wide and was edged by a natural marble and limestone strata 1.5 m thick. The floor of the harbor was loose stone and sand. In the process of hand boring, Brückner's team found ceramics (dated from 100 B.C.E. until late antiquity), glass, charcoal, marine and lagoon fauna, olive pits, grape and melon seeds, and seaweed. Datable objects were mostly from the Greek period but also the Roman and late antique/Byzantine periods. By means of C_{14} dating and pollen analysis, which give a useful chronostratigraphical framework (Brückner 1997a&b), and with insights from geophysical soundings (Stümpel et al. 1994, 1997), scientists discovered the bottom profile of the harbor during the period 4000–400 B.C.E. The harbor began to fill in about 1000 B.C.E. and was half full by the change of the era, with 14 m of sediment in all and the bottom lying 13 m below modern sea level (fig. 115 in Brückner 1994, 1996). It seems, therefore, that the lion on the north side of the harbor is not in its original position but has been moved, undermined by the sea. So, too, the south lion — if in its present position 2000 years ago — would have been under water (Brückner 1996).

4. The small northeast harbor was studied by Brückner's team in 1997 (Brückner 1998). Because of its location, it would have been the first to suffer from the encroaching sedimentation of the river, perhaps as early as 100 C.E. K. Weber pointed out to me traces of a possible bath complex at the edge of the hill next to this harbor, such as almost always accompany Greco-Roman ports, and which I surmise could have been supplied by the Stephania aqueduct. Tuttahs (1998, figs. 53 and 57) denies this possibility. Alternatively, this area could have drawn on karst springs within the limestone of the hill.

Water Supply and Drainage   Roman Miletus was mostly built on the Balat Formation of limestone, itself a poor producer and storer of water. This layer is a poor aquiclude in which water circulation depends on cracks in the inhomogenous stone, underlain by a hydrological unconformity. This may help to explain the need for long-distance waterlines to supply this area, but we also know that it was the Roman custom to add an aqueduct when a major bath complex was built for a city. Indeed, much revision of the water and drainage system took place in the Roman period. Parts of aqueducts were replaced. Details of the Roman system are in Tuttahs 1998.

The system drew on two watersheds (Fig. 5.6) that tapped the karst of the Stephania plateau south and east of Yeniköy and Akköy. This plateau incorporates large natural caves, which serve as reservoirs for rainwater, gathering it from the permeable and porous limestone. Below Stephania, springs emerge in the scree along the edge, where they can be tapped for the urban water supply. A huge reservoir on the Stephania aqueduct may be from the Byzantine period, and Tuttahs thinks it probably replaced one of Roman times. The Stephania line was free-flow and yielded approximately 1,680,000 m³/yr (diagrams of the system, figs. 53 and 54, in Tuttahs 1998). If the large reservoir for the Stephania aqueduct near Balat is Byzantine, the builder may have been Hesychios Ilustrios during the years 481–565 (Tuttahs 1998, his fig. 57, a plan of the aqueduct and its geographical

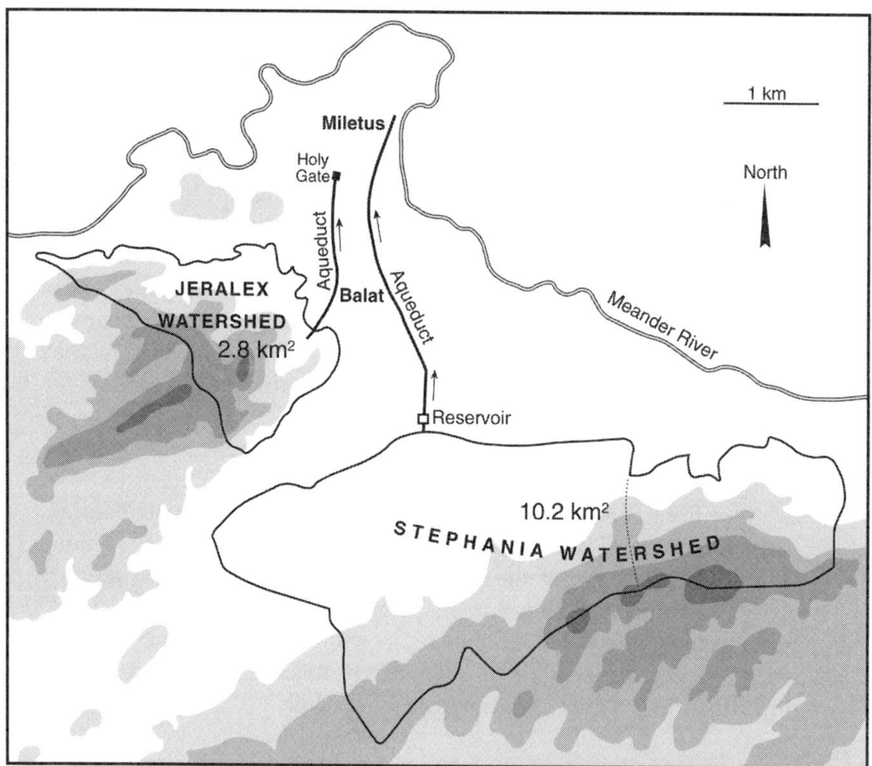

FIGURE 5.6. Map of the two main watersheds of Miletus and their respective aqueducts. The site of new Balat and the present course of the Meander River are also shown. The reservoir is possibly Byzantine, of the fifth century C.E. (Tuttahs 1998, Abb. 53, by permission; redrawn by Langford)

setting). Other springs and wells on slopes to the west, south, and east of Akköy are still flowing, although six of thirteen springs shown by Wilski at the beginning of the twentieth century were dry in the summer of 1990. Schröder thinks that the water deficit from the now-dry springs is likely to be modern, and I agree; the most likely reason is the removal of the tree cover, which led to erosion of the soil that had formerly stored the surface water.

The second watershed, west of the first, is on the slopes between Yeniköy and Akköy where landslide layers (scree) of the youngest limestone lie 35–40 m thick (Tuttahs 1998; fig. 53), a situation we have also seen at Delphi.

Given the favorable hydrogeology of the plateaus above the city, Tuttahs estimates that the two aqueducts could have supplied a population of 40,000 during the Roman era, although these sources do not supply so much water today. I agree that the aqueduct supply would have been enough for drinking and bathing for the partial population housed in the monumental Roman core of the city, excluding Humei hill, theater hill, and the southern district on and around Kalabak hill (Tuttahs 1998, fig. 52). In my opinion, however, his figures, which attribute 600–650 m³/yr (with a maximum of 940 and a minimum of 320 mm/yr) to inner-city water sources, seriously underrepresent the contribution of cisterns and wells to the urban water supply, and hence the number of people who could be supported by them. Miletus's cisterns each held 40–139 m³, with 26 m³ being a generous one-year supply per person. Thirty to fifty percent of surface water derived from the precipitation of 364,000 m³/yr is likely to have been saved in cisterns. If 50% were saved, cistern water could make a substantial contribution to the water supply, serving as many as 550 persons per hectare.

Other water storage elements were found in the hinterland. Several basins along the route between Didyma and Miletus were seasonal karst "seas," that is, products of this karstified limestone layer, like miniatures of Lake Kopias in Greece. Water in these seas could have been tapped as an additional supply for Miletus and the fields around it. Wilski (1899–1900) mapped 70 water tanks— presumably manmade—hidden in the 200-km² area of maquis south of Miletus, an area mostly unknown to the modern inhabitants (published in Wiegand and Krause 1929).

Tuttahs (1998) added to our understanding of the urban water system with his discovery of four water towers: one in the nymphaeum (begun under Titus in 79– 81 C.E. and completed under Trajan in 98–117 C.E. according to inscription), another at the west gate of the South Market and two more built into the Holy Gate. These towers functioned as settlement basins and distribution centers—part of a general rebuilding of Miletus (Fig. 5.5).

Tuttahs's investigation (1997, 1998) of the sewage channels at Miletus is an important contribution to our understanding of Roman water engineering at Miletus. Except for the nymphaeum aqueduct, the drainage system is more visible within the city than the supply system. A single sewer collected rainwater and waste water, which today is called a mixed system. The main sewers of the core city are presented in Figure 5.5. A very large sewer runs south to north from the main plaza to the Lion Harbor, draining an area of 160,000 m² or 16 ha. A second major sewer runs east to west from the South Market to the Theater Harbor, the stadium, the Athena Temple, and an unknown western termination. Both of these lines lie just under the surface of the ground. The lines are large, like the tunnels at Agrigento, and follow the Hippodamian street system. On the surface, manholes (visible near the Athena Temple) reveal their courses. Other smaller sewer lines, which run to the nearest edge of the peninsula, are assumed to drain the residential areas. These lines were adapted to the terrain. Holding basins were constructed above or in the ground, either cut into impermeable rock or, when laid in porous rock, the lower part plastered for water retention. Related to the sewers are the public latrines.[10]

The large supply and drainage system at Miletus has been disturbed enough by ground movements and later occupation that we must be critically cautious about accepting what is visible now as exactly the Roman or Greek arrangements. Not only the abrupt changes caused by earthquakes, but also the more gradual rotations of horsts might change the terrain significantly as far as water supply and drainage are concerned (Ortloff et al. 1983).

The water system worked at both local and regional levels because the ancient engineers understood the concept of water balance (Tuttahs 1998). The system was improved during the Roman era by means of better technology. For instance, the high water tanks at the ends of pressure aqueducts allowed for maintenance of the head over longer distances, settling of suspended solids, regulation of pressure, and division of the water to the various users (Ortloff and Crouch 2001).

The flow regime of the valley (Tuttahs 1998: fig. 121) could also be harnessed for irrigation, but ancient irrigation has not been studied in the Meander Valley and has been mentioned little in the ancient literature. During the Greco-Roman period, it is likely that the new, marshy land near Miletus was used for grazing, as at Syracuse.

Some geological changes in the Meander River valley may be specifically dated to the Roman period. The process of filling in the broad shelf of the river plain reached and extended past the city's peninsula during the Roman era (Altenhöfer 1974; Brückner 1998; Tuttahs 1998). The result was swampy, flat plains composed of hundreds of cubic meters of sediment, described in the first century by Strabo (XII.8, 15–17) as "friable, crumbly, and saline." Because Strabo was raised in the city of Nysa (now Sultanhisar) farther inland on the Meander river, his evidence is compelling. Pausanias wrote (VIII. 24.5) that the Meander River flowed through plowed land in the mid-second century C.E., making a continuous delta between Miletus and Priene, which he attributes to clearing the upper part of the valley, so that the resulting erosion had turned the sea into dry land. As at Ephesus, the delta front has been moving westward at the rate of about 1 mile per 250 years during our era (calculated from Brückner's 1996 data). Delta enlargement has been interrupted in modern times by building reservoirs, extracting water for irrigation, and controlling erosion (Brückner, personal communication).

No major dams or reservoirs are known from antiquity on the Meander River (Tuttahs 1998, citing personal communication from dam expert G. Garbrecht). This is unlike the situation at Ephesus, where dams were built to control the river's actions (Strabo XVI. 1.24). As early as the third to fourth centuries C.E., there were problems at Miletus with the water supply, with silt deposits in harbors, and with changing positions of swamps. The changing water situation was probably one cause of the decline of Miletus. Insufficient wealth was available to cope with the siltation around Miletus because in Roman times she had lost her economic and cultural dominance to Ephesus.

During the late antique and Byzantine periods, the delta-building activities of the Great Meander River were duplicated along the Little Meander River. "Because of superalluviation due to human-induced erosion in the hinterland, where easily erodible rocks outcrop," the process has been unusually rapid at Miletus (Brückner 1998). From just below Miletus, the river's course, controlled by faults and erosion, is entirely post-Roman.

The environment became swampy-terrestrial during the fifth through seventh centuries C.E. Partial sea walls protected the city against the Meander River. From the eighth century on, as the harbor silted up, the city was subject to Arab raids. Although there was still a Byzantine bishop at Miletus in the fourteenth century, Christianity was under increasing pressure from Islamic neighbors after the seventh century, and later from invading Turks. The city fell to the Turks in 1327, triggering a new period of renovation and construction. Mosques, as well as several baths and a new aqueduct system to supply them, remain from this period.

Modern Miletus Study

When nineteenth-century scholars became interested in Miletus, which was stranded 9 km from the sea, the ports were at least 2–3 m under silt because most of the ancient site was flooded during the winter months each year. Philippson and Grund studied the area in 1906–16, noting that swamps occupied most of the area from the ancient Priene-Myus coastline to the modern coastline, and there were at least four lagoons. Dunes, now visible only near Lade, then formed a separate zone between the outer beach and the inland tidal flats. Malaria became a scourge as salt water turned brackish and then sweet (Schröder 1990), which was ideal for mosquitoes. The former large bay is now a broad river plain (Emlyn-Jones 1980, Brückner 1996, 1998). Snell (1963) estimated that since 300 B.C.E., the river had moved 76 m^3/km^2/yr or 0.16 acre ft/mil^2/yr, but more recent work does not verify these figures. The best maps of this process are in Brückner (1997a&b his fig. 2; locations of cores are shown in his 1998 article, figs. 6a, 6b).

Powerful floods in 1945–46 moved the river's mouth from the south side of Lade into old north channels, debouching into the coastal lagoon, Varina Gölü, at the northwest. The 1955 earthquake (Brückner, personal communication) enlarged this lagoon, but subsequently it filled in rapidly (Evans 1971).

In spite of drainage projects for farming, constructed in the last half of the twentieth century, the plain at Miletus is still dry in summer and flooded in winter, resulting in an annual rhythm of sedimentation and delta building. A big flood at Miletus occurred in early 1978 when 231.5 mm (90 inches) of rainfall in January and February contributed to the annual total of 944.4 mm, far above the annual average of 604.1 mm. (Charts of the fluctuation of water levels with variation in rainfall are given by Tuttahs 1998: fig. 123.) Flooding is still a major problem for archaeologists to contend with. When I first went to Miletus in August 1985, the monumental district was under a foot of water. Modern appreciation of the ancient city is hampered by flooding and by the great size and treelessness of the site (Eder 1994).

Miletus is an ideal site for interdisciplinary and international teams. Excavators at Miletus have a history of fostering interdisciplinary research. Under the direction of Dr. W. Müller-Wiener until 1988, our team of Turkish hydraulic engineers and geologists and an American urban historian were permitted to analyze the site. Since 1988, the work in geology (Güngör and Alkan 1995, Schröder 1990; Schröder and Yalçin, 1991, 1992, Brückner 1993, 1994, 1995a&b, 1996, 1998), geophysics (Stümpel et al. 1997), chemistry, metallurgy (Yalçin 1993), building conditions (many papers on archaeology), and water management (Tuttahs 1998) has been carried out mostly Germans and Turks.

Strabo suggests the tectonic potential of the area with his reference to "fountains of boiling hot water" in the Meander Valley (XII 8.15–17); this was borne out by geologist Brinkmann (1976: fig. 52) who locates four thermal springs in the lower sections of the two Meander valleys. Yet there has been little study of tectonics in the local area. General works (Ambraseys 1978, 1996; Ambraseys and Finkel 1992) are no substitute for the kind of specific information we have found at Delphi and Ephesus. We need to correlate the findings of the sedimentologists and tectonic experts with the human response to the physical and historical changes in the valley and delta.

To date, I have been able to learn of only the following earthquakes at Miletus: 60 and 23 B.C.E., 17, 18, 49, 150, and 262 C.E., with a swarm of earthquakes between 359 and 445. Several other earthquakes may have affected Miletus: in 304–3 C.E. in Ionia; in 198 B.C.E. centered at Rhodes but felt at Didyma; in the area of Lycia, Caria, Rhodes, and Cos in 155 C.E. (Ilhan 1971; Ambraseys and White 1996). A village had long stood in the ancient ruins of Miletus, but was so severely damaged in the earthquake of July 16, 1955, that it was rebuilt about 5 km to the south at the present location of Balat (Altunel 1998; Emlyn-Jones 1980; see Fig. 5.6. for location of Balat). Indeed, of all the land forms, alluvium is the most susceptible to earthquake damage.

Further study of Milesian geomorphology could add to our knowledge of the origin and development of landforms, the relation of landforms to the underlying geologic structure, the history of geologic change as recorded in landforms, and the relation of local tectonics to local geomorphology. Yeats et al. 1997: 139–40 write that "just as tectonic landforms can form instantaneously during an earthquake, the processes of degradation can be catastrophic as well." A local perturbation such as a new fault scarp formed across a stream, a stream capture, or a major landslide may set off a sequence of events in which thresholds are crossed and the drainage system undergoes major change. The two Meander valleys give much evidence of such change in the last 2500 years, with good correlation between geological events and human history.

Priene

One Community on Many Sites

Because of its involvement in the story of Alexander the Great, Priene has received attention out of proportion to its small size. Visitors to the site are captivated by its intimacy and by the vistas within the site and from it to the sea and across the plain (former bay) to Miletus. The site has been studied and restudied historically and archaeologically, but it has received less geological analysis.

Priene's history begins millions of years before human occupation, when sea-shells fell to the bottom of the proto-Aegean, hardened and crystallized, and finally rose to form a marble mountain (Mt. Mykale). The Meander River was the major geographical factor during the last 3000 years, forcing the abandonment of the site more than once, because of the advancing delta, which cut off the city's access to the sea, making life unhealthy and economically precarious. Rivers are stronger, longer lived, and more determined than people. In the twentieth century, the large bay between ancient Priene and Miletus was drained for farming—the sea becoming dry land. Geologists can date these processes, give or take a few dozen million years, but they can also visit the site today and determine the approximate date of quarrying that provided the materials for monumental buildings and house walls 2300 years ago.

Priene was an Ionian colony, founded by Thebans sometime between the tenth and seventh centuries B.C.E. (Botermann 1994; Strabo xxii–xxiv). The first Ionian settlement along the bay was likely at Söke dating from about 1000 B.C.E. Geological investigations place a second settlement during the period 700–500 B.C.E. three-quarters of the way from the present site to Söke, 8 km northeast of the present location. Earthquakes contributed to the ruin of the old site (Aksu et al. 1987), although historian Akurgal (1995) believed that siltation was the cause of the destruction; both may be correct. At some undetermined time, because of the prograding delta and possibly because invasions in the seventh and sixth century devastated this early settlement (Landucci Gattinoni 1992), people moved farther west of Söke. The city and port had to leapfrog along the coast every couple of centuries (Fig. 5.1).

The residents moved 2 or 3 km to their port, Naulochos, which then lay farther southwest of Söke (Botermann 1994; Berchen 1979; Altunel 1998). This location of the city has not been discovered, either because no one has looked in the right place or because the site is covered deeply by alluvium, or both.[11] The stettlement again had trouble with the advancing delta front at the end of the sixth century. The early site of Naulochos was swallowed up by the prograding delta (Brückner 1998), so the population had to move west again.

In the late sixth and early fifth centuries, Ionia was ruled by the Persians, who set up local princelings as their governors. Most likely there were three-way struggles among the Persians, Greeks, and indigenous peoples over control of the territory. Pressure on Priene from the much bigger city of Miletus (Berchen), to the south across the bay, was intermittent, but we do not hear of any armed conflict because Miletus was so much larger.

Priene had been involved with other Ionian cities in the anti-Persian independence struggles of the fifth century. Persian conquest in 494 completed destruction of that site (Botermann 1994). The final location of the port seems to
have been in one of the two embayments west of the archaeological site, but
neither has been excavated (Naulochos in Fig. 5.1).

This new realization of the location sequence at Priene means that reevaluation is necessary of all earlier theories as to why no remains from before 350
B.C.E. have been found on the present site. A geological explanation seems to
distort the data less than earlier theories. Most scholars (Botermann 1994; Demand
1990; Wiegand and Schrader 1904) have unwittingly assumed the stability of the
landforms in this valley, whereas the work of Altunel (1998) and Aksu et al. (1987),
agreeing as they do with the findings of Brückner (1997a&b) and Kraft et al. (1985),
show that pre-350 B.C.E. Priene was located not on the present site but farther
upstream during the archaic era. Since the move to the final site happened several
centuries before Strabo wrote about the town, it is not surprising that the details
were lost from the historical record.

During the fourth century, conflict between the Persians on the one hand
and the invading Macedonians on the other strongly affected the town's existence
and development (Botermann 1994; Arrianus 24.5; Landucci Gattinoni 1992; Demand 1990; Berchen 1979; Badian 1966).[12] The former Persian governor, Mausolus, had planned a new town and begun the Temple of Athena (Marasco 1986)
in the decade following 350 (Koeings 1993). Constructing Priene on the new site
was strategically useful. From here, Alexander's forces dominated communication
routes along the Ionian coast. The new city-state incorporated Naulochos and
guaranteed the citizenship of some people who lived at the port city, but persons
who were not citizens of Priene were expelled from the port into the countryside
(Badian 1966); farmers—probably indigenous people—of Mirseli and Peliei had
to live in their villages.

To facilitate the settlement of citizens in the new Priene, Alexander granted
tax incentives and other privileges, about 323, so that elite families were encouraged to settle here (Sherwin-White 1985 Appendix). Antigonos was appointed governor in 334 (Botermann 1994), with a garrison that stayed at Priene until the
lawsuits about land changes were settled. Alexander determined Priene's boundaries and asserted his ownership of formerly Persian royal lands (Badian 1966;
Marasco 1986).

Alexander used Priene to further his grand conquest scheme, probably leaving
in 323 when the Temple of Athena, which he had paid to finish, was dedicated.
Antigonos managed imports and exports from Priene, especially war materiel: iron
weapons from Miletus, charcoal for campfires, provisions such as dried fruit and
dried meat, olive oil, and wool. The new town served both as a point of surveil-
lance over Greek conquests in the area, with its own garrison of Macedonian
troops, and as a collection point for tax money. Thus Priene was a gateway city
"in command of connections between the tributary area and the outside world . . .
[sited] towards one end of its tributary area; and heavily committed to transportation and wholesaling" (Burghardt 1971: 269–85). Guarding Alexander's line of
supply and forwarding money and food for the army were major tasks in the 330s,

a time of drought in Greece, the Aegean isles, and Egypt. The Persian attack on the Aegean isles in 335 had made the related famine worse (Camp 1982, n. 17; Garnsey 1988; Bintliff 1992). Grain was so scarce that pirates made good profits preying on grain ships (Marasco, 1986).

The safety of the governor, soldiers, and first citizens of Priene was assured by the new ramparts, which enclosed the city loosely in the common classical and Hellenistic pattern, taking advantage of defensible locations along the periphery, regardless of the street pattern within. "An inscription record[ed] the public subscription undertaken to finance the fortifications, [and] attributes the renewal of freedom explicitly to Antigonos" at the end of the fourth century (Crowther 1996) Priene was decreed a *free* and *autonomous* city-state—a paradox, because the great sovereign and the council of the "company town" were engaged in a balancing act, each necessary for the recognition of the rights and privileges of the other.

The Edict of Alexander and three documents of Lysimachus were carved on an anta (decorated pillar) of the Temple of Athena in 281 (Landucci Gattinoni 1992; Marasco 1986; Sherwin-White 1985; Crowther 1996, citing Pullan's 1869 excavation notebooks). The inscriptions set out the boundaries of the Priene city-state (Marasco 1986; Badian 1966; Demand 1990; Botermann 1994) and the rights of her citizens.

Construction on the Final Site as Related to Geology

The stratigraphy at Priene is as follows, beginning with the topmost modern layers:

> Cover series of schists interstriated with marbles 600–213 million years old).

> Dark grey-black marbles grading into mica schists at acropolis level.

> Platform-type carbonate rocks (limestone with active karst):

>> Marbles in thin and thick layers, forming terraces.

>> Breccia from collapse of caves in limestone, later metamorphosed into marble.

> Core of gneiss granites.

> Former large bay filled with rich alluvial soil.

Fourth-Century Construction   During the Hellenistic period, monumental construction went slowly because Priene was small. Over 350 years were spent in glorifying the town, as indicated by the dates of ornamentation of the council chamber (*bouleterion*), administrative offices (*prytaneion*), Athena Temple, and stoas (Rumscheid 1994). Koenigs (1993) notes three construction phases for the urban core:

1. Fourth-century: city planning and beginning work on the Athena Temple, the square agora, and its altar, construction of the main street and cross streets; and allocation of house lots.
2. Third century: construction of the council chamber, market, early ver-

sion of agora sanctuary, and earliest enclosures of the agora in the forms of the old north hall north of the main street and the three-sided hall to frame the large southern part of the agora.

3. Last half of second century: construction of the agora sanctuary temple with its small north hall, then the new north hall of the agora.

The temple of Athena and the plan of Priene (Fig. 5.7) have usually been assigned to Pytheos, court architect of Mausolus, and his sibling-successors Idrieus and Ada as the result of "a long-standing passion to ascribe work to a particular known name rather than to an anonymous master" (Koenigs 1993). Was there a master plan from the beginning? Koenigs concluded that extension and rebuilding of the agora was not one idea, but a gradual transformation. Priene was adjusted to its setting in ways that derive from and refer to the immediate geography. The town's earlier experiences with the marauding river induced the planners to select the moderate inconveniences of climbing up and down from the terrace to the valley rather than be subject to further flooding. Set on a terrace shelf along the southern foot of Mt. Mykale (today Samsun Dag), the late-classical/Hellenistic site rises from 20 to 90 m in two slopes, then to the acropolis at 350 m, and then up another steep slope to the ridge 350 m higher still (Fig. 5.8). The site faces south, and the streets are oriented to the compass north-south for the narrow lanes and stepped streets, and east-west for the wide major streets. The street grid, although fairly regular, had to be stretched a bit because of the steepness of the site. The result was an enjoyable tension between the regularity of geometry and the constraints of topography. In the fourth century B.C.E., the kind of urban design called the Hippodamean grid—in English a checkerboard, in German a "screen"—was popular. Hippodamus of Miletus (see Miletus case study) employed a nuanced grid with smaller blocks and more frequent intersections near the center and larger intervals farther out, adapting the grid to the specific site in spite of the difficulties (Botermann 1994; Crouch 1996)—a "combination of separation and connection which the plan alone cannot/did not express but is evident in the built form" (Koenigs 1993; Berchen 1979).

Priene was well designed as a setting for human life. The earliest surviving houses at Priene date from 330s, others from the early third century B.C.E. Houses at Priene (Fig. 5.9) were grouped regularly together in assigned lots, with party walls. Individual houses varied a bit in size; Bonnet (1920) called them "monotonous," whereas Koenigs (1993) describes them as similar but not identical. They were all made from the same materials, to solve the same social problems, and with the same concept of the courtyard house as their model. In their similarity to one another and in their variation around an upper middle class norm, they are similar to the fourth-century housing at Olynthus, which was a similar elite Greek enclave amid indigenous peasants. Houses at Priene are oriented to the south, although public buildings are not, whereas sanctuaries open to the east. In this climate, a southern orientation is an intelligent plan that minimizes heat and cold, welcoming the winter sun and shielding the interior from the summer sun. The mountain on the north side protects the houses from blasts of the north wind,

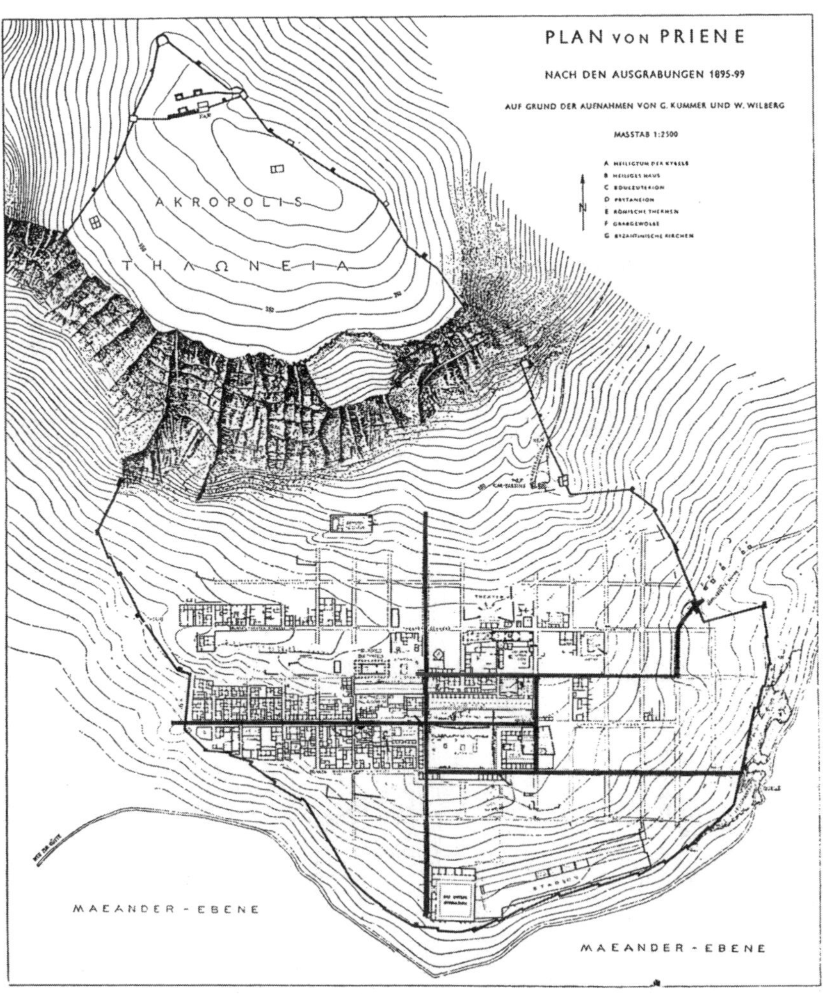

FIGURE 5.7. Grid plan of Priene. Major roads are shown in heavy black lines. The central NS axis is a stair street. West Gate (site of the flow control device of Fig. 5.12) is the only entrance for wheeled vehicles. To the right are the two streets with stairs at their east ends, the southern one leading to a spring. The agora (Fig. 5.10) lies between Spring Street and West Gate Street, its west end abutting the NS axis. Houses flanking West Gate Street are shown in more detail in Figure 5.9. Above the grid of the residences is the Demeter Temple. Near the NE angle of the rampart one contour line higher than and east of the Demeter Temple are reservoirs fed by pipelines from springs higher on the mountain. A zigzag path climbs the cliff face at the right, leading to quarries shown as a set of curved stripes. The path reaches the plateau of the Acropolis where a cross-in-square at left is a large reservoir and the triangular point of the plateau is set off with defensive walls. At the bottom of the plan, a buried fault, which runs parallel to the stadium (stadiom), is covered by alluvium labeled "Maeander-ebene" (Kummer and Wilberg, 1885–89, by permission of the German Archaeological Institute, Istambul)

PRIENE SECTION

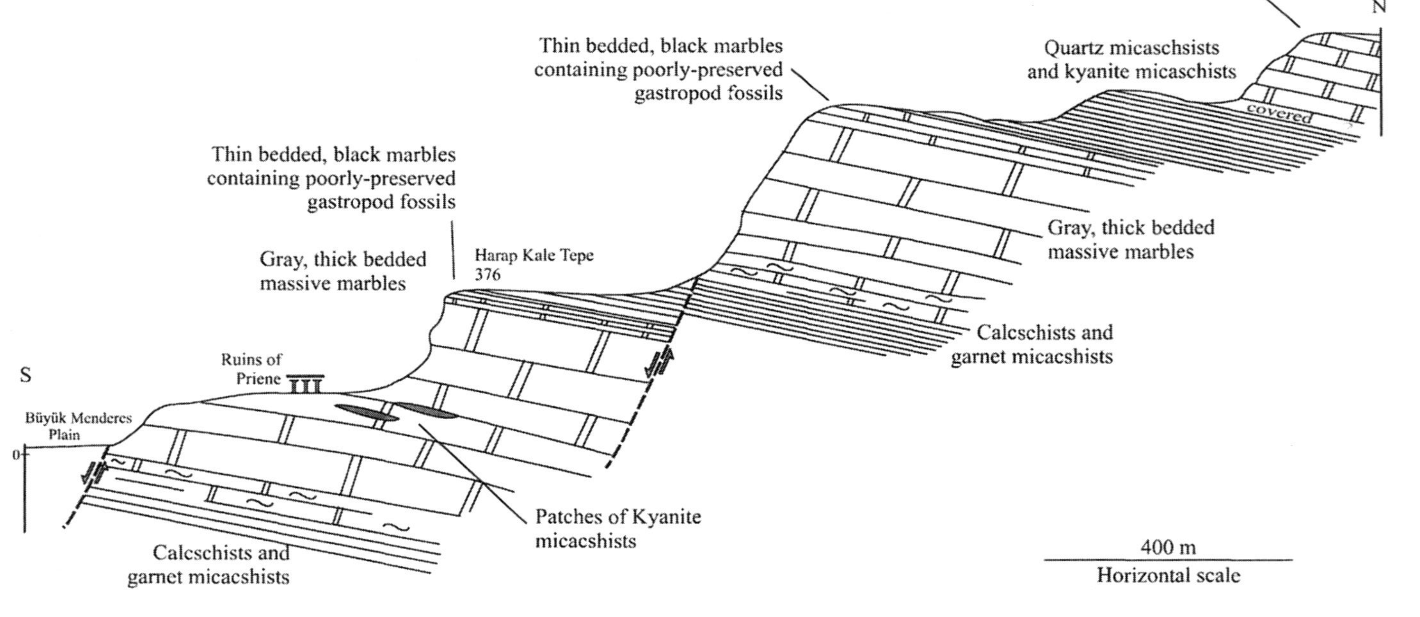

FIGURE 5.8. Geological section of Priene. The town's center is indicated by the three columns. Varieties of marble are suggested by block patterns in the layers. Local builders preferred the gray, thick bedded massive stones comprising the terrace and mountainside under and behind the town. The alternation of marbles with schists facilitated the appearance of springs (Güngör and Alkan 1998).

FIGURE 5.9. House plans flanking the main street to West Gate. A drain runs westward down the main street, fed by nearly a dozen neighborhood drains, and another big drain has survived in the stair street at right where the connecting drains from house cisterns and latrines are shown as *abtritt* (lower right). Courtyards are HOF, reception rooms are OECUS, and south-facing porches are PROSTAS. Different building periods in the houses are indicated by different line qualities. (Wiegand and Schrader 1904)

while the arrangement of domestic service rooms on the west side protects inhabitants from the west (offshore) winds (Bonnet 1920).

The variation in house plans may indicate the same long building period that we see also in the public buildings and the agora (Koenigs 1993) or may have accommodated economic variation among the residents, or both. The idea of the Priene courtyard row house as "democratic" seems to have been introduced by Wiegand and Schrader (1904), touted by Hoepfner and Schwandner (1985), and quoted uncritically in Botermann (1994). Four factors mitigate against this being a "democratic" site as the word is understood now in the United States: Judging by (1) the size and quality of the houses, (2) the small population for which they were designed, (3) the exclusion of sailors from the port and farmers from the villages, and (4) Priene's position under the sovereignty of Alexander and his successors, the city was hardly "democratic" according to our meaning of the word. At Priene, any perceived "democracy" is that of an elite of the classical era, as described approvingly by upperclass scholars in the nineteenth and twentieth centuries—democratic in the sense that a country club that elects its officers and abides by Robert's Rules of Order is democratic—but certainly not open to all the *demos.*

Houses along one side of a block had party walls and backed up to houses on the opposite side of the block and were separated by very narrow (or no) alleys with covered drains between the house rows. These drains are still visible in the vicinity of the West Gate and immediately west of the agora. The existing houses, noteworthy examples of the domestic architecture of the time, date from the late fourth century to the beginning of the third century B.C.E. (Demand 1990; Botermann 1994). The city blocks or *insulae* were approximately 20–48 m wide (Koenigs 1993, n. 6) (in round numbers, 120 × 160 Attic ft, each equal to 29.46 cm), usually with four large houses of 30 × 80 Attic ft, about 900 ft^2 (100–110 m^2) plus partial second stories, over each block (Botermann 1994).

The public buildings of Priene were often isolated, such as the Temple of Athena on its ridge or the stadium at the lower edge of the city. At the center of the town was the square agora, the only square plaza known from the Hellenistic era (Koenigs 1993: n. 9) (Fig. 5.10). It was the largest such surface within the ramparts, on an artificially flattened terrace of rock, below the hill of the Athena Temple. Its functions were political gatherings and public administration, food marketing, and social interchange. The Priene agora was unpaved, but the empty space was regularized and elaborated with memorials, statues, fountains, and halls along the edges. The agora and surrounding governmental and religious buildings at the center of the plan occupied two half-insulae north of the main street and two whole ones to the south.

Originally, one could enter from all along the periphery, as in the agora at Pella in Alexander's home area of Macedonia, which dated from the end of the fourth century (Koenigs 1993, fig. 1). The Priene agora was connected with the Temple of Athena by a stairway at the west that climbed up to the altar east of the temple. From the center of the agora, stairs led down and up the hill. The form of sloping hillside streets piercing an agora is known also at Pergamon from the mid-third century. Coming up the hill from the south, one would encounter first the edge of the agora, then an altar, then the crossing of this axis way with the main street, and finally the northern portion of the central plaza, with houses beyond. When the old north hall was added in the next building phase, it cut off the axial view, and the street no longer continued to the north. Later, another stair along the eastern side of the Bouleuterion led uphill, its downhill axis-way becoming a third set of stairs southward to the stadium. This street variation introduced a lively but subtle dissymetry within the regular plan.

Topography influenced the location of the Theater of Priene, which is set

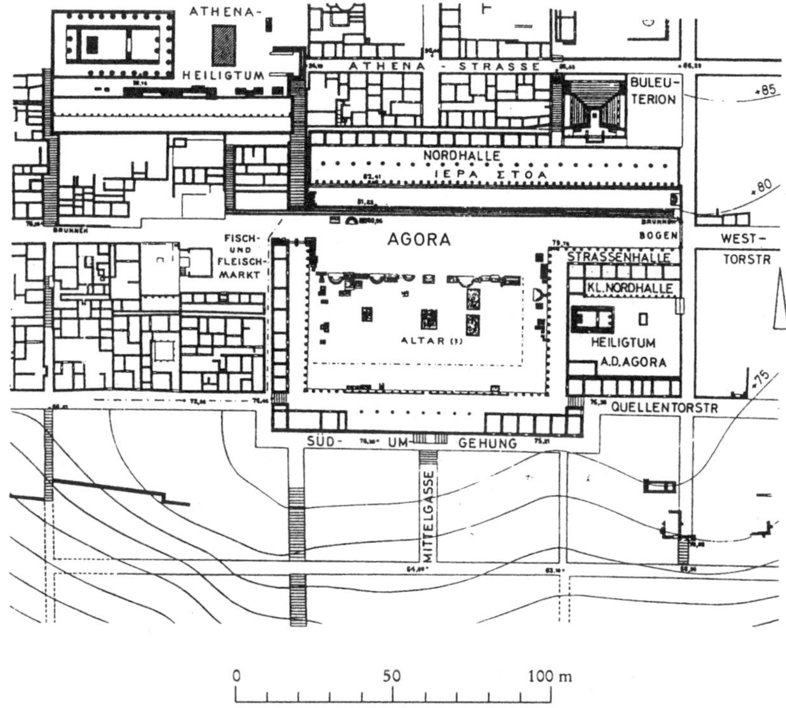

FIGURE 5.10. Agora plan with the arrangements of the late second century B.C.E.
The main (West Gate) street separates the agora proper from the terrace and
north hall above, which is set into the hillside. The agora is flanked on three
sides by porticoes and shops, which cut off the earlier stair streets up to the
Athena Temple and the council chamber (buleuterion). Many house walls of the
residential areas survive, as well as parts of the agora sanctuary (heiligtum) at
right. (Koenigs 1993, courtesy German Archaeological Institute, Istanbul)

into the hillside northeast of the agora. Begun in 332–31, according to an inscription of Apelles, its builder, it epitomizes Hellenistic theaters (Robertson 1964).

In addition to ramparts and paved streets, the infrastructure of the city boasted
an abundant piped water supply, serving not only the houses and temples, but also
the upper and lower gymnasia. Excess water and wastewater was carried away by
a full set of drains (Ortloff and Crouch 1998).

Third-Century Construction   Construction of the new town continued in the third
century. Regularizing and aggrandizing the agora began, and a three-sided col-
onnaded hall was set along the edges after an east-west retaining wall along Spring
Gate Street (Quellentorstr.) was in place to support the buildings. The stoas echoed the north hall along the uphill side of the main street. The new arrangement
changed the earlier loose connections with adjacent areas to a more regular but

asymmetrical balance, with the meat and fish market to the west of center and south of the main street, and to the east the council chamber on the uphill side of the main street balanced by the agora sanctuary downhill. About 6 m of rock had to be removed from the market site to reduce it to the agora level, but a spur of the rock was left as a "tribune" with exterior stairs for making public announcements. This enlarged the flat area of the agora, where the hard surface (now covered by a thin layer of soil) precluded paving.

Expansion of the agora south of the main street affected the street network. A stair street that had led from the lower edge of the city up to the Athena Temple was now interrupted by the portico of the west hall of the agora. The new agora hall cut off the vista that had been one of the glories of the fourth-century city. Alteration of the streets around the agora was convenient for the expanded commerce of the city. West Gate Street was the chief connection to Priene's hinterland, to its port, and to nearby cities, because it was the only route into the city for wheeled vehicles. In contrast, the eastern gates opened into steep streets unsuitable for carts.

After this construction, the center of the city had achieved its characteristic form, with major entries from the northeast and northwest, a secondary entry from the south, and other modest doorways piercing the north, east, and south walls. With these halls, the center of Priene began to approximate the "outdoor room" that we see in several large porticoed courtyards at Miletus in the later Hellenistic period and at Rome in the Imperial Fora of the next period. Indeed, a key example of the "Ionian agora," for example, according to Pausanias, is the one found at Priene.

The agora halls were devoted to banking of customs' duties and other monies generated by trade, which provided the main support of public services and was commemorated in fourth- to second-century decrees regulating commerce (Marasco 1986).

Architectural changes in the accessibility of the council chamber, however, imply that publicly debated self-government had become less important during the third and second centuries. The king of Cappodocia gave money about 150 B.C.E. to rebuild the north hall and extend it across the little plaza, which stretched across the south facade of the council chamber, thereby separating the public outside from the councilors within. The agora was completed between 350 and 100 B.C.E. at the latest.

Second-Century Construction   The fairly rapid construction of Priene in about two centuries is evidence not only of local wealth but also of the ability of the citizens to attract donations from outsiders, such as Alexander and the Lydian king. Costs for a typical set of urban buildings, being apportioned over several centuries of construction, were quite affordable (Hopkins 1978b: 70). The costs were usually borne by expected donations from local dignitaries, but sometimes also by imperial donations, such as the help Ephesus received after the earthquake of 30 C.E. (CIL x, 1624). The Athena Temple was substantially completed during the second century B.C.E., when a copy of the Athena of Phidias from the Parthenon at Athens was donated by Oropherne to ornament the temple (*Blue Guide: Turkey* 1983).

The Theater was remodeled before the middle of the second century B.C.E. (Botermann, 1994).

Priene's history is typical of gateway cities that grow at precipitous rates and then slacken off, even to the point of stagnation (Burghardt 1971). Unfortunately, sedimentation of the valley was a force too implacable to be resisted permanently, and Priene's political importance waned as its geographical troubles increased.

First-Century B.C.E./C.E. *Construction*   Priene added few buildings during the Roman and early Christian eras. The west end of the Athena Temple was finally finished during the time of Augustus; thereafter, the temple was used for the double cult of Athena and Augustus. The bath near the theater was built in the Roman Imperial period, rather awkwardly over the upper gymnasium. Because by this time malaria was a problem in the river valley, people may have clung to Priene's terrace site because it lifted them above the range of the mosquitoes.

Fourth-Century C.E. *Construction*   Later at Priene, an early Christian or Byzantine church was built into the western part of the upper bath area. Gymnasia, baths, and stadia were common locations for early churches, replacing shrines to the emperors. Under the Byzantine empire, Priene had a bishop, indicating the maintenance of its population until that era. The site was abandoned after the twelfth century.

Geology

Resources and Sedimentation   Priene is situated midway between Izmir at the center and Bodrum at the south of Ionia. Western Anatolia consists of three tectonic belts: the Izmir-Ankara zone to the north, the zone of Menderes metamorphics, and the zone of Lycian nappes to the south (Güngör and Alkan 1993). Neogene and sedimentary rocks overlie these three belts unconformably. Like Ephesus to the north, Priene is built on marble, whereas Miletus to the south is situated on Neogene clayey limestone (compare their stratigraphies). In the first century C.E., the Roman writer Strabo (XXII–XXIV) recorded Priene as lying 40 stades—a Roman stade was 607 English feet—some 7.5 km, from the coast, whereas today the site is twice as far from the sea.

Slow geological processes formed the base of the city. After millennia during which the carbonate rock of the later Mt. Mykale was laid down then metamorphosed into marble and raised into a mountain, earthquakes added their sculpturing forces, and the little terrace of Priene emerged at the foot of the mountain high above the skirting Maeander River (Berchen 1979). Fractures lie mainly along the lines 15°N/85°S (Güngör and Alkan 1995). The mountain face north of the city terrace is a weathered fault plane (Roberts 1988).

The Menderes metamorphics of Mt. Mykale at Priene consist of a core of gneiss granites and a cover series of micaschists below, with platform type carbonate rocks above (Fig. 5.8). These metamorphic marbles occur in thin and thick layers, forming the terraces of Priene and the mountain behind it. On the highest part of the acropolis, dark gray-black marbles crop out, grading into micaschists. Two marble quarries utilized their marble for the buildings of the city: one quarry was on the acropolis near the front (south) edge and the other outside the city wall to the northeast, as mapped by Güngör and Alkan (1995) (Fig. 5.8). The steplike appearance of the rock wall separating the acropolis from the city proper suggests that human quarrying activities may have contributed to the present configuration, just as the narrow stairway leading up this rock wall to the acropolis is of human construction.

The necessity of building in local materials because of the cost of transport was an advantage for Priene, given the excellence of the local marble used for temples, public buildings, house walls, and street paving (Koenigs 1993). Stone blocks used in the city ramparts and those observed in situ in the quarries are very similar. The final effect was a city strikingly unified in its appearance and subtlely blended with its environment (Fig. 5.11). Neither the limestone from the very top

FIGURE 5.11. View from the agora up to podium of Athena Temple and the quarried cliff behind. Blocks from a fountain lie in the grass near the observer and another is tucked into the foot of the stairs ascending the hill. The agora buildings, the podium (retaining wall) of the temple, the temple (not visible in this view), and the cliffs behind are all of the same light gray marble that darkens with age. (Photo by Crouch).

of the mountain nor the schists interstriated with the marbles were used for building materials in any appreciable amount. Bricks are not common, but terra-cotta was used for pipes and for roof tiles.

The hinterland changed drastically in historical times as the Meander River rapidly silted up its valley. Indeed, the coastal plains of western Anatolia are characterized by alluvial formations. The inland mountains that ring the central plateau of Turkey have been rising noticeably during the past 2000 years, contributing to the loosening of materials for the coastal rivers to carry away as well as increasing the steepness of the river's slopes. This combination meant that a river might deposit as much as 4 m of materials in one year (Gage 1978). Similar coastal changes have been noted at Smyrna and Ephesus (Berchen 1979) and at Miletus (Schhröder and Yalçin 1992; Tuttahs 1998).

Problems with the Little Maeander River may have favored the shift from an agricultural to a commercial economy (Berchen 1979). There was very little agricultural land available for Priene, since the broad river valley now visible at the site was, during Greco-Roman times, a bay of the sea. Because of the lack of farmland, the centuries-long rivalry with Samos over coastal territory continued into Roman times (Berchen 1979; Sherwin-White 1985), with both polities wanting to dominate the coastal plains on the mainland nearest to the island. The value of the coastal plain would have increased if it were not subject to malaria. Trade developed at least partly to overcome dietary constraints caused by the small local supply of grains, which ordinarily formed the major share of the Greco-Roman diet.

Priene continued to be habitable as long as its inhabitants could make a living from grazing, from fishing, and from lumber and other forest products, such as locally important seeds, barks, flowers, leaves, resins, aromatic oils, fungi, and galls, not to mention tree crops, such as olives, carobs, nuts, and other fruits (Thirgood 1981). The Mediterranean has 500 fish and shellfish species, 120 of economic importance (Hughes 1994), but access to this resource was jeopardized when the delta separated Priene from the sea.

Hydrogeology   Water supply and drainage were strongly linked to the karst geology. The fractures from tectonic activity contributed to the development of karst systems at and above the site, notably springs, dolines, and canalizations such as the still active karst caves just above the lower gymnasium (at 20–35 m elevation). Springs were mapped at the beginning of the twentieth century by Wiegand and his colleagues, and checked again by our team. We noted small, active dolines in the limestone along the ridge high above the acropolis. Most of these springs are located toward the top of the mountain. Their waters were collected into a reservoir on the acropolis (Fig. 5.7) and then led in pressure pipes down to settling chambers inside the ramparts but above the residences and public buildings of the town. From the settling tanks positioned 10 m higher than the Demeter Sanctuary, the water was carried by pipelines to lower buildings. Miniature karren, another karst feature, occur on the slope below the Demeter Sanctuary at 125 m, which was watered by its own spring from the mountain. In the city, springs and seeps were tapped by the houses adjacent to the mountain wall, notably those along the street north of the Athena Temple. Fountains supplied by piped water stood in the agora, along West Gate Street, at the council chamber and near the theater. Close to the bottom of Spring Gate Street was an animal watering trough, and another still occupies a spot about halfway down the street behind the Athena Temple. Karst water systems based on these same sources still provide water to the modern village of Güllübache adjacent to Priene and to other local users.

Duplication or redundancy was a basic principle of water management in the Greco-Roman world. Therefore, the people of Priene made duplicate provisions. In addition to the known water system elements listed previously, Dr. A. Alkan of our Turkish team discovered a second waterline from the mountains that runs along the northeast side of the ravine separating Priene from the mountains behind; this waterline appears to be directed to an unrecorded reservoir that he and I discovered in the northeast corner of the ramparts, about 150 meters lower than the water chambers mentioned previously. This reservoir was also fed by a spring that C. Ortloff and I discovered just within the rampart but some 40–50 m higher than the reservoir. From the workmanship, I think it is a fairly late construction, no earlier than late Roman. These discoveries are now being verified by archaeological investigation. Altunel (1998) writes that the reservoirs of Priene, in existence since Hellenistic times, were repaired later with Byzantine cemented masonry.

Pipelines and drains have been found at Priene, both following the pattern of the streets. With so much rock so close to the surface, laying the pipes deep into the earth or in a network separate from the streets was cost-prohibitive. Drilling wells into solid rock was equally difficult, so wells are not common here. However, approximately 25% of the houses—those farthest from the slope—do have rainwater cisterns, cut shallowly into the rock or set into soil that had accumulated from flaking off the mountain or from wind-blown dust. Some houses supplemented the municipal piped supply by tapping directly into the mountainside for water. Private cisterns were the usual Greco-Roman provision for duplicate supply; their scarcity here points to the copious piped water supply and the stony site. The once-abundant drainage from the stadium area near the south rampart was directed toward what appears to be a karst shaft that has been formalized with masonry. Local people assert that water from this shaft communicates with a spring in the house just below and outside the archaeological site (Rumschied personal communication).

C. Ortloff and I (Ortloff and Crouch 1998) have studied the long channel that drains the southwest part of the city and exits the city under the west gate. We have analyzed a flow control device with no moving parts that kept the exit drain clear of debris (Fig. 5.12). The device indicates a keen understanding of the nature of turbulent flow in this area, and it provided a simple, economical way to control this flow.

Tectonic Events and Processes The Great Meander River valley (graben) runs EW from Aydin (ancient Tralles) to Magnesia, where it turns to the SW past Söke and Priene on its north side, and Miletus on the south, to the coast. This is one of the most tectonically active areas of western Turkey (Appendix A). Earthquakes

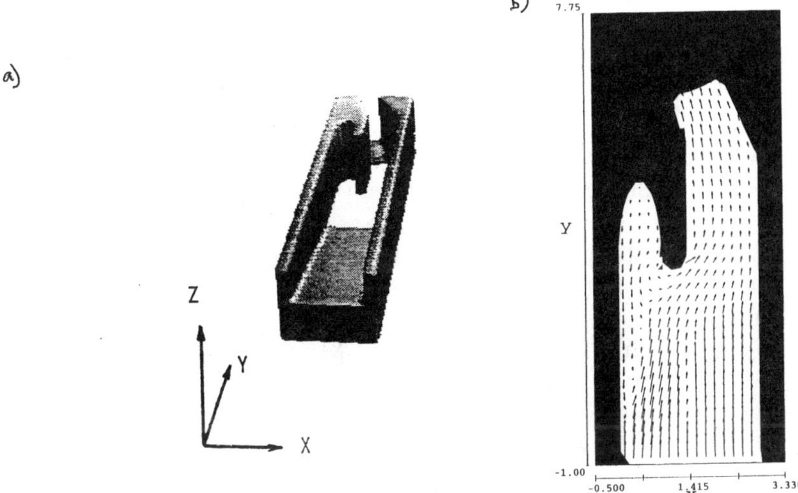

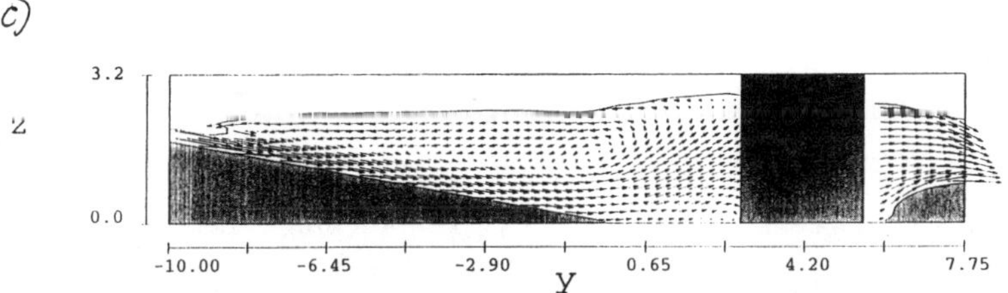

FIGURE 5.12. Flow control device made of stone, discovered under the west gate at Priene, where it receives rain runoff and wastewaters, possibly laden with debris. The device, without moving parts and needing almost no maintenance, churns the waters and debris, directing them outside the city and preserving the western end of the city from flooding. The device has not been dated, but it seems integral to the west gate, which was part of the fortification system of the second half of the fourth century B.C.E. (a) Model of the channel with sloping bottom and narrow exit at right, dead-end channel at left. (b) Flow diagram seen from above. The dead-end section caused a vertical eddy (diagrammed in the change of arrow patterns) that pushed water and debris to the right through the narrow channel. (c) Flow diagram seen from the side. The changing surface of the bottom of the control device caused a horizontal eddy (again represented by arrows) that interacted with the vertical eddy to entrain debris and carry it outside the city through the channel (Ortloff and Crouch 1998).

at Priene have been studied by Altunel (1998), who uses visible damage to buildings at the city's core to assess the severity of the event. These damages include fractured, dilated, and collapsed walls, rotated wall blocks, rotated and broken columns, and fractured and offset floor blocks. Damages occurred within a NE-SW corridor (his plan fig. 3B) where right-lateral oblique normal faults or dilational fissures are visible. This corridor is roughly parallel to the northern margin of the graben, indicating a smaller fault with its slip antithetic to the major fault system. Perpendicular cliffs left by the quakes are shown in Figure 5.11.

In addition, tilted or toppled walls and rotated blocks indicate shaking. Altunel also noted collapsed house walls, columns that were toppled, cracked, or broken, and cracked walls (i.e., the eastern wall of the upper gymnasium) and floors (gymnasium and church). Large chips broken off the bottoms of the column drums of the Athena Temple bear witness to the shaking and shear stress they underwent before falling over, and according to Altunel, suggest that the temple was damaged by more than one quake. He lists earthquakes in 26–25 B.C.E., 68 C.E., 238, 244, 262, 1653 (level IX), 1891, and 1899 (another level IX), and 1955. Earlier quakes of the fourth and second centuries B.C.E. led to repairs or new construction, he concludes, but unrepaired damages to the monuments seem to have occurred as late as the thirteenth century C.E. after the town was abandoned. Quakes may thus have contributed to the final demise of the city, as they did to the preclassical city on the earlier site and to the destruction of the Athena Temple (now partially re-erected).

Conclusions: The Interaction between Man and Nature

The geology and hydrogeology of Priene interacted with the desire of the Hekatomnid Dynasty for increased status, the need of Alexander the Great for a supply base, the desire of descendants of the original settlers of Priene to occupy their traditional elite status in the Hellenistic world of western Anatolia, and the capabilities of excellent builders working with intelligent patrons. These factors came together to produce not only a viable city but also a delightful one. Each site has its own list of actions and reactions between the setting and the settlers. At Priene, we have prime examples of solar orientation—the city backed up to the mountain for protection from the northern wind—and of employment of a sophisticated grid.

Citizens of ancient Priene gambled that the river would not destroy their livelihood as seafarers and middlemen, but they lost out in the Roman period as the sea gradually receded from the foot of their mountain terrace. Not until the twentieth century draining of the swamps along the river did agricultural wealth make possible new villages which were kept modest by the plague of mosquitoes (and hence malaria) until pesticides brought that nuisance under control.

The site has been long abandoned, after an active occupation of about 700 years. Priene may be compared with Morgantina, a hill town in Sicily, where after a century or so on the Cittadella site, the town was moved to the present archaeological site and endured there for about 450 years. Or Priene may be contrasted with Argos, which has actively occupied the same site for close to 3000 years. Like Argos, Priene had a mountain hinterland of both nutritional and economic significance. A forested mountain was an ideal resource during Imperial Roman times when there was an acute shortage of wood in the Mediterranean area (Hughes 1994). Each household during Greco-Roman times required one to two tons of firewood per year, and to burn one kiln of lime for mortar required 1000 donkey loads of juniper wood (Wertime 1983: 446, 450, and table 1, Energy Costs of Pyrotechnology, p. 451). Between the early fifth century B.C.E. and the mid fourth century, the present site of Priene was left vacant. This fostered new growth of timber in mountains above Priene, which in turn helped to make the site attractive to Alexander the Great and his successors. Macedonia and Thrace, which still had big timber reserves by 313, according to Theophrastus, could also have supplied his expedition, but Priene was much closer to the action (Thirgood 1981: 38)

Another geographical feature, the meandering river with its annual load of silt and gravel, dominated the human history of Priene. The river forced the abandonment of urban sites more than once, because of the flooding and the marshes — and their malarial mosquitoes — that formed in the advancing delta. The first and last sites were subject to the same siltation process. Priene's most ancient site was bypassed by the delta no later than the Geometric period. Then a possible archaic site southwest of that was engulfed and finally destroyed in the Persian War at the beginning of the fifth century B.C.E. The late classical (modern archaeological) site was again bypassed by the aggrading river in the first century B.C.E. The advancing delta cut off the city's access to the sea and made life not only unhealthy but also economically precarious.

Meanwhile, continued harvesting of trees on the mountain slopes above the city had interfered drastically with the karstic water system within the mountain. Faced with much less water, no trade, earthquakes, and swampy land unfit for crops, the people of Priene moved away. The latest chapter of the story of the large bay between ancient Priene and Miletus has changed a process into an event. From Roman times, the bay had been filling through delta building, eventually becoming a swamp. Finally, as recently as 1920–50, the swamp was drained for farming, completing the process of the sea becoming dry land.

Ephesus

Introduction

Many ruins of Greco-Roman cities in Ionia are found in areas long-since poor and depopulated. Political instability and environmental change seem to be the major causes of this change in urban pattern (Walker 1981). Disappointing sources for learning about this landscape are the so-called topographical studies in the archaeology literature. They report what ancient authors wrote about geographical features and settlements, and compare those opinions with modern archaeological data. Since the geography of the valley has changed drastically in historical times, the ancient literature does not apply directly to modern conditions. At Ephesus, physical change relates to the natural frequency of catastrophes such as earthquakes, climate variation, and sedimentation. Modern geographers and geologists, fortunately, can give us physically based topographic information from their surveys and sampling. After a brief description of the historical sequence at Ephesus, we survey the physical geography, the petrology as it related to defensive walls, tectonics and sedimentation as they relate to the location of ports and changes in settlement sites, and karst as it relates to water supply and drainage. Finally, we summarize the human response to the geology in a discussion of topography and some features of urban design. The Ephesus Chronological Table (Appendix A) shows that geological events and constructed changes in the city are interrelated.

Description and History

From south to north, the main features of the Ephesus settlement area are the Bülbül Mountain, with faults along both faces, the graben (valley) occupied by the Roman city, the horst (upthrown block) of Panayir Mountain, and the wider valley toward the hollow of Artemis and the hill of the crusader castle with modern Seljuk to the east. Occupation of the Ephesus site stretches from about 3000 B.C.E. through Turkish times of the mid-nineteenth century C.E. to the present. As a result of relatively frequent shifts in the shoreline and occasional earthquakes, settlements have stood on several different places in the area. Figure 5.13 is a map

———————————————————————→

FIGURE 5.13 (opposite). Geological map of Ephesus, with locations of quarries, faults along rivers and along the bases of mountains, and settlements (A–T) over time:

A. North end and east side of Ayasoluk. Copper Age settlement from the first half of the third millennium B.C.E., followed by Hittite settlement called Apasas (second millennium) and then Mycenaean trading post in late second millennium. Lydians lived here when the Greeks arrived during the period 1000–700 B.C.E. This is also the site of a Byzantine-Turkish castle from the eighth to the fourteenth century C.E. located above Mycenaean remains. Seljuk, indicated by double line with grey dots, grew up on the rest of the hill and on the plain to its east.
B. Ionian Greek settlement tenth–eighth century B.C.E., using an early harbor related to the Artemis shrine.
C. The sacred way from the Artemis shrine skirted the east side of Panayir Dagh (mountain) and extended to the goddess's birthplace at Ortygia (modern Arvalia) and back again along the west side of the mountain. C near H is the probably location of Smyrna, partly under the Hellenistic agora.
D. Possible sites of Koressos settlement of archaic times.
E. Lysimachus's city of about 300 B.C.E. and later.
F. Late Hellenistic expansion under Roman rule.
G. Roman and early Christian city, mid-fourth century C.E., developed on recent sediments.
H. Roman harbor, still visible because of vegetation differences, although silted up.
T. Byzantine-Turkish settlement of later first millennium C.E. to the present.

(Vetters 1998, redrawn by Langford)

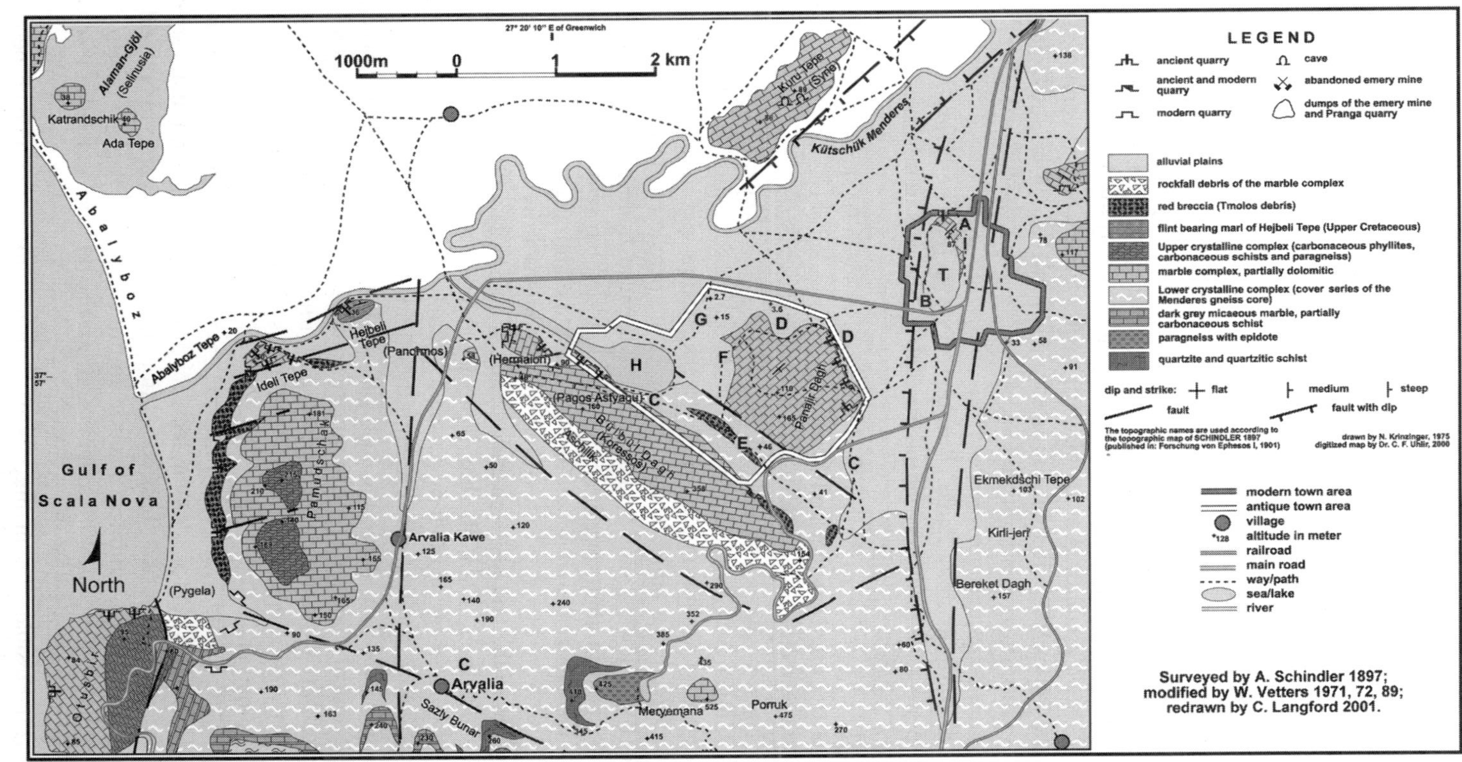

27° 20' 10" E of Greenwich
1000m 0 1 2 km
LEGEND
ancient quarry
cave
ancient and modern quarry
abandoned emery mine
modern quarry
dumps of the emery mine and Pranga quarry
alluvial plains
rockfall debris of the marble complex
red breccia (Tmolos debris)
flint bearing marl of Hejbeli Tepe (Upper Cretaceous)
Upper crystalline complex (carbonaceous phyllites, carbonaceous schists and paragneiss)
marble complex, partially dolomitic
Lower crystalline complex (cover series of the Menderes gneiss core)
dark grey micaceous marble, partially carbonaceous schist
paragneiss with epidote
quartzite and quartzitic schist
dip and strike: flat medium steep
fault fault with dip
The topographic names are used according to the topographic map of SCHINDLER 1897 (published in: Forschung von Ephesos I, 1901)
drawn by N. Krinzinger, 1975
digitized map by Dr. C. F. Uhlír, 2000
modern town area
antique town area
village
altitude in meter
railroad
main road
way/path
sea/lake
river
Surveyed by A. Schindler 1897;
modified by W. Vetters 1971, 72, 89;
redrawn by C. Langford 2001.
Aleman-Gjöl (Selinusia)
Katrandschik
Ada Tepe
Abalyboz
Abalyboz Tepe
Küru Tepe (Syrie)
Kütschük Menderes
Gulf of Scala Nova
North
Hejbeli Tepe (Panormos)
Ideli Tepe
Pamudschak
(Pygela)
Arvalia Kawe
Arvalia
Sazly Bunar
(Hermaion)
(Pagos Asfyagui)
Bülbül Dagh (Koressos)
Aschük
Panair Dagh
Meryemana
Porruk
Ekmekdschi Tepe
Kirli-jeri
Bereket Dagh
37° 57'

of these locations and of the geology, with alphabetical order indicating chronological order:

- (A) The earliest settlement on the north end of Ayasoluk near the later site of the medieval castle (151 and 155 in Fig. 5.14), dated to the Copper Age (first half of third millennium B.C.E.). Additional prehistoric remains have been found about a half mile east of the present south entrance to the site, near Luke's Grave (14). [Parenthetical numbers in the text (14) refer to buildings mapped in Fig. 5.14.] Hittites called their second-millennium settlement Apasas (Marchese 1986, citing Gulltekin and Baran 1964, and Mellink 1961, 1983). When the Mycenaeans from Greece established a trading post in the late second millenium, they also chose the hill, then an island or peninsula, which rose high enough to furnish good visibility to lookouts. At the time, the hill was taller than it is today because the base was lower; alluvium 3–25 m deep now disguises its stony flanks (Brückner 1997a & b). The height helped to protect the foreigners, lifting them above dangers from winter flooding or possible enemies.
- The Mycenaeans buried their dead farther south (153) on the same hill. The Little Meander and Marnas rivers provided access to customers in the hinterland.
- Tradition dates founding of the shrine of the Great Goddess to the Minoan period of the second millennium B.C.E. (Pausanias VI.2.4f); the goddess was assimilated to the Greek Artemis and then to the Roman Diana as the ethnicity and culture of the people changed.
- (B) Following after Minoan and Mycenaean trading posts of the late second millennium, Ionian Greeks founded ten settlements along the west coast of Anatolia during the tenth to eighth centuries B.C.E.[13] Lydians were living at Ephesus when the Ionians arrived (Pausanias 7.2.8). Before the end of the archaic period, Ephesus gained dominance over the villages of Marathesion, Neapolis, and Anaea of the coastal plain in the direction of modern Kusadasi (Marchese 1986). The sanctuary of the great goddess called Artemis drew both Greeks and indigenous peoples from Asia Minor, the Near East, and the wider Mediterranean area. Physical evidence of the sanctuary comes from no later than the eighth century B.C.E. (Bammer 1986–87). During the sixth century, the sanctuary became a huge temple[14] rebuilt several times during the Greco-Roman era (1).
- (C) A Sacred Way (7 in Figs. 5.14, 5.17) was important for all ancient versions of Ephesus, as the focus of processions and as the main street for many other uses. It extended from the Artemision around both sides of the central mountain Panayir, to the village of Ortygia (modern Arvalia) in the coastal mountains southwest of the city, and it was revered as the birthplace of the goddess.
- (D) Possible sites of the city of Koressos during the archaic period. Ancient literary records of the Ionian foundation myth associated it with

FIGURE 5.14. Plan of ancient Ephesus with numbered buildings. The city lies between Bülbül Dag and Panayir Dag (mountains). Important for our purposes are

1. Artemis Temple
4. Meter sanctuary
8. Quarries
10. Meander Gate
11. Rampart
16. Bath built in its own quarry
17. Fountain house of government center
21.–22. Roman government center
32–55. Embolos Street
50–51. Slope houses
55. Library of Celsus
61. Agora (built over part of Smyrna site)
65–66. Street leading to Lysimachus harbor
67. Serapis temple

74. Karst cave made into shrine
75. Theater
76. Palace above theater
77. Byzantine walls
78. Late Roman version of street to stadium area
79. Hellenistic bath
80. 85–87. Early Christian buildings 4th–5th centuries C.E.
83–87. Street to Roman harbor
91–98. Harbor bath, gymnasium, temple complexes of second century C.E.
102–106 Stadium and bath
151 Medieval castle over Mycenaean trading post.
156. Aqueduct and reservoir rebuilt in Byzantine era

(Courtesy of the Austrian Archaeological Institute, Vienna)

the Koressos Gate (107) (Rogers 1991), and this has been verified by archaic relics (Engelmann 1991; Karwiese 1985).

- Archaic graves (Karwiese 1995, Karte 2), but no archaic settlement, were found in the saddle area where the Roman government center stands. Another archaic settlement was the old village called Smyrna (not the Smyrna at modern Izmir) at the mouth of the creek that drains the valley between the two city mountains. The water of the bay was very close to the old Smyrna site about 500 B.C.E. (Brückner 1997a & b; Langmann and Scherrer 1993). Excavations found an archaic street under the Greco-Roman agora, flanked by graves and tumuli (Thür 1989; Rogers 1991; Scherrer 1999).

- About 200 years later, Lysimachus, successor of Alexander the Great, founded a new city about 300 B.C.E. in the valley between Panayir Mountain to the north and Bülbül Mountain to the south (Strabo XIV 1.21); the settlement occupied about 100 hectares (Fig. 5.13: E). To defend his town, Lysimachus built typical Hellenistic defensive walls, extending 4.5 km (Seiterle 1964) along the ridges of the mountains (Hueber 1997). After Lysimachus, from the third century B.C.E. through Byzantine times, the area where the Tetragonos Agora (61 in Fig. 5.14) was later built was a commercial center of the settlement, and the port was located somewhere off the northwestern edge of Panayir Mountain (Kraft 1998) (See Lysimachus Harbor in Fig. 5.15).

- (F) When Asia Minor came under Roman control about 130 B.C.E., not only did new buildings and building types appear at Ephesus, but also new areas of settlement extended the city to the west and northwest. New temples for the cult of deified emperors (30, 98), new civic buildings (21, 22), dwellings (50, 51), long-distance waterlines (Fig. 5.16), fountains (17, 33 in Fig. 5.14) and sewers were added throughout the Roman period. The baths (79 and 92), theater (75), and stadium (104) replaced and amplified earlier Greek arrangements. Particularly in the first century C.E., there was a notable increase in monumental architecture that utilized marble for veneers and cornices, with niches for marble statues; these were paid for mostly by local "first citizens" but also by Roman proconsuls of western origin. Reconstruction of Ephesus was simultaneously an economic stimulant, consolidator of Roman power, and divisor between social groups in the city (Walker 1981; Rogers 1992, 1991). (H) The city enlarged by both spreading out and increasing density, with apartment buildings and houses rising on many of the slopes (Vetters 1998; Pliny: V.29). Winning land from the sea by filling in former swampy areas, the Romans obtained sites for a wealth of new buildings from the late first century C.E. on. New port facilities ("Possible 1st C. A.D. Harbor" in Fig. 5.15) were necessary, as the old ones were either left high and dry or covered over by the delta-building river.

- In the third century C.E., the final size of the city was some 280 ha, with a population of 20,000–30,000 male citizens plus their families and slaves, as well as uncounted foreigners (Hueber 1997), for a total

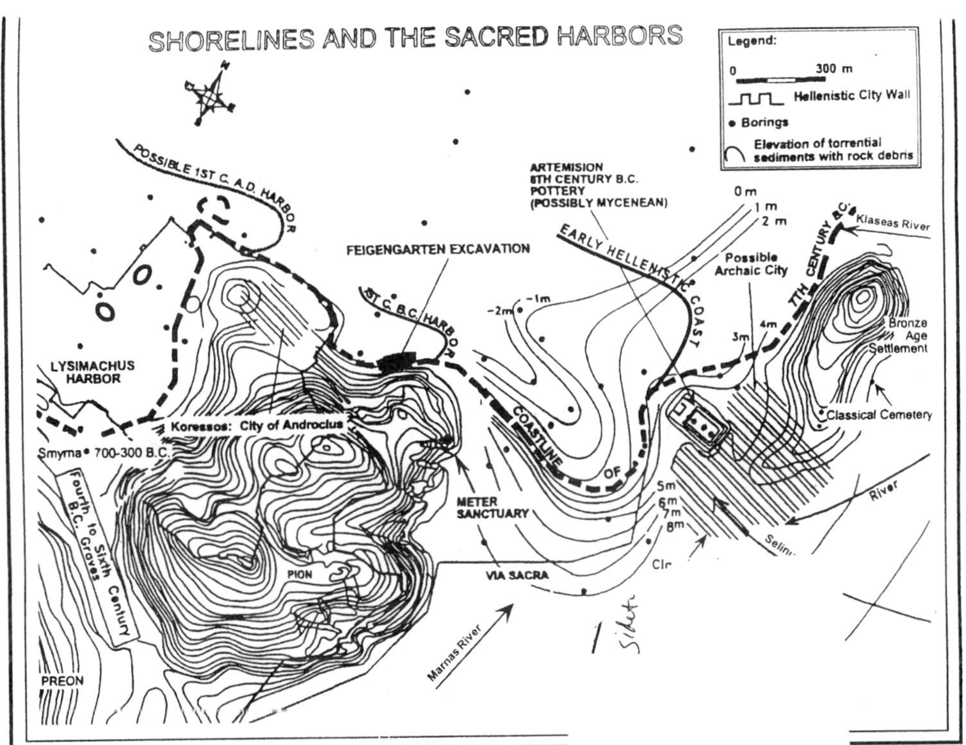

FIGURE 5.15. Paleogeographic reconstruction of the harbors at Ephesus (Kraft, Brückner, and Kayan 1999). From right to left:

Heavy dashed line, seventh-century B.C.E coastline; Artemis Sanctuary and probable settlement from the seventh–fifth century B.C.E.; Ayasoluk on far right, designated as "Bronze Age Settlement;" at left, Pion, the ancient name of Panayir Mountain; Via Sacra along the east side of Pion; the road turns west between Pion and Preon (Bülbül Mountain), past Smyrna, to the village of Ortygia in the coastal mountains (not shown); Smyrna was another archaic settlement, overlapping the site of the later agora, at the mouth of the creek that drained the valley between the two mountains. Creek mouths often served as ad hoc harbors in the archaic period; The site was originally drained by three rivers, the Klaseas/Kaystros (Little Meander) from the NNE, the Selinus from the E (from the springs at Sirínçe), and the Marnas from the SW; a possible river harbor lay between the Via Sacra and the Meter Sanctuary. During the sixth–fifth century B.C.E., the harbor seems to have shifted to the north edge of Pion, possibly west of present stadium and bath area, to serve the city called Koressos; By the late fourth century B.C.E., the Meander River's delta had advanced from the NE, silting up the Artemis harbor. The Selinus and Marnas rivers later became tributaries of the Little Meander River, uniting at an unknown place beyond the early Hellenistic coastline, in the former bay; Lysimachus built a new harbor to the west of Pion. His new settlement was built above the fourth–sixth-century graves in the valley between the two mountains. Roman power, during the second century B.C.E., necessitated a new government center in the valley between the mountains, at the south end of the rectangle marked "Fourth to Sixth Century B.C. Graves." Apartment buildings for the increased population rose along the north slope of Preon (Bülbül Mountain). The Roman town also expanded along the return route of the Sacred Way (Via Sacra), on the western edge of Pion. In the first century C.E., the port was at the NW edge of Pion, as discovered by Burckner and Kraft's shoreline investigations. A new harbor (Fig. 5.21), not shown here, was built in the second century C.E. to the west of Lysimachus's harbor, which had filled in. The progression was sea, then swampy land, then land solid enough for large public buildings.

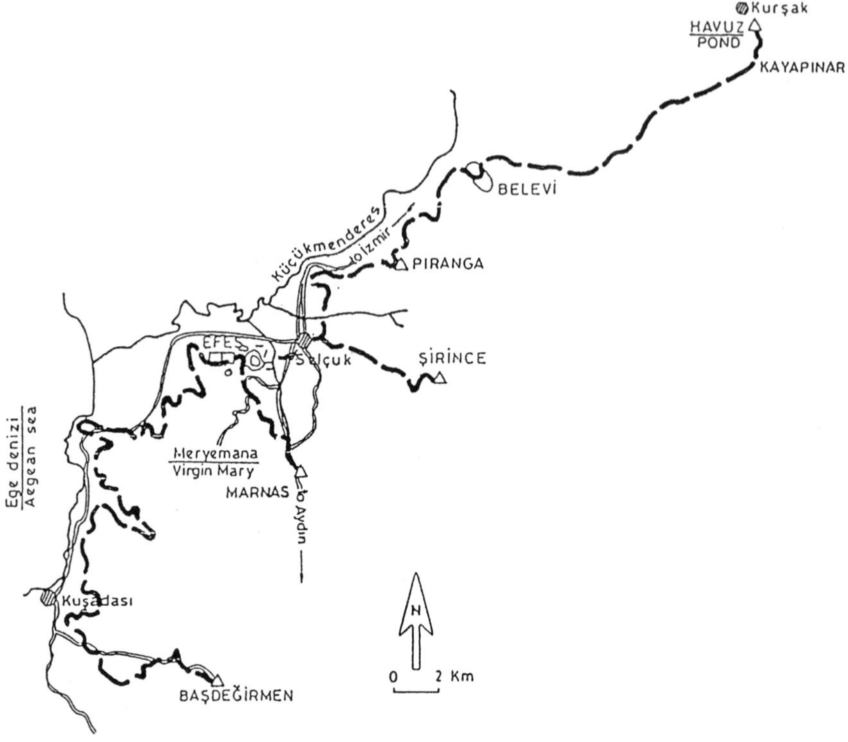

FIGURE 5.16. Long-distance waterlines at Ephesus ("Efes" in Turkish). Major lines came from north, east, southeast, and southwest. The northern line ran from the springs at Kayapinar (Kaya Springs), with additional water from springs near Belevi and Piranga. Sirinçe springs to the east supplied the Byzantine aqueduct to the center of Seljuk and probably supplied the line running through the stadium. The Marnas line came from the direction of Aydin (ancient Tralles), paralleling the Marnas River, and another (newly discovered) comes from the mountains directly south of the city, perhaps from springs near the Virgin Mary's shrine. A double line comes from the SW, from springs at Basbegirmen (ancient Kenchiros). The Kütschcük Menderes (Little Meander) River passed Ephesus on the north, running SW to the Aegean Sea (Özis 1996, reprinted by permission).

population of possibly a quarter of a million. During the Imperial Roman period (first–third century C.E.), the harbor as seen in Fig. 5.14; see also Fig. 5.21.

- (G) Christianity, from the mid-fourth century on, reused temples and imperial shrines as churches as paganism died away (Fig. 5.14). By this time Christian buildings stood on new land to the north and west of the earlier sites. Destructive earthquakes in the third quarter of the century necessitated other changes. Earthquake debris was utilized to increase the stability of the new land, and the materials of the monu-

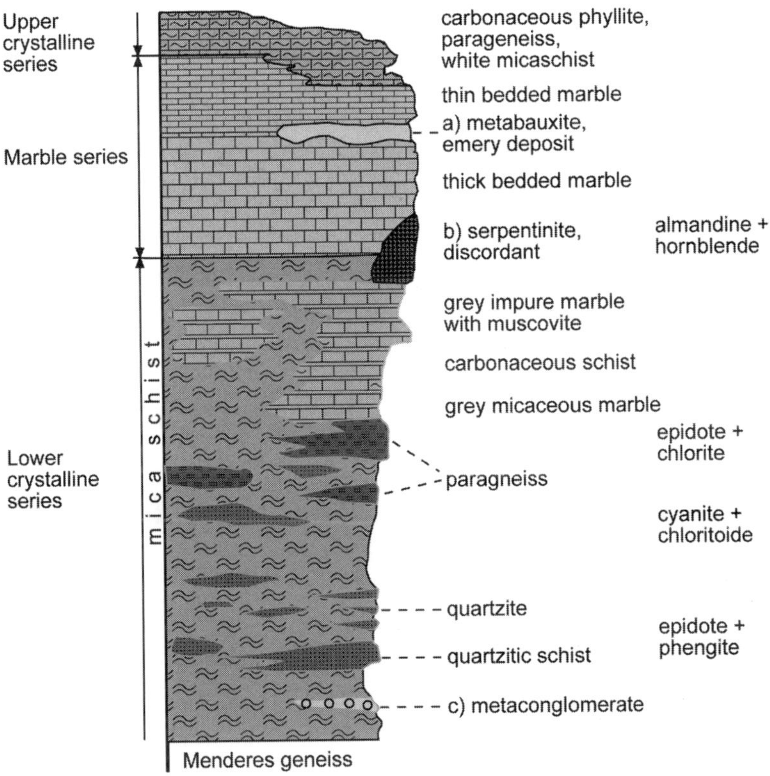

a) found at Ala Dagh
b) one small outcrop near Pranga quarry (in the NE of the map in Fig. 5.13)
c) one outcrop within the Kirkindsche brooklet

FIGURE 5.17. Geological column at Ephesus (Vetters 1998; redrawn by Langford).

mental buildings of the old centers were reused in later buildings in the new areas. The Hellenistic-Roman city between the two mountains continued as a residential area but lost its status as a government center. In the so-called Byzantine quarter to the northwest, a bishop's palace (81) vied with the new governor's headquarters for prominence.

• In the Byzantine period, Ephesus was a minor player, but with a great enough population to warrant a fine new church to St. John. Modest pilgrimages continued to the older shrines connected with the Virgin Mary (95), who was honored by both Islam and Christianity. The castle hill seems always to have lacked an ample water supply, so the early Christian and Byzantine buildings on the hill needed large cisterns. In spite of occupying different places in the overall site from time to time, the city was known as Ephesus to its stubborn inhabitants until the Turkish conquest.

- (T) The medieval castle (151) was useful in the military struggle for domination of the area. Arabs and Turks from the east and south were countered by the Byzantines and later by Franks and other Europeans from the west and north. The castle was equipped with a large vaulted cistern, so that it could withstand sieges (Schindler 1906). After the Turkish conquest in the fourteenth century, a large new mosque (170) was built near the site of the Artemision, utilizing stones from the old temple. Turkish Seljuk expanded east of Asayokuk.

The Relationship of Physical Geography to Geology

To see Ephesus in its regional setting, we must, with Vetters (1985, 1998, 2000; Kandler et al. 1995), include the hills on the north side of the Little Meander River, beginning at Alaman Hill near the coast and continuing over the road from Izmir to Seljuk along the eastern mountains. The eastern edge of the study area is the mountainous area called Kirkindsche with the town of Sirinçe some 6 km east of Seljuk. One kilometer to the south lies Ala Mountain (664 m tall) with the shrine of Meryemana (the Virgin Mary's House). On the west, the coast lies beyond the mountains along the south bank of the Little Meander River.

The Little Meander River valley ("Kütschcük Menderes" in Fig. 5.13, now called the Kaystros River), runs NE-SW in a former bay of the sea. Arising from springs at an altitude of 1300 m, it drains a basin of 2300 km^2 along its 150-km-long course.[15] The Little Meander is a fairly shallow braided river 0.5–1.8 m deep (Vetters 1985). Tributaries from the north (Torbali), northeast (Belevi), east (Sirinçe), and south (the ancient Marnas River from the direction of Aydin) run in fault valleys or along discontinuities. Some of the rivers are self-damming: the deposition of sediment at points of confluence tends to develop the valleys into chains of marshes—a process like that of the Great Meander River near Miletus. These tributaries deposit mostly fine sand, which over many millennia has created a plain as much as 10 km wide and 60 m thick, extending into the sea for about 5 km offshore (Westaway 1994). The mouth of the river has been constrained by the patterns of Mediterranean currents and wave regimes. The Little Meander River and the Selinus stream from Sirinçe are perennial, whereas other tributaries, such as the Marnas (modern Derwend), are dry during the summer because the springs that fed it have dried up (Strabo XVI.1.24; Snell 1963; Philippson 1892; Grund 1906a; Kraft 1998; Brückner 1986; 1995b, 1997a & b, 1998).

The denuded marble mountains in the area surrounding Ephesus exhibit sheer vertical and perpendicular walls with bare rock, as do the two mountains, Bülbül and Panayir, on the lower slopes of which the city was built (Fig. 5.14). Inland to the east, the river valley is enclosed by smooth mica schist and gneiss hills (Vetters 1998). These hills are much weathered, rounded, and covered with pine forests, contrasting with the bare marble slopes of the higher mountains (to 600 or 1800 ft), which seem even taller because of their seaside location, deep valleys (400–600 m), and steep slopes. Farther inland are taller mountains, from which the winter precipitation contributes to the summer water supply at the coast.

Quarries   Studies of quarries at Ephesus yield more data about local stone resources than any of our other sites. Therefore, we will give numerous details to indicate what could be learned elsewhere.

The stone was the result of millennial processes of deposition and compression (Fig. 5.13), the focus of the studies by geographers and geologists over more than a century. For instance, archaeologist H. Vetters initiated and his son geologist W. Vetters carried out (1985, 1988, 1994, 1996, 1998) studies on the origins of stones used in the Slope Houses. More than one hundred varieties of local and imported stone were used in these houses. Their structure, veneers, and mosaics displayed high technological development and sophisticated craftsmanship, because pieces 0.5–2.0 cm thick could be cut by saw. Many minerals and stone types, especially certain high-quality marbles, are from the Ephesus area. That there were so many types and qualities of stone here was a great aesthetic and economic advantage for Ephesus.

This stone was used for construction, decoration, and tools. Stone was being exploited for tools, and possibly for houses, in the third millennium B.C.E. (Runnels 1981). By the Hellenistic era, decorative stones were being imported from all over the Mediterranean—Tunisia (Simmithus), Egypt (Syene), Greece (Pentelikon), the islands (Euvöa), and Italy (Liguria)—and stones from the Ephesus area were exported as well. One was a green-pink marble from Teos (Türk et al. 1988) along the coast northwest of Ephesus,[16] from whence stone could easily be brought by water to the metropolis. The stones used here are either metamorphic or non-metamorphic, in several subtypes. Metamorphic stones include the marble series and the group of mica schist, quartz and quartzite, serpentinite, paragneiss, and metaconglomerate. Nonmetamorphic stones include marly hornstone, limestone, breccia with marble components, and recent sediments.

The stratigraphy at Ephesus is as follows, beginning with the topmost modern layer:

> Metamporphic series (600 million to 248–213 million years old):
>> Marble series in two varieties:
>>> Upper: thick-bedded, coarse-grained marble with some dolomitic intercalations.
>>> Lower: thin-bedded (15–30 cm) marble under metabauxite.
>> Mica schist, quartz and quartzite, serpentinite, paragneiss, and metaconglomerate.
>
> Nonmetamorphic series:
>> Marly hornstone, thick-bedded beige to clear yellow.
>> Limestone.

Red breccia with marble components in a dense red, chalky limestone matrix, found between Panayir and Bülbül Mountains, and in the karst tubes of Panayir Mountain.

Sand and silt-forming alluvium.

The location of quarries and attention to their products give important insight into the topography of Ephesus and begin to explain its economic importance. There were three groups of quarries and a few scattered ones (Fig. 5.13). The oldest quarry was in the mountains near Belevi. A small quarry was located on the north side of Ayasoluk Hill, with at least one on the southern extremity of one of the hills on the northern side of the Meander, not far from the modern village of Hejbeli. A second important group lay on the northeast side of Panayir Mountain. The most extensive cluster was on the north and west sides of the hills from the Roman harbor to the sea and south toward Kusadasi, and a few scattered quarries lay in the southern mountains.

The marble series was a favorite building stone during Hellenistic and Roman times. It occurs at Ephesus in two varieties. The lower series, 150 m thick, is mostly thin-bedded (15–30 cm) marble with metabauxite above, and the upper series is thick-bedded. The lower marbles are gray to whitish-gray, fine- to coarse-grained calcite marbles. The grain size of marbles correlates with the amount of pigmentation, with white marbles being coarse grained and having finer pigmentation. The thick-bedded upper marble series is coarse grained with some dolomitic intercalations such as in the quarry NE of Belevi (northeast of the area mapped in Fig. 5.13), where the dolomite is a distinct layer within the marble. The color varies from clean white through yellow to grey. Because fossils are mainly lacking in the local schists and marbles, as a result of their metamorphism, the rocks are difficult to date; they are possibly as ancient as the younger Paleozoic (Permian) era nearly 600 million years ago, but at the latest from the younger Mesozoic (Cretaceous) era, 248–213 million years ago (Dürr et al. 1977). Studies of the stable isotopic composition of marbles (Herz 1988: 305–14 and Wenner et al., 1988: 325–38) show that Ephesian marbles were similar to those from Naxos and Paros.

The quarries on Panayir Mountain closest to the city were mostly slope quarries (north and east sides of Panayir Mountain on Fig. 5.13), with the dip of the strata to the southwest (section in Hidiroglu 1988), which meant that extraction from the northeast had to work against the slope. [Compare Dirkin and Lister's (1983) study of how an ancient marble quarry in Crete was worked with Nenci's account (1979) of quarrying at Selinus, Sicily.] The physical evidence from tools and wedges in the Panayir quarries seems to be Hellenistic, from about 300 B.C.E. on, but some of the quarries were reused or newly developd in the second through fourth century C.E. for church buildings (W. Vetters interview 1998; Bessac 1988). W. Vetters estimates that the ancients removed more than 132,000 m^3 from the eastern quarries of Panayir, whereas the two small quarries on the north slope yielded about 1600 m^3.

Most of the quarries that H. and W. Vetters found were in ancient times on or adjacent to either the bay of the sea or rivers that were at least seasonally available for shipping. From the Belevi quarry, water transport moved the great blocks of white marble that were used to build the older Artemis Temple (1) of the archaic period (mid seventh to early fifth centuries B.C.E.), but when the temple was rebuilt in the fourth century, new transportation arrangements had to be made because of the seaward growth of the plain (discussed subsequently). At the Ionian quarry of Beylerköy, the use of rafts to transport marble is evident from blocks found in the river (Türk et al. 1988, with map).

There are two ancient quarries on the westernmost hill of Bülbül Mountain, from which grey to white, flecked or banded marble was cut to build the rampart along that mountain. These ramp quarries 20 m deep with variable slopes yielded 20,000–60,000 m³ of stone; there is a modern quarry on the same hill. One ancient quarry on the next hill to the west, Hejbeli, yielded a thick bedded beige to clear yellow marly hornstone as well as a chalky limestone. Hornstone, a quartz similar to flint, could be used for tools, which were also made from obsidian and flint. Tools of stone, though dominant in prehistoric times, continued in use through Greek times at least (Runnels 1981) because they were cheaper than metal.

Another four or five quarries were grouped on Ideli Hill, a western outlier of Bülbül Mountain, which W. Vetters calls a "complete hill of marble." Three of these quarries are shown in Figure 5.13. This group is one of the largest quarry complexes in western Anatolia, the center of the Ephesian stone industry, producing 250,000 to 300,000 m³ of stone. Marble blocks cut from layers 0.7–1.3 m thick weighed 2.4–7 tons. Smaller blocks were cut for the rampart along the ridge, since dragging huge stones upward was not easy. Other blocks of this stone were used in foundations, for example, in the theater. This quarry appears to be the source of the stone used in the basilica of the government center (Güngör, personal communication).

Six more large quarries, reactivated in modern times, were cut in the western slopes of Pamudschak Mountain, near the sea (Fig. 5.13). All are at least 40 × 40 m, but the stone does not lie on the surface and so the quarries have to be mined. Total yield for this group was 70,000–100,000 m³. The four on the northern side of the river produce red breccia (used for 500–600 years beginning about 300 B.C.E. for foundations and repairs), which incorporates marble elements in a dense red chalky limestone matrix. Breccia is also found within the city between Panayir and Bülbül Mountains, and it fills the karst tubes (discussed subsequently) of Panayir, an independent formation of the same date as sedimentary breccia, like the Timolos breccia near Priene and Miletus. Other quarries here yield gray-white marble.

Another four quarries are on the peninsula near Kusadasi (called Otusbir on Schindler's map); they produced 12,000 m³ of marble. One isolated quarry in the southern mountains at Kurt Kaya (Schlinder 1966; Atalay 1976–77) tapped a layer of excellent marble 20 m deep. This gave 1400–2000 m³ of stone in blocks of 1.5–2 m³, and a second underground quarry here, like the latomies of Syracuse in Sicily, yielded large-crystal, translucent marble, some 6,600–11,000 m³ in blocks of 2 × 0.7 × 1.5 m. This stone had to be hauled 300 m or so down the mountain and then by boat or raft on the river to Ephesus, a total distance of 11 km. Apparently, the ancients considered this effort worthwhile for high-quality marble for special buildings.

In addition to marble, on Ala Mountain in the area south of the city, where metabauxite is bedded with dolomitic calcite marble, there is an ancient quarry or mine for emery, a variety of corundum. It was used in antiquity—at Naxos, for instance—for smoothing and polishing hard decorative stones such as porphyry and granite. The next nearest emery quarry is 15 km to the south.

Both the Romans and the Greeks before them had considered quarrying a noncommercial venture (Fant 1988: 58). About 100 C.E., however, the emperors developed a Mediterranean-wide and protifable trade in marble that lasted well into the fourth century. Marble was a good symbolic material because, Fant thinks, it was expensive, imported, and unnecessary because there was plenty of good local stone such as peperino (a pyroclastic gray tuff found in Italy) and travertine (a calcium carbonate rock deposited at hot springs), that was widely distributed. Fant's list of imperial quarries does not include any at Ephesus, perhaps because the owners here were people of local rather than imperial eminence. From Ephesian quarries alone, Vetters estimated that 500,000–600,000 m^3 of marbles was extracted. Compare this number with the 25,000 m^3 of stone quarried for the fifth-century B.C.E. rampart at Miletus (Schhröder et al. 1994). Even considering how much rebuilding went on at Ephesus, there was plenty of marble for shipment to other Roman cities and for earning a healthy balance of trade for the Ephesians.

Aside from the archaic temple of Artemis, the ramparts of Ephesus are the earliest use of stone in construction (11). Most of the ramparts date from about 300 B.C.E. and from the Byzantine period (sixth–twelfth century C.E.) (77) (Seiterle 1964; Ozylgit 1991). The walls were built as part of the imposition of Hellenistic Greek culture in the area at the end of the fourth century B.C.E. They were possibly paid for by Athenis of Kyzikos (Koenigs 1993 citing *IvE* 1441), who was most likely a rich merchant honored with citizenship in the new city. The Hellenistic walls utilize the topography of the site, placed along ridges and around heights that posed danger if they fell into the hands of enemies. The Byzantine walls (77) followed them for much of the circumference, but also enclosed the new areas on the plain northwest of Panayir. Ramparts were built by the Turks about 1400 to enclose the hill of Ayasoluk where the church of St. John and the medieval castle now stand.

Tectonics and Sedimentation   Earthquakes affected mostly the southern part of the Meander plain, where the important faults of the system lie between Ephesus and the former island of Syrie (Fig. 5.13) (Vetters 1998). Location of earthquakes at Ephesus relates to graben formation. The most easily visible fault is revealed in a sheer scarp immediately south of the tourist entry on the south side of the archaeological site, along the road that leads up to Meryemana, that is, on the north side of Bülbül Mtn.

By Westaway's calculations, clusters of seismic activity (coupled with creep) occur at approximately 250-year intervals in this region. Because the last cluster was in 1650–70, in 1994 Westaway wrote that the area was overdue for major shaking, although there had been five earthquakes over magnitude 6 in the region during the twentieth century, one on the Söke fault close to Priene. All of these recent quakes involved south or southwest extension of faults (Fig. 5.13) (Westaway

1994). In addition landslides that shook the slagheaps of the marble series were worsened by discontinuity-related slippage between the marble and the underlying mica schist (Vetters 1998). Many quakes affected storage of silt along the still-active plane of motion in the river valley, where the behavior of water relates also to the boundaries between permeable and impermeable rock.

Because of the action of the African plate pushing beneath the west Asian plate (Fig. 2.2, discussed subsequently), the numerous faults in the region, and the prevalence of alluvial soil, which liquefies when shaken, Ephesus is subject to destructive earthquakes. Recorded quakes necessitated rebuilding of many urban structures. The written records are more likely to mention governmental and religious buildings, but excavation has shown repeated damage to residential, commercial, and recreational buildings as well. A summary of important earthquakes and reconstructions must include the following: those of 60 B.C.E., followed by the first efforts to Romanize the city architecturally (Rogers 1991); another of 23–30 C.E.; others less well documented may have occurred between 28 and 49 C.E. (Ambraseys 1971), necessitating the rebuilding of the agora (61) and moving the Marble Street (between 52 and 72) eastward and uphill; an earthquake of 150 C.E. or thereabouts; and a destructive quake in 262 C.E. Major quakes and fires from 350 through 455 coincided with the religious conversion of most of the population to Christianity (cf. Soren 1988). Particularly bad quakes in 358–66 ruined the city when bursting of the high waterline caused water to rush down the stair streets to the agora (Hueber 1997).

To rebuild after the latter catastrophe, Ephesus received imperial help. The grateful city rebuilt the shrine of Hadrian (41) and rededicated it to the emperor Theodosius as the new founder of Ephesus. The many rebuildings of this shrine suggest that the design was less robust than necessary for a spot where so many quakes were felt. The Slope Houses across the street also needed frequent renovation, probably due to the fact that the layer of clay on which they were built was leaning against a strongly tilted plane of rock. The seam between the two materials was lubricated by the groundwater, making an unstable base. The nearby Library of Celsus (55) was also ruined, then later rebuilt as a nymphaeum. The quake toppled the northern retaining wall of the theater (75), which was repaired about 400; this may be when the large reservoir behind the analemma went out of use. Other earthquakes as late as the seventh century have been invoked to account for the fact that some dead persons were buried between the slope houses and Kuretes Street in the shops along the street; ordinarily bodies had to be buried outside the city, except at the crossroads (Embolos) where the Octagon and other tombs from the first century B.C.E. still stood (52).

Thus, responses to the geological setting included repairs after earthquakes. Quarrying operations were affected by several changes along this coast: tectonic movement, eustatic (worldwide) changes in sea level, and the erosion and deposition cycle (Rapp and Kraft 1994). Additionally, decisions about water management and urban design were made in response to tectonic events.

Sedimentation—More Recent Geological Processes Hydrogeology interacted with tectonic overturn (thrust), liniation (parallel lines formed by alignment of minerals or by tectonic movement), folds, and other geological formations to change the setting of Ephesus. Subsidence along the Aegean coast correlates with the uplift of inland mountains, causing regional erosion and consequent sediment loading of the waters; Westaway (1994) asserts that erosion furnished only one-fifth of the sediment in the river. As the inland mountains are technically elevated and the river conveys fragments of the mountains to the sea, the river cuts deeper into its bed, a geological process known as rejuvenation (Ortloff et al. 1983). Elevation of the mountains is the result of the African plate plunging slowly under the Anatolian section of the Eurasian plate (Fig. 2.2) (Hsü and Ryan 1972), but it is also related to the reduction of load as the mountains erode. The slow pressure tilts the whole stone basis of the area and produces not only vertical shear near planar normal faults, but extensive heaving, which thrusts up large horsts such as Panayir Mountain. There can be rotation of the coastal blocks, as we have seen in discussing Miletus. Geophysicist Westaway (1994) has noted the coupling of surface processes and isostatic compensation, and thinks that plate-based uplift accounts for more of the local sea-land changes than does eustatic sea-level variation or human intervention. "Human activity has caused some increased erosion and sedimentation in historic time," Westaway (1994) writes, but not enough to distort the sedimentary record over time. His figure 2 shows coastal progradation, offshore sedimentation, and the fault zones of both Maeander valleys. The result of these interactive processes is a prograded coast at each river mouth. Geographer Kraft, personal communication, questions this reasoning.

The main fracture systems of the region (Fig. 5.13) are the NS fault along the Marnas River valley to the east of Ephesus and the NE-SW fault along the Little Maeander River valley—considered by Westaway (1994) to be among the most active normal faults of the area. In addition, an EW fault through the saddle between Panayir and Bülbül is known mainly from geophysical analysis but is not on the standard regional maps of faults; this morphological (not tectonic) fault is, however, quite significant in the local building history. Of these faults, the oldest is the NS set, but the deepest is the Meander Valley system at 300–400 m (Vetters 1998). Bülbül Mountain has one fault along its southwest side and another along its southeast end. Panayir Mountain has fault surfaces along its northern face, and it is somewhat closer than Bülbül to the major NS fault of the area (Vetters 1998).

Tectonic and sedimentary forces interact. Each bay along the Little Meander River is a down-drop graben controlled by local tectonics (Rapp and Kraft 1994: 69–90). The valley of each tributary has both a local relationship with the sea-level curve and its independent history. Surface features were quite different during the Ice Age (fig. 1 in Bammer 1986–87), with land bridges connecting the later islands of Samos and Chios to the mainland north and south of the mouth of the Little Meander River.

The most continuously active geological process at Ephesus seems to be not faulting but alluviation, which has left this city—a major port for over 1000 years— to become a land-locked village by the fifteenth century and 500 years later a tourist site 7 km inland. Sedimentologists have more recent dates for the history of the valley than the petrographers cited previously because their study material is newer. Recently, Kraft et al. (1999) have studied aspects of the alluviation of the Ionian coast. Combining the coring investigations of the sediments of the Little

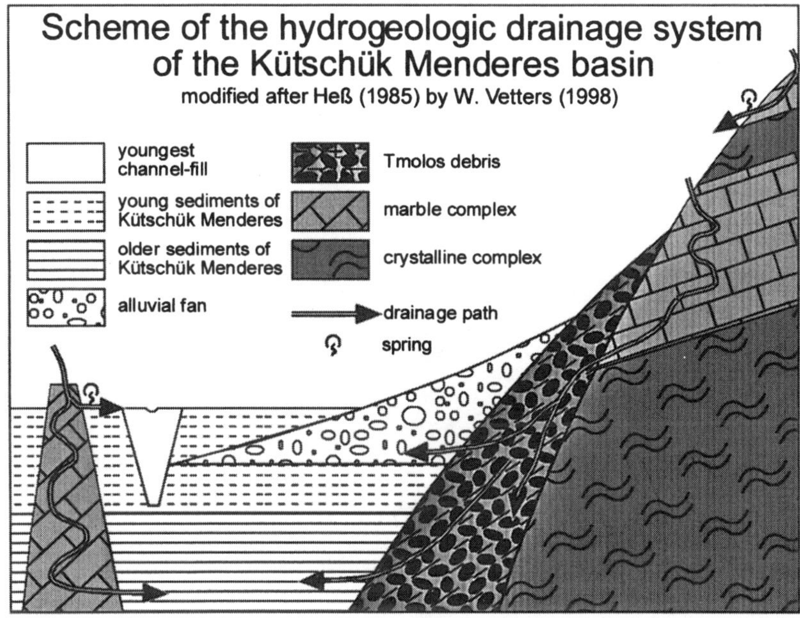

FIGURE 5.18. Sketch of alluvial deposit layers along the NE side of Panayir Mountain, containing both hard (stoney) and soft (sedimentary) materials. Arrows show paths of water in karstified marbles (Kraft, Brückner, and Kayan 1999).

Meander River valley by Brückner (e.g. 1996, 1997b) with Kraft's 25 years of geomorphological investigations of shoreline changes in the Aegean (e.g., Kraft et al. 1985) and Kayan's intimate knowledge of the coast of West Turkey (e.g., Kayan 1991), this team has come to understand that the late Holocene sediment units record continuous and rapid environmental and geomorphic change in the last 3000 years (Figs. 5.18, 5.19).

The Little Meander River has transported enough sediment to fill a bay 10 km wide by 15 km long, raising the level of the flood plain as far as 70 km inland. Since 3000 B.C.E. alone, the river has transported billions of cubic meters of sediment (Snell 1963). Present fluvial sediment loads are carried at the same rate for the Little Meander Valley near Ephesus and Great Meander Valley between Miletus and Priene (Westaway 1994, his table 6, derived from Aksu et al. 1987), though the latter drains an area more than six times as large as the former. Selective coring at different places in the deltaic plain has recovered much information about ancient shorelines. The late Holocene sediments are a "natural archive" of geological information (Brückner 1996), available to enrich archaeological and literary evidence (Rapp and Kraft 1994: 69–90).

Dating of fossils using C^{14} indicates a slow accumulation of sediments during the later part of the Holocene sea transgression, considerably before our period. Flooding of the valley when the ice caps melted about 18,000 B.P. brought about a change of boundary between the rivers and the mountains and along the valley

FIGURE 5.19. Stratigraphy in the Ephesus agora, from Roman at the surface to archaic Greek at the bottom, masked by seepage from the water table. The sounding is about 4 m deep. (Photo by Crouch)

between Ayasoluk and Panayir. The ancient shoreline along the Little Meander River, when sea transgression was at its maximum, was inland of Belevi (now 16 km inland) (Brückner 1997a&b, 1998). The mountain streams carried debris downhill to fill in the valley so effectively that only a few lakes along the northern edge remain from the ancient bay. Kraft et al. (1999) have done over 100 corings in this area and have compared them not only with each other but also with their earlier work, for instance, in the Troy area. Their provisional conclusions are set out in their map of the changes in harbors (Fig. 5.15).

Two deposition patterns are evident in the Ephesus valley: In the first, extending from 3000 to 1000 years B.P., the area was a prograding, multiple distributary coastline with changing channels protruding into a very shallow marine bay that was at the same time filling with silt. The natural process of delta formation accelerated in the second period, after 1000 B.C.E., when the sea level dropped some 2 m (Kayan 1988, 1991, 1996; Westaway 1994: his table 8).

Between 300 B.C.E. and 700 C.E., the delta bypassed in turn the sites of the archaic, Hellenistic, and Roman harbors (Fig. 5.15). Every two or three centuries, delta formation required shifting the port westward. The river continued to deposit silt against the western foot of Panayir, which extended more and more into the bay (Fig. 5.18). From Roman times on, the prograding delta front with its multiple and ever-changing channels carried alluvium through the shallow marine embayment to the western coastline.

There is no evidence of sandy coastal barriers until after 1000 C.E. Sands were dumped where the force of the river met the waves of the sea rather than in the adjacent landward floodplain, which became a backswamp region. The western shoreline moved 600–800 m per century or a total of nearly 7 km during the last two and a half millennia—a very rapid development in geological terms—but not at a steady pace. Delta growth between 750 and 300 B.C.E. was less than 1 km per century, 5 km during 300–100 B.C.E., and 2 km during 100 B.C.E.–200 C.E.; but since 700 C.E., the coast has changed little (Brückner 1997a & b; Westaway 1994). Reasons for the slowdown may have been depopulation, which reduced the harvest of trees on the slopes, which in turn reduced erosion, or a lull in uplift and earthquake activity, which also reduced the amount of loose material available for erosion and sedimentation, or both. Vetters suggests (personal communication) that downwarping of the valley surface from tectonic activity is also likely. Whatever the trigger, the wider and deeper bay took more sediment to fill than had the narrower river valley inland.

The interaction between river sands and the wave-dominated coast eventually produced the present lower delta barrier accretion plain that has much standing water. This was ideal terrain for malaria, which increased from the fourth century C.E. (Brückner 1997a & b), so it is not surprising that depopulation took place, as commented on in many travelers' accounts of the eighteenth and early nineteenth century (Foss 1979).

These physical changes affected the development of the Roman Imperial and early Christian city in the northwest quarter of Ephesus, which was built on land that did not exist in Hellenistic or early Roman times. The northwest edge of the Olympian complex filled in between 400 B.C.E. and 130 C.E., as Kraft's maps of coastal change show. The boreholes made in and near the Olympian/Church of Mary area have pinpointed the edge of solid land (F in Fig. 5.13) (Brückner 1997a & b). Just at the northwest edge of the Olympian (at boreholes 7 and 10), sea sediments and fossils, datable to the early sixth century B.C.E., were found at 16 m below the present surface. The accumulation of river sediments was enormous in this area through the third century. The residents also dumped a great deal of earthquake debris here (Brückner 1998), probably after the major earthquakes of the first and fourth centuries C.E. The Hadrianic Olympian, for instance, was so affected by the fourth-century quake that the vaulted substructure had to be filled in to provide better support for what was left (Karwiese 1985). At both this Olympian borehole and at number 6 to the northwest, evidence of earthquakes has been found. Yet borings to more than −30 at the site of the Harbor Bath just south of the Olympian showed it to be an erosional remnant, not a mound of earthquake debris. These data about the shoreline are far from intuitively obvious, so Brückner concludes that the location of the land's edge here is a problem complicated by numerous earthquakes, requiring paleoecological, archaeological, and historical data to solve.

The pattern of sedimentation after about 1000 C.E. differed from the earlier one. The Little Meander River fed sand into the wave-dominated bay farther west than in the first pattern, dropping its sands into the lower delta barrier accretion plain. Progressively younger and more western shores were built by the beach ridges along the western edge of the delta. Thus, at the rate of 6 m per century, the delta prograded while silts filled the bay. Progradation was independent from the formation of low-lying swampy marshes, swamps, and bifurcating channels that occupied the zone between the solid land and the beach ridges. More recently, a similar process has taken place at Miletus.

Human Use of the Site and Resources

Humans characteristically respond to their environment by finding ways to use it. We will discuss briefly the kinds of water systems set up at Ephesus and the chronological changes in harbors as examples of interaction of humans and their settlement site.

Karst Process and Water Management In spite of increasing aridity in the area during the period studied, the karst terrain provides especially good subterranean sources of water, making possible the survival of cities. The karst channels collect the usual 700 mm of rain per year within the mountains, where the waters gradually percolate to lower surfaces and reappear later at surface springs. Because of its karst geology, the Ephesus region can be expected to have many springs. Yet on-site springs to support the ancient population who lived at high elevations in Ephesus have been invisible to excavators during the past century (Brückner 1986), partly because the tops of the mountains are bare as a result of drastic erosion. The field work of Ortloff and Crouch (1998, 2001) has verified the presence of springs all around the peak of Panayir Mountain; we found constructed catchment basins in several places just below the peak. Further, because numerous ancient springs around the foot of each mountain have been covered by several meters of alluvium (Kraft interview), they are now invisible.

We have additional evidence for Ephesian karst. First, in an ancient inscription, the chief priestess Claudia Trophime (92–94 C.E.) wrote the following:

> The Pion drinks unnoticed from the air the rain into his belly
> and lets it fill its flanks just to the largeness of an ocean wide.

Inschriften von Ephesos, 1062,
as translated by S. Karwiese[17]

Pion is the mountain at the center of the city, now called Panayir. Drinking the rain "unnoticed" refers, I think, to the disappearance of rainwater into the cracks and crevices of the mountain, where it fills the empty spaces within the rock. The amount of water so stored could be compared to a wide sea. From here the water came out at resurgences on the lower flanks of the mountain, months or even years after a particular rainfall.

Second, archaeological and historical evidence for karst exists. On the northern side of Bülbül Mountain (71 in Fig. 5.14), higher than the Serapeion Temple and to the west of it, is a cave some 15 m deep by 2 m wide (Miltner 1955) with a carefully leveled floor. In early Christian times, its entrance was established as a chapel of St. Paul, with frescoed walls (being studied in 1998). This is likely to be a karst cave, because the marble mountain is a suitable lithography for a cave spring and because it was customary to protect sources of water with shrines, as it has been since remotest antiquity (Fahlbusch 1982). Similar cave shrines are found on the south side of the Acropolis in Athens (Crouch 1993, Chap. 18).

Third, we have observational evidence. On the southern side of Panayir Mountain, above the bath on the north side of the government center, what appears from below to be a large water channel comes directly out of the hill above the water-using rooms of the bath. Examining the channel will be a task for speleologists because of the steepness of the hill. It is normal for baths to be situated at sources of water, like the stadium bath at Priene.

Ephesus was supplied not only by karst springs within the site, but also by copious springs in the hinterland that were tapped for the long-distance waterlines, which have been studied for many years by Professor Ü. Özis and his students (Özis 1997) (Fig. 5.16). The known aqueducts tapped springs such as those in the eastern mountains, where in the late first century B.C.E. an Augustan setting was created for the spring on the mountainside high above Sirinçe. Water from a seam between marble and mica schist was captured in a 12-m-long reservoir that held (and still holds) the water long enough for it to drop its sediment load before flowing into the aqueduct to Ephesus; the aqueduct is still visible along the valley between the village and Seljuk. In addition, a deep-lying aquifer from the same area may have been the source of springs such as those at the Artemis Temple. The former island of Syrie (modern Kuru Tepe) (Fig. 5.13), which deflects the course of the Meander River to the north of Panayir, lies in the path of this aquifer. It is a horst, part of the natural system for storing water in the valley. Although no physical remains, which are now probably buried under as much as 25 m of alluvium, have been found of an ancient system for delivering Syrie's water to Ephesus, only detailed excavation could exclude the possibility that it was a water source comparable to the artesian springs of Pisma and Pismotta at Syracuse.

Between the two mountains of the settled area, a deep layer of weathered mica schist still serves as an aquifer (Fig. 5.13: E). This aquifer collects surface and underground waters from the two hills, and drains under the agora. Wells in houses on the hills tap this underground flow. One well with a hand pump in the house at the corner of Marble Street and Kuretes Street was still producing ice-cold, delicious water in the hot summer of 1985, and another across the street in the Slope Houses was used by workmen doing repairs in 1998.

Forest clearing during the last century has made karst increasingly visible, but it has reduced the amount of available water percolating through the channels. As at Delphi, slope debris (scree), fallen boulders, and quarrying debris facilitated the drainage of surface waters through fractures in the limestone into deep aquifers.

Locating and utilizing the water supply (springs and seeps) within the site was essential. As the society became more sophisticated, dams and weirs were employed to avoid flood damage, to improve the harbor, and probably to supplement natural rainfall for irrigation. The Romans combined supplies from springs, local aquifers, natural reservoirs, and artesian springs of the area to supply water for the estimated population of 300,000, supplementing those sources with long-distance aqueducts and on-site cisterns and reservoirs.

At least four to six long-distance water supply lines were built to bring additional potable water to the city for fountains and baths (Özis 1995). The long-distance lines originate in copious springs to the north at Kayapinar (*pinar* means spring), where there is still a small lake and a picnic resort; from the east in the mountains near the town of Sirinçe; from the southeast beyond the House of Mary; and from the southwest where lines from two sets of springs are gathered to constitute the longest line from Basdegirmen Pinarlari. The Marnas line and the Basdegirmen lines overlap along the north side of Bülbül Mountain (Fig. 5.20), supplying the Slope Houses and the large fountain house and distribution center south of the government center (18) (Ortloff and Crouch 2001). Similarly, the Sirinçe and Kayapinar lines overlap along the east-northeast foot of Panayir Mountain. The exact connections between the visible water elements within the city and the long-distance lines outside it are lost under several meters of alluvium in the river valley and along the slopes, except for the fountain house in the government center.

On-site water came from springs, such as those on Panayir Mountain at the level of the palace above the theater and from the seeps that wet the area from the peak of the mountain to the palace; we found the whole ground surface on the west side between peak and palace green with moss after a hot and dry summer. Wells, cisterns, and reservoirs were ubiquitous. A huge reservoir, previously unknown, was found behind the left front wall of the theater during the process of dismantling the upper wall while we were at the site in September 1998 (75).[18]

Perhaps the most significant finding of our on-going research into the water system at Ephesus (Ortloff and Crouch 2001) is that Roman engineers had more hydraulic and geological knowledge than is evident in the writings of Vitruvius (late first century B.C.E.) or Frontinus (first century C.E.) (Fig. 5.20). Additional information will be published after further study of the water system with the Austrian Archaeological Institute.

Harbors　Shifts in port location constituted the human response to geologically rapid changes in the pattern of relationships between river and shore from the seventh century B.C.E. to the seventh century C.E. The Ephesus harbor of the seventh century B.C.E. was located near the Artemision, south of the present castle hill (Fig. 5.15). The first harbor was shallow, but adequate for Greek keelless boats. By 600, the bay had contracted to a shallow harbor. It was filled in by 407 B.C.E. (Brückner 1997a and b) or became a swamp by 300 B.C.E. and went completely out of use by about 200 B.C.E. (Kraft 1998) at the very latest.

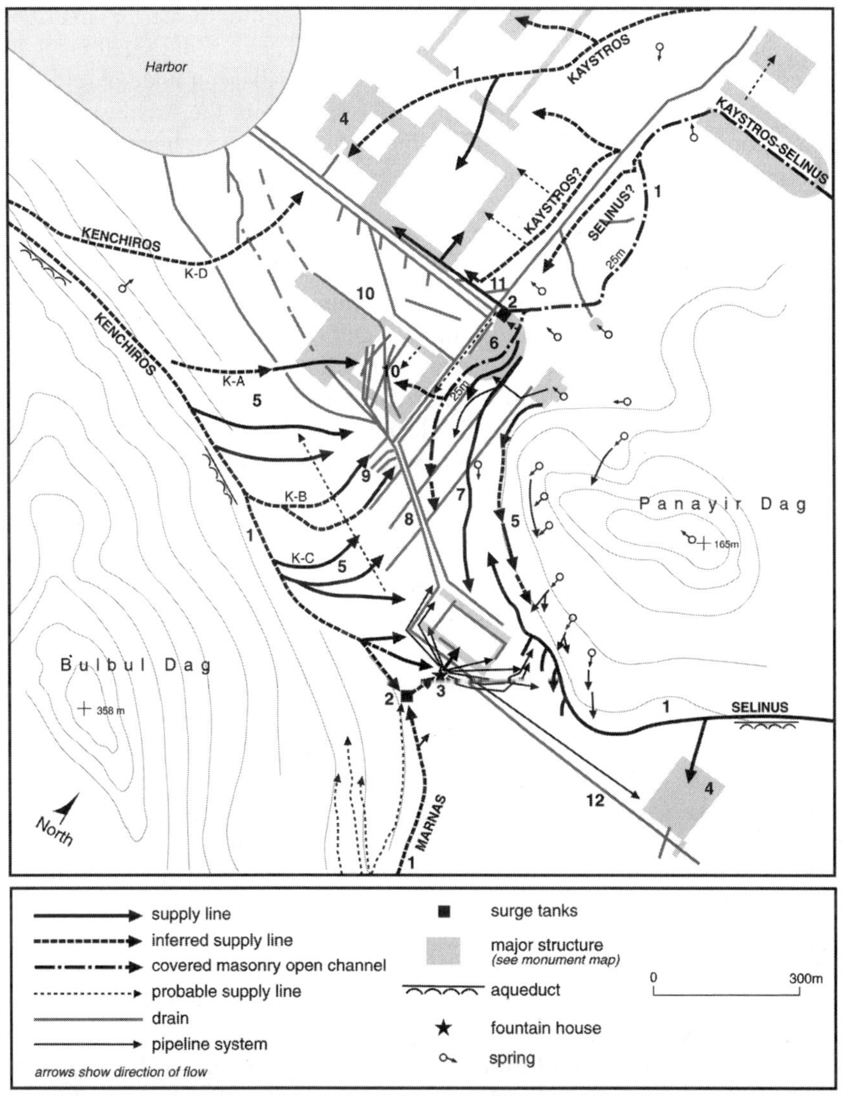

FIGURE 5.20. Roman water supply and drainage system at Ephesus, based on extensive field work and computer modeling (reprinted from Ortloff and Crouch 2001). Supply lines provided water for public display in fountains and baths, supplementing traditional springs, cisterns, and wells. Drainage of used water and wastewater was by gravity flow to the harbors.

During the same archaic period, an informal berthing place for small boats had probably been located at Smyrna, between the banks of the small creek that drained the valley between the two mountains. This village, a node of settlement from the eighth century B.C.E., stood about 14 stadia from the Artemision. In the swampy conditions of the valley at that time, boats would have been the preferred mode of transportation from Smyrna to the shrine's harbor, except for religious processions.

Another harbor was located along the northern side of Panayir Mountain during archaic times (Fig. 5.13: D) to serve the Koressos settlement, whose very name, according to Karwiese (personal communication), implies a harbor and some altitude — perfect for a settlement on a cliff overlooking a port. At this time, Panayir Mountain was a peninsula with water on three sides. Alzinger (1970, 1972) thought that a fifth-through-fourth-century harbor was tucked into a recess at the northwest corner of Panayir Mountain (Knibbe and Alzinger 1980). The Koressos residents are said to have made a cable seven stades long connecting themselves to the Artemis sanctuary across the Marnas River (Karwiese 1985). During the sixth century B.C.E., the Koressos harbor was rapidly silting up (Kraft, et al. 1999), but the western flanks of Panayir Mountain were still dropping steeply into the bay with only enough level border for the circuit road.

When a flood at the end of the fourth century destroyed the old town near the Artemis Temple, Lysimachus moved the people from that area, and possibly those from Koressos, to the valley between Bülbül Mountain and Panayir Mountain. The new settlement, named Arsinoe for the ruler's wife, was a half-hour's walk (2 km) to the south of the Artemis area, but on higher ground in the valley between Panayir and Bülbül Mountains and with abundant water. Lysimachus built a port close to the west side of Panayir Mountain, perhaps where the port of Smyrna had been (Kraft 1998) (Figs. 5.13: E, 5.15: Lysimachus Harbor). By then, old Smyrna had been buried by 5 m of sediment. The supposed Hellenistic port was closer to the mountain than the subsequent Roman port. Hypothetical river ports on the island of Syrie would also have had to struggle with the shifting course of the Meander River during the last three centuries B.C.E.

Drainage from the two mountains into the creek may have helped to keep the Hellenistic port open. The size of the sewers under Kuretes Street and under the agora, some big enough to walk in, speak to the volume of drainage waters that had to be handled. Early Roman times saw the Hellenistic port west of Panayir filled in by river sediments and urban drainage (Brückner 1998).[19] We catch a glimpse of the ongoing battle between river and society in the description by Strabo (XVI 1.24) of the efforts made about 100 B.C.E. to build a dam and channel to the sea.

By Roman imperial times, the edge of the delta had moved several hundred meters west of the Hellenistic port, and a new harbor was needed (compare Fig. 5.21 with Fig. 5.1 in the Miletus study). To maintain the harbor in the first century of our era and later, the local government appointed an official to direct the works: dredging and constructing offshore islands with the accumulated material, constructing moles and other harbor structures, and erecting diversion dams (barrages) — possibly useful also for irrigation — in the river and its tributaries.

FIGURE 5.21. View of the Roman harbor from the palace above the theater. The change in vegetation clearly reveals the round center of the harbor and the long, narrow channel leading out to the late Roman and Byzantine location of the river. (Photo by Crouch)

In the second century C.E., a new harbor basin (Figs. 5.14, 5.15, and 5.21), warehouses, and other buildings were constructed approximately 200 m to the west of the Hellenistic harbor. Roman efforts to protect the harbor and control its use were memorialized in a large inscription now in the Seljuk museum (Bammer et al. 1974). Since there were at least five locations of ports at different times, we must be careful not to attribute either Strabo's description or the lengthy Roman inscription to the wrong harbor. The Roman harbor was kept open for at least 300 years. In spite of ongoing physical difficulties, this harbor was one of the two greatest centers of international trade on the Ionian coast, generating the wealth that made possible the long struggle against the river.

Zabehlicky (1991, 1996) made the first excavations of the Roman harbor since Bendorf's in 1898. Kraft (1998) and Brückner (1996) (Fig. 5.15) provide additional geographical and geological information.[20] The harbor, today a swamp, lies in a flood plain 100 m wide (Fig. 5.21). Between the harbor and the flood plain in Roman times was a mole 4 m wide, forming an east-west bank. Quays of stone outlined the basin for ships. Divers found the bottom at 6.3 m toward the eastern end, but shallower ponds now occupy the western section. Ceramics found under the plaster coating south of the mole and an inscription of 100–130 C.E. date the use of the harbor to the early second century C.E. Zabehlicky speculates that the architectural development of this port was impelled by the concurrent change from road to sea transport in the second century.[21]

Later physical evidence, such as amphoras from the fourth to sixth centuries, made in workshops at Ephesus, an anchor, and a statue base of Aphrodite, as well as travelers' letters (Foss 1979), indicate that the port was in use through 650 C.E. at least (Meriç 1985). The Ephesians struggled to retain access to the sea that enriched them by trade, provided seafood, and connected them to the wider Greco-Roman world, but the reduced population of early Christian and Byzantine times was not able to overcome the river's pattern of deposition. Shipping moved westward to a site called Panormos, possibly in one of the southern bays of the valley as on Schindler's map (1906; see also the topographical sketch in Karwiese 1985). The seaport of the Turkish era after 1300 was at Alaman Gölü at the north-west edge of the valley next to the modern coast. This is one of two former sea areas, now shallow enclosures, that have persisted since 100 B.C.E. (Schindler 1906; Brückner 1997a & b). Hellenistic watchtowers to the NW and N of this area suggest the location of the shoreline of that era (Meriç 1985). A paved way, constructed in the Turkish era, of debris from ruined buildings at Ephesus, ran through the swamps to connect this port with the settlements upriver at Ayasoluk. As recently as 1872, H. Kiepert (1906) sketched a map of the region showing the possible location of the Panormos harbor at a "lake" enclosed in a bend of the Maeander River approximately three-eighths of the way between the old Roman harbor and the coast.

Conclusions: Topography and Urbanization

Political change and geological processes and events combined to destroy and rebuild Ephesus many times on sites within its core area. A fine example of human–site interaction in the river valley is the processional route (Via Sacra), which must always have run on ground lying above the active river, probably at about 6 m above sea level in the seventh century B.C.E. (Fig. 5.15). The route went from the Artemis sanctuary, across the valley, to the east side of Panayir, to the east end of the valley between Panayir and Bülbül Mountains, through that valley, past Smyrna, westward to the village of Ortygia in the coast mountains, back to Smyrna, along the western flank of Panayir, through or near the Koressos village, and back across the valley to the sanctuary of Artemis (Kraft 1998).

The vistas from place to place within the site began with natural forms, enhanced by an architectural vocabulary that related single buildings to streetscapes. Not only was the city fabricated from stones quarried on and near the site, but the buildings stood on specific sites formed by the geology and "edited" by the lengthy geological processes that took place at Ephesus. The Varius Bath (16) of the government center, for instance, was set into the angle formed by quarrying the stone to erect it. Built form echoed natural form at the theater (75) and the slope houses (50, 51), both of which extended up hillsides. The many fountains brought the waters of the surrounding mountains to the city and enlivened the major streets and public plazas. Local stone also depicted the important people of the city in statues set up in many public places. The city of Rome is said to have had as many statues as living people, and Ephesus as the first-ranked and richest city of Asia doubtless had similar numbers of marble and bronze dignitaries.

Ephesus flourished because of the agricultural wealth of the valleys plus the grazing, seafood, timber, charcoal, and stone resources of the outlying areas. For a long time, there was a substantial balance between the needs of the population and the productivity of the place. The change in balance is indicated in the use of lumber: In 407 B.C.E. Ephesus built wooden ships for Spartan war use (Plutarch, *Lysimachus*, vol. 3), but by the second century C.E., Ephesus had to import large quantities of timber, as noted in the list of abuses at the harbor (Keil 1959: 242–46; Meiggs 1982). Later still, drastic change and turbulence in the political system, and the division of the eastern and southern Mediterranean littoral from the northern and western, interrupted the flow of trade. Centuries of deforestation, which contributed to silting up the whole bay, coincided with the political problems. The introduction of the malaria mosquito was the final factor that reduced the splendid late antique city to a small medieval and post-medieval village. As is true in most human situations, the observed changes had multiple causes (Vetters 1997; Keil 1922–24b; Schneider 1997).

With the completion of this chapter, we are ready to compare our findings from the ten case studies to see what similarities and differences in their geology reveal about the patterns of urban development in the Greco-Roman world.

FINDINGS AND REFLECTIONS

6

Comparisons of Cities

A brief reprise of the geological aspects, organization, physical constraints, and appearance of each city will remind us of their common and unique features. Then we can compare them by groups linked by research questions (Table 6.1).

Sicilian Cities

Agrigento is built on two ridges of 120 and 320–390 m, setting generous limits not yet filled by the modern city. A plain extends from the lower ridge south to the sea. Vistas were provided along contours and across elevations. Grouping the public buildings on stony ridges, with temples above and below and government structures along the west side, made economic and aesthetic sense. Landslides provide important clues to the nature of the hill the city is built on, and they correlate with occupation of various parts of the site. Additionally, the water system shows unexpected correlation with the families of discontinuities in the stone rather than the surface grid of the streets (Ercoli and Crouch 1998; Crouch 1989).

Morgantina stretches along a ridge about 600 m in elevation. The agora most clearly reveals the interface of urban design and geology. Sanctuaries and fountains were the focus during the fifth century B.C.E. In the third century, modest but elegant new architecture (theater, great steps, flanking stoas, fountains, sanctuary) combined with pragmatic engineering as framework and connector between points of observation. Morgantina had one aqueduct, from the springs that later supplied Aidone. The site has numerous springs although some are now dry or give less than 1 l/s. Yet, during the third century B.C.E. when the population was at its maximum, the aquifer was also at maximum, and higher springs were fed from it more amply than at present. Improper management of water resources likely hastened the demise of the town after the Roman conquest. At the turn of the era, the shift from small rural towns to great landed estates as centers of population affected Morgantina strongly. Deforestation of the hills and mountains for fuel and

TABLE 6.1 Site Comparison by Features and Duration[a]

Sites	Date Occupied	Culture[b]	Karst	Springs	Clay	Silting	Building Stone/ Quarry	Earthquakes	Landslides
Agrigento port & plains	14th c. B.C.E. present	pre-G, G, R, post-R	+	+	+		+		+
Argos near coast & mountains	Neolithic– present	pre-G, G, R, post-R	+	+		+			
Corinth near coast & mountains	Neolithic– present	pre-G, G, R, post-R	×	+	+		+	+	
Delphi mountains	1600 B.C.E.– 19th c.	G/R+	+	+			+	+	+
Ephesus port	3rd M. B.C.E– present	G/R+	+	+		+	+	+	
Miletus port	9000 B.P.– present	pre-G, G, R, post-R	+	+	+	+	+	+	
Morgantina hill town	Neolithic– 1 c. B.C.E./ C.E.	G/R		+	+		+		1 (modern)
Priene river terrace	ca. 800 B.C.E.– 12th c. C.E.	G/R+	+	+		+	+	+	
Selinus port & plains	7th c. B.C.E.– early 1st M. B.C.E.	G/R+	off site	+		+	+	+	
Syracuse port	mid 2nd M. B.C.E.– present	pre-G, G, R, post-R	+	+	+	+	+	+	

a. + means the feature is present; × means there is some indication of feature.
b. G, Greek; R, Roman.

building materials could have resulted in desiccation, with climate change a related factor. Occupation by the Hispanii (Spanish veterans) who replaced the Hellenized Sicilians after 211 B.C.E. coincided with a negative water balance. Negative hydrogeological phenomena—collapse and landslide—known at Morgantina from ancient times have recurred because of modern pumping of the aquifer and dumping from modern excavations.

Selinus occupied three promontory heights of 30 m, 31 m, and 29 m separated by two rivers. From the city southward to the sea, there were fine combinations of vistas, articulated by splendid, widely spaced groups of temples. The promontories and slopes were united in the sixth century plan. The central acropolis preserved the bar-and-stripes plan of seventh–sixth century. Development was limited neither by extension, which was more than ample, nor by topography—but by time. Stasis in topographic change marked approximately the first 300 years of Greek settlement. The Selinuntines constructed river walls to live in balance with nature. From 300 B.C.E., however, the river was more of a challenge, with increased sedimentation and probable flooding. The location of Selinus, so close to rival Carthage and so far from centers of Greco-Roman life, made it impractical to devote major resources to keeping its ports open after 200 B.C.E. War, geopolitical change, and malaria attacked the site at once, after which Selinus was never more than a village. The occupation of Selinus was effectively terminated by an earthquake in the twelfth century C.E.

At *Syracuse* geological factors and human history interacted at key moments. In the eighth century B.C.E., the site was chosen for its river access into the hinterland and for the small bays at the north end of Ortygia island to be used as ports. From the earliest settlement on Ortygia island, the Hellenistic and Roman city expanded onto the mainland, which slopes from sea level to 60 m, halved by the line of limestone cliffs where the major archaeological park is located. A conscious effort to provide vistas of the harbors and sea in many directions characterize all periods of development. Natural underground channels were transformed into aqueducts and later into cemeteries. Utilization of the on-site limestones for construction and clays for ceramics shaped the city, but the quarries were later damaged by earthquakes. Erosion and deposition along north shore of the bay and in Large Harbor, coupled with earthquakes, eventually changed the shape and location of the ports. Earthquakes struck in the first, fourth, twelfth, sixteenth, and late seventeenth centuries C.E., the latter completing the wreckage of the bridge from Ortygia to the north. Also noteworthy were the submergence of the edges of Ortygia by 3–4 m after the Roman period, and the quiet changes of the erosion of hillsides, silting up of swamps, and action of salt air on the stones of monuments.

All in all, the Sicilian cities yield a 50% success rate for the foundation of durable cities in resource-rich settings. Syracuse, the dominant city during the classical period, was still the largest and most successful in the twentieth century. Agrigento, its ancient rival, still rejoices in its monumental past and persistent beauty, but the city lacks an active port to connect it to modern international trade. The agricultural resources it can draw on are not as significant now as they were in Greco-Roman times. Opening of the new university at Agrigento specifically oriented toward archaeological study may improve the economic climate. The relative inconvenience of access to Selinus and Morgantina has helped to preserve them. The increase of interaction between Sicily and North Africa may improve the economic position of Selinus, but can do little for Morgantina on its central highland ridge.

Greek Cities

Argos was built between a 276-m mountain and a plain sloping southward to the sea and eastward to mountains across the valley. The mountains on the west and north limited the city in those directions, and the river to the north, northeast, and east was a moving barrier, sometimes flooding dangerously. At Argos, careful analysis of karst systems helped to explain both present and past water supply patterns. Perched on its ledge above the best agricultural land, the city looked up to the steep, fortified Larissa Hill, crowded public and private buildings together near the linear water sources at the foot of Larissa, and used the lower Aspis Hill as a sanctuary. Although never as regular as colonial Greek cities, the street pattern of Argos has persisted through Greek, Roman, Byzantine, and Turkish rule until now, when it is closer to the stimulating density of Rome itself than to the spacious cities of the Ionian littorals or the magnificent cities of Sicily.

Corinth extends down nine terraces from Acrocorinth to the Gulf. Although it had the largest city area in mainland Greece, the city seems never to have filled all nine of the terraces at once. Terrace edges were the major physical features that determined how Greek Corinthians perceived their world and how they occupied it (Engels 1990). Geology and architecture were melded, for instance, at the Lerna bath, located where a major road from the harbor met a terrace with karstic water supply—a successful building location and plan which was in use for a thousand years. Builders filled in the hollow of the central agora-forum area, bracketing it with fountains, regularizing it with stoas and colonnades, and displacing houses and workshops toward settlement edges. Greeks combined diagonal streets with a grid; Romans emphasized their new grid, but rebuilt monumental sites following the Greek design principles of isolation and angular views. Even severe earthquakes have not been able to dislodge the surviving residents, because the site has both agricultural and trade location advantages.

At *Delphi* the central sanctuary lies south of Mt. Parnassos in a cup of scree, between a karst aquifer and a flysch aquaclude. Channels within the rock feed karst systems in limestone and numerous small springs situated over flysch. The site is notable for the number and even distribution of its springs; the water supply was visible in many fountains, some still flowing. The precarious balance between imposed weight and internal pressures within the rocks surrounding the sanctuary was broken at strategic moments in the history of the site, either to protect it by avalanches or to wreak destruction in earthquakes and rockfalls. The steepness of the slope and the instability of the scree necessitated frequent rebuilding. The strong impact of the physical setting is inescapable, but equally strong has been the perseverance of human activity. The layout of the site preserves both the enclosure formed by the arc of mountains uphill and the exposure to the 200° view toward the Gulf of Corinth. The second focus of the site (Marmaria) lay on a terrace below the modern road and to the east, well watered by springs in that area.

The three Greek cities were built on slopes related to imposing mountains; they had good access to building materials and plenty of water. Argos and Corinth were active traders, and Delphi occupied an important trade position in the eighth century but later became mainly a religious center. Corinth was the most regular and geometric in plan, although not as regular as colonial Greek cities. It was not so densely occupied as either Argos or Delphi. Earthquakes have been rarest at Argos, of medium frequency at Corinth, and both frequent and devastating at Delphi.

Ionian Cities

Miletus spreads out on low hills and the alluvium connecting them. Easy access by sea and a river route to inland customers drew traders to Miletus in the second millennium B.C.E. The Cretan/Carian settlement of that time and an Ionian colony in the early first millennium B.C.E. were stimulated by nearby iron and coal deposits, a wealth of forested mountains, and fertile valley soil. After several centuries of successful trade and colonization, Miletus was destroyed by the Persians in the early fifth century, then rebuilt according to the Hippodamean plan. The city flourished in late Hellenistic and early Roman times. The mercantile city used several harbors to connect with the Mediterranean world. Street and plaza systems focused on the harbors, a fine combination of architectural regularity with geographical irregularity. A new political and physical organization in Roman times added large buildings, such as sports facilities and baths (Voigtländer 1985), and refurbished harbors (Brückner 1998) to the old core. The grid was adapted to the new monumental complexes. The settlement history of Miletus is similar to that of Ephesus because their geological and political histories are parallel. It is hard to tell whether the different locations of settlement at Miletus are the result of geological processes alone or whether the data sets are incomplete.

Priene is situated on a sloping terrace that falls abruptly into the valley to the south in front of a mountain rising abruptly to the north. Limited by its small terrace, the city was intended only for upper-class families, with sailors living at the port and farmers living in villages. The differences between the geometry and topography of the central grid set up aesthetic tension. The axes of the plan were streets oriented EW crossed by NS stair streets. Greek design principles of isolation and angular views are evident, especially in the placement of the main temple above and at an angle to the agora. Priene's climate fostered outdoor life, necessitating a variety of open spaces: the agora, temple precincts, theater, and stadium, as well as the paved streets and public stairways. Like Argos, Priene had a mountain hinterland of nutritional and economic significance, especially for timber resources. The city boasted an abundant piped water supply from the springs of the karstified marble mountain. Excess water and waste was carried away by a full set of drains (currently being restudied by H. Fahlbusch et al.). The Meander River, with its heavy annual load of silt and gravel, dominated the history of Priene, forcing the abandonment of the site several times. By the first century B.C.E., the final site was landlocked by the aggrading river, and Priene is now more than 8 km inland. The site also suffered from frequent earthquakes and was abandoned after an active occupation of 700 years.

Ephesus is the most complex site we studied—physically, historically, and institutionally. It was built on and around two low mountains. Shifting locations of settlement in the Ephesus area reflected the changes in the river valley over time, when alluvium collected in different places along the lowest slopes of the mountains. During periods of prosperity and dense population, the houses spread up the slopes. From 300 C.E. on, the city occupied new land northwest of the older city. The urban design of Ephesus was strongly related to the underlying geology, with the urban armature (MacDonald 1986) of the streets relating to the topography. Vistas began with natural forms and materials, but were modified by the Roman architectural vocabulary unifying buildings, ensembles, and streetscapes. Of all our sites, Ephesus is the most architectural in its vistas.

Both monumental and vernacular buildings related to their settings by the use of materials. For instance, the bath of the government center is set into the angle formed by quarrying the stone to erect the bath. Enduring prosperity was made possible by both long-distance trade and an ample water supply, signified by the many fountains that delivered the waters of the surrounding mountains to the city, thereby enlivening major streets and plazas. The negative challenge of water came from centuries-long alluviation of the valley, shifting the shoreline so that the port had to be moved several times. Intermittent earthquakes affected the development of the city.

The three cities of Ionia had different settings; a terrace at Priene, the lowest slopes of several little mountains at Ephesus, and low hills of a mature karst terrain at Miletus. Yet they also had similar settings: rapidly prograding river valleys. Indeed, Priene and Miletus were located on the north and south sides of the same valley, and Ephesus in the next valley to the north. Priene was small, the other two very large. Miletus exemplifies the Hippodamian plan and Priene utilized this kind of plan, but Ephesus shows only a small sector developed with such regularity. Destruction came from earthquakes and the river at Ephesus, from the river at Priene, and from river valley changes coupled with political attrition at Miletus. Quarries have been studied at Miletus and Ephesus, not Priene. In surviving structures, Priene is the most Greek, Ephesus the most Roman, and Miletus an architectural and cultural mixture (Greek, Roman, Byzantine, and Islamic). Sources of wealth were trade, government, and religion at Ephesus and Miletus, but government was the source of wealth at Priene during the fourth and third centuries B.C.E. and thereafter forest products.

Comparison of All Ten Cities

The ten cases may be considered a series of experiments in urbanization in the Mediterranean area (Sheets and Grayson 1979), and our studies a series of interdisciplinary investigations. Our initial hypothesis that these cities exhibit similarities based on geology has been verified.

Three cities (Agrigento, Morgantina, Selinus) were set on ridges; three on terraces (Corinth, Delphi, Priene); one at a valley ledge between rivers and mountains (Argos); and three (Syracuse, Ephesus, and Miletus) on coastal plains. Of the latter group, Syracuse and Miletus, and probably Ephesus, took advantage of offshore islands for their earliest Greek settlements. Five (Selinus, Corinth at its ports, Ephesus, Priene, and Miletus) had sedimentation problems during the period studied.

Medium-size Argos and Morgantina, and small Priene, were the most severely constrained by their topography. The large cities, Ephesus, Miletus, Corinth, Syracuse, and Agrigento, had wider physical limits and more wealth to pay for development and alteration of the terrain. All shared a common design vocabulary sensitively related to particular sites, similar building materials, and the traditional Greco-Roman architectural forms. Other common features were the karstic water elements with their necessary holding tanks and drains.

All of these cities combined public buildings (plazas and religious precincts edged with stoas and graced with temples) with districts of densely packed houses oriented to the south and fabricated mostly from locally quarried stone and mud brick. Streets and public plazas were enlivened with fountains. The Italian/Roman pattern of interspersing shops in the main-street facades of houses is not common in any of these ten cities, so far as their residential districts have been excavated to date. Most of these cities were notable for both their natural setting and the beauty of their urban arrangements. Only Ephesus and Miletus relied for effect on their architecture and urban planning more than on their geographical settings, yet their locations beside harbors meant that in antiquity their settings were enlivened by views of ships and water, and their economic success meant wealth to be spent embellishing the city.

Shifting shorelines in all three regions forced the harbors to build new facilities at frequent intervals, necessitating matching shifts in settlement location. Argos went through this process in Mycenaean times, long before the period studied here. Like Delphi, Argos of the Greco-Roman period depended on its coastal port. Only Syracuse has a port that still functions today. The other ports silted up in this order: Priene's near-Söke site, about 1000 B.C.E.; Agrigento, fifth century B.C.E.; Selinus, second century B.C.E.; Priene's last site, first century B.C.E.; Corinth in Roman and Byzantine times, fourth–eighth century C.E.; Ephesus and Miletus, eighth century C.E. Only Morgantina and Delphi are completely inland, dependent on roads to connect them with the world.

The 10 sites have similarities and differences in stratification and materials. All four Sicilian sites have clays and calcarenite. Morgantina adds cemented sandstone and flysch, Selinus adds clayey sand, and Syracuse adds silt and basalts. In Greece, all three sites have karstified limestones. Argos adds alluvium, Corinth adds conglomerate, sands, and sandstone (poros), and Delphi adds flysch, siltstone, mudstone, sandstone, breccia, red shale, and schists. Because of folding, Delphi's geology is the most complicated of our ten cases. Of the Anatolian sites, all three have karst, with both Ephesus and Priene being marble based, whereas Miletus is on clayey Neogene limestone bedded with sandstone and tuff. Conglomerate, varieties of quartz, gneiss, and micas, serpentine, and hornblende enrich the stratigraphy at Ephesus. At Priene the mountain's core is gneiss with coverings of mica schist, limestone, and marble. Thus the stratigraphy varies, though all have karst or karstlike processes.

The effects of catastrophes on the cities suggest that I was overly optimistic (Crouch 1993) in claiming that the Greeks understood how to maintain their environment for future generations. This may have been true for small towns like Morgantina, but even at Morgantina the new geological studies link the city's demise to the lack of water, which was probably worsened by deforestation. We may come to realize that the greatest threat to human survival is not lack of food but lack of fuel. Will we realize this in time to harness the sun, wind, and tides for our energy needs? As Mumford (1961) pointed out, these sources are less efficient but free and infinitely reusable.

What happened at these ten places, however, is not only a matter of human choice but also of implacable natural processes and events. The true human history of these sites and of our own is a comedy if the residents learn about the reality of their physical setting in time to do something about it, but a tragedy if they do not.

7

General Conclusions

When we began this study, we hypothesized the following: (1) similar geology fosters similar urban development, but not every urban difference has a geological basis; (2) geological differences are likely to result in developmental differences; and (3) geology forms an active backdrop to human actions. We assumed that (1) the growth of classical cities was in response to the setting with which the inhabitants interact, (2) tectonic, sedimentary, structural, and engineering aspects of the geology affect how the city developed and its durability, and (3) hydrogeology defines the nature, amount, and location of water resources.

At the end of the study, we have verified that implacable geological forces affected each city. Specific interactions between the local terrane and the residents of these ten sites during the Greco-Roman period can be documented. We have struggled with problems of scale and interpretation, and with which methodologies to use in weighing evidence. We have wondered which links between phenomena and process are the crucial ones. In our research, we have used both the optimism and the informed skepticism that Dincauze (1987) called for in her parallel work of reconstructing paleoenvironments. Temptations in research can be subtle, such as "equating what is observable with what is significant" (Snodgrass 1987). We do not claim to be exempt from these problems.

New information does not of itself produce better science or better human history. The contributions of oversimplification to the scientific discourse continued unabated in the last half of the twentieth century, with Wittvogel's *Oriental Despotism* (1957), which based urban development and the growth of the state on irrigation, or Vita-Finzi's concept (1969) of one period of sedimentation for the whole Mediterranean area. The attentive reader may have noticed similar examples of arguing ahead of the data in this present work, in spite of our efforts to be moderate and truthful.

We apply scientific and engineering data to the sites in question, and we offer the results as models for investigation rather than exhaustive studies. This study has teetered between the modest samples of archaeology and the definitive mathematics of computer modeling, attempting to integrate the tangible, semitangible, and intangible into a vivid story told in historical language. Even with fuzzy data sets and imperfect theories, we have created models of how these ancient cities functioned in their geological settings. "Crude and rough estimates are much better than pure guessing and idle speculation" (Hassan 1978: 88). Indeed, "incompleteness, ambiguity, and complexity" (Snodgrass 1987: 41) are inescapable in historical investigation.

Findings to Date

Human Decisions about Location, Resources, and Vistas

Geological elements form both the landscape and the building materials of a site, and determine its natural beauty. All ten sites have been susceptible to human alteration, the enhancement as well as the exploitation of the environment. Builders had to weigh the natural advantages (see Table 7.1) of each site against both natural and human deficits. Landslides and earthquakes as at Agrigento are the most dramatic problems, but long-term changes in the relationship with the local resource base as at Argos, Morgantina, and Priene are just as implacable. Because long-range outcomes are still difficult to foresee today, we can only imagine the difficulties of forecasting in a nonscientific society. Still, the lively oral traditions of these cultures mitigated against the total ignorance of long-range consequences.

For much of the 1400 years we are studying, there was balance between the needs of the population and the productivity of the city-state, until a drastic change in the political system and the division of the eastern and southern Mediterranean from the northern and western Mediterranean coincided with deforestation, local climate variations (Fig. 7.1), silting up of the bays, and the scourge of malaria mosquitoes to reduce splendid late antique cities (Miletus, Ephesus, Argos, Corinth, and Agrigento) to small medieval and postmedieval villages. Some sites were abandoned completely (Morgantina, Selinus). Others suffered from earthquakes that recurred even in the twentieth century (Corinth, Delphi). Urban populations continued most successfully at Syracuse because of its great harbor, at Agrigento because of its defensible hill and fertile plains, and at Ephesus-Seljuk, which continually reinvented itself.

We understand the effects of people on land in more detail now than we did in the nineteenth century. More and higher-quality publications in the last 20 years have dealt with relationships between humans and their physical setting—most often, what people do to the land, not vice versa. The changes in thought patterns about our environment are mirrored in the topics of research, beginning with Hutton's descriptions of the earth's stones (1788), continuing through the nineteenth century with the controversies over the age of the earth, rising to a high point with Semple's masterful synthesis of geography and history in the Mediterranean Semple (1931), and continuing in the twentieth century with innovative analysis by Doxiadis (1972), Ambraseys (1978), Rapp and Gifford (1985), the specific geoarchaeology of Zangger (1993), and the expansive history of Hughes (revised

TABLE 7.1 Human Decisions about Settlement Location

Site	Locations	Dates Flourished	Resources	Decision	Vistas
Agrigento	Hilltop, ridge with slope between	6th c. B.C.E.– 6th c. C.E.	Grain, sulfur	Lack of competition; rich soil	Scenic, architectural
Argos	Valley edge	8th c. B.C.E.– ca. 400 C.E. +	Wood	Trade route, water, soil	Scenic
Corinth	Terraces[a]	750 B.C.E.– ca. 550 C.E.	Pottery, shipping, government	Trade routes	Scenic, architectural
Delphi	Terraces	9th c. B.C.E.– 1st c. B.C.E.	Oracle	Oracle, neutrality	Scenic
Ephesus	Coastal hills plain	mid 7th c. B.C.E.– end 4th c. C.E.	Stone, shipping, government	Trade, rich hinterland	Architectural (some scenic)
Miletus	Coastal hills	8th c. B.C.E.–present	Iron, furniture, wool, philosophy	Ports	Architectural
Morgantina	Hilltop ridge	5th–1st c. B.C.E.	Ceramics, grain, etc.	Route dominance	Scenic
Priene	Terrace	6th c. B.C.E.– 4th c. C.E.	Wood, government	Access	Scenic, architectural
Selinus	3 ridges at coast	7–5th c. B.C.E.	Grain	Lack of competition; rich soil	Scenic, architectural
Syracuse	Island & coastal plain	6th–end of 3rd c. B.C.E.	Shipping, culture	Ports, trade	Scenic, architectural

a. Terraces began to be built in the eastern Mediterranean between 1200 and 900 B.C.E. according to Wertime (1983: 446).

1994). Significantly, the twentieth century concluded with interdisciplinary studies by the archaeologist Higgins and geologist Higgins (1996) and by Kraft, Brückner, and Kayan (two geographers and a sedimentologist; 1999). Humility before the wonders of our planet now fosters utilization of the most sophisticated sampling device for this research—a team of scholars from different fields of expertise.

Geology Determines City Form

The geology of each site affects the topography (and thus the placement of buildings), the water supply and drainage, and the availability of materials for construction (Table 7.1). The problems associated with building on limestone at Syracuse are quite different from those of expanding onto new sedimentary land at Ephesus. Where builders had marble, at Priene, their buildings are lighter structurally and

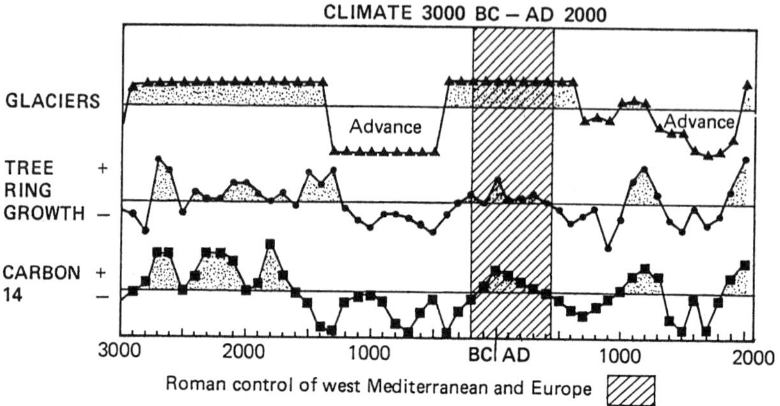

FIGURE 7.1. Diagram of climate 300 B.C.E.–2000 C.E., comparing the advance and retreat of European and North American glaciers, the rate of growth of bristlecone pines, and the fluctuating levels of C^{14} in the atmosphere. The trees and glaciers have matched phases, whereas the C^{14} reflects solar radiation and hence climate warming. A warm period coincided with the flourishing of the Roman Empire, 200 B.C.E.–400 C.E. (MacKenzie 1972).

more elegant aesthetically than those made of calcarenite at Selinus. Managing the ample water from rain and springs at Delphi was quite a different challenge from managing ample river water at Miletus. Quarrying, port development, and irrigation and drainage systems, as well as buildings, altered the landscape.

Changing political circumstances, such as those at Priene and Morgantina, were embodied in the physical form of these two ancient cities. The very pattern of the streets and fabric of the buildings was different in the seventh century B.C.E. compared with the fourth or second century. Similarly, the social organization of the polis and its distance from centers of power or integration into the empire are evident in the surviving structures.

Greco-Roman culture had assigned functions to particular buildings and spaces. The proper amount of governmental and personal income to allocate to construction was one of the issues local and imperial governments considered. We would have liked to study not only the building materials but also the energy costs of construction, using the methods demonstrated by Abrams (1994), who analyzed the cost of construction in terms of man-hours per each type of building task and then compared relative man-days per building type. This work is beginning to be done by others, but the building-type applications of his methods have not been used for classical archaeology (but see Kuznetsov 1999 and his bibliography).

Resource Management

In the management of resources, we see the confrontation of natural constraints with human needs and practices. Water, which is essential to life, is one of the most obvious limits on human plans. Karst, the common geology of these sites, offers visible and hidden water to settlements, making year-round residence feasible. These sites need intense research by karst geologists. Karst interrupts the urban fabric, as in the collapse of the Lerna bath at Corinth, provides ample but mysterious water through underground channels at Argos, and supplies numerous fountains in all these cities. Most sites have ample evidence of karst, but even Morgantina's sandstone and clay strata exhibit karstlike behavior. All sites had springs with enough drinking water for initial settlement, and all found additional springs close and copious enough to supply increased populations. The only exception may have been Morgantina, where the hinterland was not rich in surface water, a lack of which contributed to the town's demise. Providing water systems was a positive political act, strengthening regimes in the sixth century B.C.E. and again characterizing Roman rule. Drains have been found at all sites. Advanced technology and wealth permitted more fountains and other amenities in the eastern Roman cities than in the western cities (Crouch 1993).

The severe drought that affected Athens and Attica in the last third or quarter of the fourth century B.C.E. (Camp 1982) is a possible example of the political effects of water. Little has previously been made of the ramifications of this drought on political activity, colonization, and foreign conquest during this time period, although the drought stimulated political changes that directly influenced the form and function of Priene and again stimulated Greek colonization in southern Sicily in the fourth century. The form of a city was often determined by its relationship to a river or the sea. Selinus and its two rivers interacted well for less than 300 years, but then the rivers imposed their will and the city faded away.

Buildings fabricated from locally quarried stone and finished with local terracotta roof tiles, with some wood for beams, shutters, and rafters, reiterated the natural setting through human manipulation of the elements. Imported stone articulated major monuments at the richest cities of Ephesus, Miletus, Delphi, and Syracuse. Quarry development paralleled urban history, as at Selinus. The rural road network was directly tied to urban streets and monuments. At the eastern end of the Mediterranean, the quarries of Ephesus were mined for stone not only for local structures but also for contributions to the international trade in decorative stone. In the third millennium B.C.E., the first known exploitation of stone was already taking place at Ephesus, though we see no evidence of quarrying at Miletus until the eighth century B.C.E. In the seventh century B.C.E., Argos, Ephesus, Selinus, and Syracuse began to utilize their stone; in the sixth, Agrigento, Corinth, and Delphi; in the fifth, Morgantina; and Priene was built out of stone from quarries adjacent to its last site, in the fourth century and later. Normans and Spaniards in Sicily and Turks in Ionia continued to mine the same quarries for their own buildings after 1100 C.E.

The simplest types of construction, adobe brick and piled up slabs of stone, are also the most easily destroyed by weathering and earth movements. Gravity-based Greek construction in stone was resistant to weather, but still subject to collapse during earth movements. Its custom design features required an educated workforce. Roman concrete buildings had two advantages: They were more durable in earthquakes, and they could be constructed by mass production, which is consistent with the Roman social organization of uneducated workers from many ethnicities and languages brought together as slaves. Embedded in the stones or mud brick disinterred by the archaeologists are engineering solutions to the geological challenges of any ancient site. We must work backward from the answers to the questions, from what they accomplished to how they went about it and what they had to know to do it (Garbrecht 1987: 97–110; Ortloff 1989, 1993; Ortloff and Crouch 2001).

Clay was used raw for mortar and flooring material. It was fired for terra-cotta roof tiles, pipes, gutters, ceramic dishes, vases, and other containers. Centers of ceramic and terra-cotta production were Agrigento, Morgantina, Selinus, Syracuse, Corinth, and Miletus for local use, and Syracuse, Corinth, and Miletus for export. At Agrigento, Morgantina, Selinus, Syracuse, Corinth, and Miletus, clay was important to both the setting and the development of the city. The sites of Priene, Ephesus, Delphi, and Argos have not been investigated for their clay geology.

All of these cities display a standard of living generally not to be equaled until the twentieth century. The great temples of the Greco-Roman world were made of permanent materials and were often rebuilt to satisfy changing religious behavior, some ultimately being converted into churches for the new religion of Christianity. The Greco-Roman courtyard house has still not completed its life span, and the Roman invention of the apartment house shows no abatement of popularity. How these cities allocated resources to subsistence, export, or display can be recovered, although not easily, from the documentary and archaeological record. The three sites in Sicily relied on shipments of grain and other agricultural products for their wealth; two in Sicily, two in Greece, and two or all three in Anatolia relied on trade and transshipment of architectural elements, sculptures, and especially high-value ceramics from Corinth, Miletus, and Syracuse. Minerals were important exports at Agrigento and Miletus. Charcoal was probably an important export at Morgantina, Argos, and Priene, whereas forest resources ranged from enough for local consumption to exportable surpluses in the case of all Sicilian cities and the two large Anatolian cities. Government was an important industry at Corinth and Ephesus; religion drew participants and traders from wide regions, especially at Delphi but also at Corinth and Ephesus and somewhat less at Miletus and Agrigento.

Human History and Geological Processes

The history of each site has been disrupted by geological processes and events. All have been affected by earthquakes, landslides, and/or sedimentation. Quakes sometimes occurred simultaneously or sequentially at scattered sites (within one decade or quarter century): 17 C.E. at Selinus, Ephesus, and Miletus; 351–365 at Miletus and Selinus; 375 at Corinth; 551 at Corinth and Delphi; 1965 at Delphi; and 1968 at Selinus. The other quakes seem unrelated in time. There were quakes at three of the Sicilian sites (one in modern times only), at two of the three Greek sites during the study period, and at all three Anatolian sites. Corinth and Delphi have recorded the most earthquakes. Nine landslides are known, all but three at Delphi; those three are in Sicily—two at Agrigento and one at Morgantina. This is probably not a complete list.

Severe storms and climate fluctuations (Fig. 7.1) undoubtedly made life difficult also, but we have too little information about them to factor them in properly.[1] Storms likely increased the effects of sedimentation, from the modest sand deposit below the Great Steps at Morgantina to the three consecutive physical removals of the entire city of Priene necessitated in less than 500 years by the silting up of the bay. The sites most affected by sedimentation follow: Agrigento (river port silted up during the sixth–fifth centuries); Corinth (serious siltation in Byzantine era, especially at the northern Lechaeon port fed by Corinth's small river); Argos and Selinus (Mycenaean and Hellenistic silting, respectively); Syracuse (affecting mainly the isthmus, in our era); and Ephesus and Miletus (severe silting of river plains, leaving them stranded after the seventh century C.E.). The Anatolian coastline changed by as much as 30 km in 3000 years because of erosion and deposition.

Long settlement by a particular culture brings about construction of the religious buildings, cafes, plazas, workshops, and houses that are normal for that group, and these are arranged in patterns considered convenient by the people. These patterns perhaps yield to the forces of nature. By developing thoughtful attention to the physical environment, cities such as Syracuse persisted for over 2700 years. The political pattern of oligarchy, democracy, regional despot, or governor imposed by Rome could not operate independently from the natural setting.

8

Physical and Intellectual Issues

Issues at Two Levels

The issues associated with this study are both physical and intellectual, as are the factors in urban development. The correlation of known data from inscriptions (epigraphy), literary references (ancient history and more recent government and church documents), evidence from destroyed and rebuilt buildings (archaeology), and modern scientific and technical findings (several kinds of engineering and subfields of geology such as seismology and sedimentology) can give a more complete picture of each city's development than does one kind of information alone. Most of the necessary site-specific studies, however, have not been done. Our problem parallels the study of global warming, where precise records of weather events have been kept for less than 200 years.

Urban elements must be studied by experts in that building type and in social expression. Ramparts need more study by historians of warfare, theaters by scholars of drama and literature, stadia by those who study the history of sports, plumbing by hydraulic, civil, and fluids engineers, and temples by historians of comparative religion. Insights into institutional and political aspects of ancient studies and the historiography of all the disciplines involved in ancient studies would be both useful and fascinating (Kardulias 1994). The benefits and difficulties of interdisciplinary research are clearer now to us than when we started.

In considering the physical setting and geological processes in the Mediterranean area, is description sufficient or should scholars strive for explanation, even if this involves theory building? "In much of art history and classical archaeology traditional practices have continued without explicit theoretical support" (McNally 1985; but cf. Preziosi 1989). The theorists of archaeology and urban history desire comprehensive and precise theories—even in the absence of enough data to make that possible. Some data may be rescued by followers of one discipline after being ignored or thrown out by followers of another. The awareness of theoretical difficulty is part of the increase in consciousness typical of the second half of the twentieth century when we began to question the nature of both knowledge and culture. When forced by the data to conceptualize and even to theorize, we have done so, and we have tried to honestly label the differences between facts and concepts.[1] We have also contributed our personal experiences in the field as evidence. It will take more than scattered individual effort, however, to build the necessary intellectual bridges between geology and history, geology and the social sciences, geology and the history of architecture, geology and ecology, and geology and hydraulic engineering.

Physical Issues

To find the meaning of what is said, done, or made, you must know the question to which it is the answer (Collingwood 1939:31). We have had to work backward from the answers discovered in the field to the questions posed by ancient settlements, from what the Greeks and Romans accomplished in urban development to how they went about it and what they had to know to do it. The entire urban fabric displayed the value system of the residents. Ancient building traditions derived from and molded the physical manifestation of the community, embodying the social values of the people in the structures of the city, their understanding of the physical properties of the stone, and their appreciation of water. Particular materials controlled the mechanics of construction and determined the visual aspects of each place.

Can we extrapolate modern data neatly to the past? This was an unrecognized problem when scientists first began to examine the ancient sites. Such interpolation requires an understanding of the physical processes and an analysis of the rate of change of how those processes affected the site, factoring in both short-term events and long-term changes. Modern geological data must be used cautiously, for it is an approximation to, not a precise description of, the ancient situation. Two thousand years of erosion, robbing of stones for other uses, weathering, climatic variation if not outright change, and socio-politico-economic differences have altered the site. The geologist will have difficulty calibrating all these influences to arrive at quasi-precise knowledge of the geological setting 2000 years ago. Although we are unable to see a true likeness of past circumstance, our imprecise understanding is nevertheless likely to be close to the truth.

We have had to leave for others the complete study of timescales. At first glance, the rapid pace of ephemeral human life and the slower-than-glacial pace of geological change seem incommensurable. Yet synchronizing human and geological timescales is difficult, not impossible. In some places, the convergence of geologic evidence with that from archaeology and literature makes possible a unified chronology, as at Delphi. The collaborative work of geologists and archaeologists can produce chronologies for other sites that are increasingly accurately calibrated (Appendix A).

Intellectual Issues

As background to the research, we have looked at the historiography of the three main contributing disciplines: archaeology, which we have traced from antiquarianism to scientific anthropological archaeology; geology, which grew from a descriptive to historical to analytical to humanistic discipline; and the newest, urban history, which is already widening to include scientific and engineering contributions.

Integration of the history of science and technology into the humanistic disciplines is essential. Human culture and even survival depended on the invention of cities, and cities depended on careful resource management, which in turn depended on a profound understanding of the geography of urban location and on clever management of resources by technological inventions, such as cisterns and aqueducts, and technological improvements such as the shift from fuel-expensive bronze to fuel-efficient iron. Human culture includes not only religion, language, and marriage customs, but also water management, farming practices, long-distance trade, and building construction.

Science and technology, the study and control of the environment, are squarely at the center of human history—not peripheral. Before 1850, science and theology were united in one discipline (natural history) and usually in one scholar. The social distress caused by the Genesis controversy was a necessary by-product of the separation of science and theology. It became safer and more remunerative for geologists to concentrate on oil and mineral reserves, the history of glaciation, and other aspects of geology that seemed benign rather than threatening to the received wisdom of the nineteenth century. After 150 years, theology-free geologists should now be free to examine issues such as urban location and other humanistic topics, even if that research bears cultural or emotional content.

The Greco-Roman world was more urbanized than any other society before the modern era. For modern investigators, Greco-Roman sites have the advantage of long duration and depth of study. The town-country symbiosis was the basic governmental and social unit, as the Greeks and Romans themselves understood. Data from individual cities makes possible an understanding of the impact of the geological setting on human life.

Human response relates only to what is known in a situation, but a significant set of factors remain unknown (Van der Leeuw 1987: 228). Geological or engineering study in particular had previously been limited at most of the sites. Further geological study will make additional contributions to human welfare as it illuminates the physical constraints on settlement, points out human adaptations and inventions regarding those constraints, and raises human consciousness about the environment.

Tangibility of Site Assets

Decisions about urban location follow the human pattern of mixing the tangible and the intangible (Table 8.1). For survival, we must recognize positive environmental patterns. When there is a minor deviation from what we expect, we have pleasure; whereas a major deviation gives us negative feelings. Penning-Rousell and Lowenthal (1986: 4, 9, 11, 17–18) assert that the choice of habitat influences survival, and therefore strong selection of certain behavioral choices meshed with

TABLE 8.1 Tangibility of Site Assets

Site	Tangible Asset	Intangible Asset
Argos	Water, renewed soil	Trade route[a]
Corinth	2 ports and dilkos, land and water	Trade routes
Delphi	Water, gas for oracle	Neutrality, oracle
Ephesus	Rich hinterland, river access, renewed soil	Emperor cult, trade routes
Miletus	Iron, rich hinterland, river access, renewed soil	Trade, oracle at Didyma, intellectual life
Priene	River/bay access, forest resources	River valley dominance
Syracuse	Bays for shipping, clay/ceramics, water	Intellectual life, trade routes

a. Trade might be classified alternatively as a semitangible aspect, because trade goods are certainly tangible, whereas the process is less so.

certain environments for many millennia: "Good habitats, as measured by the features that contribute to survival and reproductive success should evoke strong positive responses while poorer habitats should evoke weaker or negative responses." They then discuss the evolution of the sense of beauty, because beauty is associated with some landscape features as a sign of excellence or value.[2]

The study of ancient cities is still producing a mosaic (where each data set is but one element of the whole) not a hologram (where each fragment contains the whole), but because of interdisciplinary work, the picture is more detailed and accurate than in the past. With each contribution to the mosaic, our story of the site is more precise. At a site like Ephesus or Delphi, enough information is available to begin a true integration of geological history with construction history and social history.

This work is not the full multidisciplinary investigation that these sites deserve, but an introduction to what could be learned. We hope that our research induces other scholars to apply geological insights to other groups of cities, such as those that ring the Swiss Alps or the group of Caribbean ports. Infusion of specific geological information can make more exact, specific, and plausible the urban history and archaeology of each site. We are keenly aware that we have not dealt with all the useful and interesting questions along the cusp between geology and human settlement. Nor have we been able to exhaust the richness of any one question. Rather, we offer questions to stimulate further scholarship. *Geology and Settlement* remains an introductory study.

APPENDICES

Appendix A: Chronologies

Dating order is a passion with *historians* as a skeleton on which to hang further layers of detail and meaning. This does not mean that every date given in these tables is equally valid or is even the same genre as another date. Archaeological and historical dates are *"fuzzy sets"* but not in the same way (Millett 1981: 527, Fig. 32.1). Nor are either of these dates conceptually the same as geological dates. *Archaeological dates* tend to have generalized names (*middle Helladic, late Bronze Age, Roman Imperial* times), whereas *historical dates* of events are nominally more precise as to year (the reign of Tiberius 14–37 C.E.). *Geological dates* relying on *stratigraphy* and *carbon dating*, especially of *fossils*, introduce vast *timescales* to compound the problem. Archaeological dating relies on archaeological findings from *site studies. Historical dating* incorporates *archaeological findings* but relies especially on *documents*. Lately, both *archaeological* and *historical dating* incorporate *scientific data*. All three disciplines are looking for *patterns* that characterize the periods in question. More precisely, subdivided geological timescales within the last 10,000 years would be of inestimable value to the historian who attempts to use geological information in correlation with the events of *human history*.

In this appendix, dates are presented for each site. Names and dates of historical periods are as follows:

Prehistoric: through third millennium B.C.E.

Bronze Age: 2900–1100 B.C.E., especially the second millennium B.C.E.

Dark Ages: 10–8th c. B.C.E.

Archaic: 7–6th c. B.C.E.

Classical Greek/*Early Roman*: 5–4th c. B.C.E.

Hellenistic Greek/Roman Republic: 3rd–1st c. B.C.E.

Roman Empire: 1st c. C.E. through 476 C.E.; thereafter, the *Eastern Roman empire* continued until 1453

Late Roman/Early Christian: ca. 350–550 C.E.

Byzantine: 6th through mid-15th c. C.E.

Medieval: 7th–15th c. C.E.

Modern period: 16th c. on

The two most detailed chronologies presented here are those of *Delphi* and *Ephesus*. At Delphi there is remarkably good correlation of historical and geological events, especially *earthquakes* and *landslides*. At *Ephesus* the painstaking work of the archaeologists to date their findings makes possible a more precise and ample list. *Chronologies* for the other eight sites are given in abbreviated form, which calls attention to the relationship between geological processes and events and the historical periods and dates of construction in the cities. Numbers in parentheses refer to archaeological maps of each site.

Agrigento

Mid-6th c.	Port at confluence of Akragas and Hypsas Rivers silting up.
2nd half of 6th c.	Stone buildings at Rupestrian sanctuary.
6th c. B.C.E.	Asklepios Temple sited in plain, at spring. Extramural sanctuaries near river port; first use of W summit (2).
6th c.-Roman era	Sicilian wheat an important export.
480–475	Water system, drains built by Phaiax (3).
480–406	Seven temples built.
5th c.	Completion of nine-gate ramparts; river walls link Gates 5 and 7.
End of 4th/beg, 3rd c.	Two temples remodeled, one built.
3rd and 2nd c.	New construction: public buildings, first peristyle houses in Sicily; sewer of lower agora.
200 B.C.E.–500 C.E. +	Port for shipping wheat, the basis of prosperity.
By 14 C.E.	Water: 13 springs, domestic cisterns, wells, and large drains.
2nd c.–1st quarter of 4th c.	Sulfur industry corroded stone; declined after 350 C.E.
3–4th c.	Fragapane catacomb.
5th century C.E.	Spring under early Christian church of Santa Lucia.
7th c. on	Population retreated to W summit for defense.
11th c. on	Cave reservoir below castle, attached to ancient Purgatorio passage system.
18th c.	Corrosion of monuments; conservation necessary.
2nd quarter 18th c.	Rural waterline extant (5).
Late 19th c.	Railroad to Girgenti set along possible EW fault.
1920s and 1950s	First excavations of the port(s).
1966	Landslide in Addolorata (Arab) quarter on W. Girgenti.
1976	Landslide under temple SE of lower ridge, along Greek rampart.
By 1980	Early Christian church of Sta. Lucia collapsed.
ca. 1995	Spelelogical investigations of passages.

Sources: Trasatti and Micceché personal communications; Pindar, Scholia XIII, 32, Pancrazi 1751–52.

Argos

50,000 years B.P.	Earliest occupation of Kephalari cave S of Argos.
10,000–6000 B.P.	Coastal settlements, later covered by 80-m sea rise.
Early 3rd millennium	Lerna Lake reached the outskirts of Argos (1).
	Springs along E base of Larissa (2) attracted settlers.
2000–1000 B.C.E.	Sea receded, changed the shoreline. N alluvial plain farmed; marshy S coastal plain used for grazing. Channel above Lerna Lake controlled floods, supplied dry years (3).
1400–1200	Possible earthquakes (2). Unstable political situation.
ca. 1000	Earthquake (?), possible volcanic action, tidal waves damaged eastern Argive plain (2).
Since ca. 1000	Delta formation along Inachos River SW of Tiryns (1).
ca. 8th c.	Wells dug in alluvial plain to reach water table.
7th c.	Quarries: first stone construction
5th c. (?)	River walls built to control flooding and erosion.
	Earthquake destruction wake across the Agora (7).
5th–4th c.	Clearing steep hillsides for olives led to erosion.
	Harbor structures built at river mouth.
3rd c.	Stone sewer in agora.
3rd–1st centuries	Erosion from overgrazing and lack of terrace maintenance (8).
	Shoreline depressed in harbor area.
Roman era	5 m of alluvium deposited in river valley above Argos.
	Roman wells dewatered the urban site.
ca. 120–130 C.E.	Hadrianic aqueducts from S and NW tapped karst springs.
552 and later 6th c.	Earthquake, epidemics, Slavic invasions.
7th or 8th c.	Church of Virgin Mary replaced Hera Temple on Larissa terrace, both utilizing karst spring in cave.
Later Middle Ages	Turkish rulers settled Albanian shepherds here to manage transhumant grazing in mountains and plains.
20th c.	Lerna Lake mostly filled in. Shipping structures (at river mouth in 5th c. B.C.E.) now visible below sea 100 m S of river mouth.

Sources: Zangger 1991, 1993, 1994; Scarborough 1994; Knauss 1996; Pariente, private communication; Van Andel and Zangger 1990a & b.

Corinth

10–5 M B.P.	Gulf of Corinth graben produced by activity along parallel faults.
Since 450,000 B.P.	19 marine transgressions made terraces along the graben.
235,000 B.P.	Old Corinth terrace.
124,000 B.P.	New Corinth terrace.
Bronze, Early Iron Ages	Villages on two central terraces.
10th c. B.C.E.	First version of Sacred Spring shrine.
750–700 on	Corinth the chief Greek trading city, ports at Kenchreai and Lechaeon
End of 8th c.	Peirene spring made into fountain; wells already in use.

after 700	Stone cutting industry
Archaic period	Quarries developed within site and at 1 mile and 4 miles E of agora.
7th c.	Sacred Spring in use.
Mid-3rd quarter 7th c.	Defensive walls around Potters' Quarter. Aphrodite sanctuary at Acrocorinth.
6th c.	Diolkos begun, to drag boats over isthmus. Spring W of Lerna formalized.
By 575 B.C.E.	Corinth ceased to dominate pottery trade; clay exhausted?
5th c.	Lechaeon the principal harbor of Corinth.
2nd quarter 5th c.	Increased water supply and improved drainage.
5th/4th c.	Greek bath W of Lerna. Diolkos rebuilt.
338	Lechaeon the operation base of Macedonian fleet.
77 C.E.	Earthquake.
Early Roman era	Acrocorinth quarried for pink breccia. Roman fill raised level of agora to build forum.
Roman era	Diolkos rebuilt.
mid 1st c.	Nero proposes Isthmus canal, begins work (finished in 19th c. C.E.).
2nd c	Aqueduct built by Hadrian from Tripoli plateau.
267	Kenchreai still active port. At Lechaeum, dredging of basins.
375	Earthquake destruction of Lerna bath, etc.
395	Upper Peirene filled in.
522 and 551	Earthquakes; ramparts and Temple C destroyed.
Last 300 yrs.	Gulf of Corinth most active graben on earth, at 11 mm/yr.
1858	Earthquake; Ancient Corinth destroyed.
late 19th c.	Isthmus canal finished.
1928	Earthquake damaged New Corinth.

Sources: Williams all listed papers; Armijo et al. 1996; D. G. Romano 1993.

Delphi

1600 B.C.E.	Minoan and Mycenaean material.
12–11th c.	Mycenaean settlement.
ca. 1000	Landslide wiped out lower villager.
10th c.	Houses E of Apollo site.
875/860	Rocky Pytho settlement. Sacred stones: Omphalos (world's navel) and Stone of Kronos, from early rockfall.
	Shrine of the Earth Mother Ge between buried Mycenaean houses and Apollo site; spring or seep still visible.
8th c.	Expansion of trade routes.
730	Earthquake; settlement abandoned. Cult of Pythian Apollo, with initial laurel bough shrine.
3rd quarter of 8th c.	Delphic oracle and divination rites recorded.
7th c.	Archaic racing area for Pythian Games located in plain near port (2).
	First wooden Temple of Apollo. Kastallia Fountain at downhill site. Earliest treasury, at beginning of Sacred Way.

ca. 675–650 B.C.E.	Wooden Apollo shrine replaced by first (Corinthian) stone temple.
End of 7th c.	Shrine or early temple to Athena on Marmaria terrace.
600	Earthquake, one of the strongest here.
596–86	First Sacred War, followed by new construction.
6th/5th c.	Quarries opened at Livadi in northern doline area.
6th c.	Terracing and early treasury buildings.
550–535	Stadium fountain (Doric) at upper W corner of sanctuary. Kerna fountain fed from adjacent spring.
548	Apollo Temple burnt.
530s	Apollo Temple rebuilt of tuff or tufa from Corinth; W and N walls of sanctuary built. Polygonal wall below Apollo Temple.
Last third of 6th c.	Kassotis Spring in the oldest, uphill location. Treasuries added to Sacred Way.
by 510–500	Temple of Apollo rebuilt with gold statue of the god.
By 5th c.	Colonization oracles come to an end.
5th/4th c.	Oracle changes: "Pythian frenzy" recorded.
480+	Earthquake of magnitude 6.3. Persian raid aborted. Avalanches required rebuilding and repairing. Stoa of the Athenians erected.
about 400	South Stoa.
373	Earthquake of magnitude 7, with landslides, rockfalls; destroyed temples and terrace walls. Kassotis Fountain now nearer Apollo Temple.
380–370	Circular tholos of pentellic marble built on Marmaria terrace.
356	Repairs begun in Apollo area; polygonal wall (Ischegaon).
348	Earthquake of magnitude 6.7 destroys sanctuary.
	Kassotis spring again moved closer to Apollo Temple.
340s	Stone bought from Corinth for new Apollo Temple.
330s	Wood bought for ceiling beams for Apollo Temple.
4th c.	Treasury of the Cyrenians, the last built.
	Lower Kastallia spring enclosed in orthostates.
334–279	West Portico built.
330–300	Gymnasium on the Marmaria terrace.
279	Avalanche and earthquake of magnitude 6.8 repulsed Gauls.
	Quake increased flow of Stadium fountain. Kerna fountain overflow led to Kassotis fountain.
	Sanctuary of Apollo enlarged. Theater built.
Hellenistic/Early Roman	Exploitation of quarry at Prophet Elias.
mid 3rd c. B.C.E.	West Portico (later a cistern) renovated by Attalos I of Pergamon.
	Kings of Pergamon completed the theater, rebuilt altar of Apollo.
210–100	Kassotis Fountain waterless — result of earthquake?
2nd c. B.C.E./2nd c. C.E.	Archaic Kastallia fountain supplemented by, then replaced by Roman Kastallia fountain.
88 B.C.E.	Temple of Apollo burnt.
60s C.E.	Nero looted Delphi of 500 bronze statues.
Late 1st c. B.C.E.	Delphi now a "museum city."

88	Temple of Apollo destroyed by fire set by invaders from Thrace. Kassotis spring and reservoir now closest to Apollo Temple, flowing abundantly.
ca. 125 C.E.	Temple of Apollo rebuilt. Water from western source brought to fountain at theater; travertine deposits produced.
2nd c. B.C.E.–5th C.E.	Town occupied sanctuary area and slope below it. Gymnasium rebuilt, hot baths added.
2nd c.	Stadium renovated by Herodes Atticus. Baths and other secular buildings built in sanctuary.
4th c.	Constantine and Theodosius plundered Delphi to adorn Constantinople.
394	Imperial edict shut down the Pythian Games.
6th c.	Apollo Temple succumbed to fire, lack of maintenance.
5–6th c.	Sacred Way repaved with old inscriptions. Great houses of the Imperial Roman period renovated. New mosaic of the late sixth century C.E. Three Christian basilicas.
551	Earthquake of magnitude 7.2.
ca. 600	Final conflagration of Apollo Temple, during Slavic raid.
996	Earthquake of magnitude 6.8.
1402	Earthquake of magnitude 7.0.
1580	Earthquake of magnitude 6.7.
1870	Avalanche, earthquake of magnitude 6.8. Fault zone reactivated. Series of 300 quakes in 4 months, with magnitudes of 5.3–6.7.
1887	Earthquake of magnitude 6.3.
1894	Apollo Temple excavations begun.
1905	Major landslide affected Temple of Apollo. Stadium fountain stops flowing.
1939	Temple of Apollo restored.
1958–59	Kastallia spring restored.
1965	Earthquake of magnitude 6.3.
1970	Earthquake of magnitude 6.2.

Sources: Pausanias X.5.9, X.9.1, and X.10.5; Mulliez, personal communication; Childs 1993.

Ephesus

Oldest fault	NS, along Marnas River valley.
After 10,000 B.P.	Sea transgression, sediment accumulation. Ancient maximum: shoreline beyond Belevi, now 18 km inland
Pre-historic	Remains near Luke's Grave (14)*
3000 B.P.	River valley a bay 10 km wide extending up to 70 km into the land.
3000–1000 B.P.	Multiple distributary coastline with changing channels protruding into shallow marine bay filling with silt.
By 3rd M	Stone exploited.
3rd M	Chalcolithic (Copper) and early Bronze Age settlement (156).

2nd M	Hittite settlement (Apasas) on Ayasoluk (151).
1600–1400	Mycenaean and/or Ionian trading post (151) N of Ayasoluk, then an island or peninsula amid alluvial plain and marshes.
14th c.	Mycenaean grave on S slope of Ayasoluk (153).
2nd M	Mycenaean and Minoan finds at site of later Artemis Temple.
Early 10th–9th c.	First securely dated Artemis cult votives.
10th to 8th c.	Ionian Greeks settled here, controlled coastal plain.
8th or 7th c.	Artemis sanctuary: first peripteros temple (1). Bay near SW Ayasoluk a shallow harbor.
8th/5th c. (?)	Settlement of Smyrna (?) at western agora site.
800–600 B.C.E.	Delta of Little Meander River transforming bay into a valley. Vegetation changed from oaks to scrub. Ancient springs at bases of mountains (now covered by alluvium).
Mid 7th–early 5th c.	White marble transported by water from Belevi quarry for older Artemis Temple.
750–300	Delta growth less than 1 km.
6th c.	Sacred Way in frequent use. Archaic Koressos settlement in later stadium area; harbor NNE or NW of Koressos Gate. Possible acropolis at site of later Macellum (100).
ca. 550	King Croesus begins huge stone Temple of Artemis (1).
6th–5th c.	Plaza where road to Ortygia crossed roads to Artemision and to Magnesia (52). Graves in government center area (21).
407	Earliest date for filling in of Artemis harbor; occurrence of malaria.
5th c. B.C.E.	Shrines on N side of Panayir Mountain (4).
4th c.	Marnas River still poured directly into the bay.
early 4th/end 2nd c.	Most active phase of sedimentation: 12–20 m deep between Syrie (former island) and Panayir's N edge.
4th–3rd c.	Temple of the Crack built.
356	Artemis Temple burned; replacement built 2.7 m higher because groundwater higher (1).
By 300 B.C.E.	Artemis harbor area a malarial swamp. Smyrna covered with alluvium and abandoned.
After 300	Hellenistic agora built above Smyrna ruins (61).
300–present	River carried down billions of cubic meters of alluvium and rock.
300	Lysimachus built defensive wall (11) on mountain heights to the sea. Magnesia Gate (10) part of this wall.
ca. 300 B.C.E.–ca. 700 C.E.	Panayir quarries (8 and 11) in service. Decorative stones imported and exported.
300–end of era	Hellenistic port W of Panayir Mountain (32).
ca. 300 and later	New settlement between Panayir and Bülbül Mountains. Shops, fountain(s) along Kuretes St. (36). Aqueduct (?) along Marnas River.
By ca. 200	Artemis harbor definitely filled in.
300–100	Delta growth of 5 km. Early versions of stadium (104), theater

	(75) with fountain house at front of theater. Houses on site of Serapeion court (67).
ca. 150–50	Main streets take on permanent pattern with monuments and fountains (36) (110) (39) (75) (79). Palace (76) on hill above theater.
ca. 100 B.C.E.	Roman control of area.
	Efforts to save Hellenistic harbor by dam and channel to the sea.
60	Earthquake, then first efforts to Romanize the city architecturally.
34 and later	Monuments on main street; new settlements W and NW (area of 80), with civic buildings (18–22) and port facilities W of Hellenistic town. Temples for emperior cult (30, 98). Apartments and houses along slopes (e.g., 50, 51).
Late 1st c.	Long-distance waterlines, fountains, and sewers.
ca. 29 B.C.E.	Government center (22 etc.) refurbished.
ca. 0	Memius Monument (32).
10 C.E.	Marnas aqueduct built. Artemision (1) repaired.
11 C.E.	Basilike stoa (21).
20s	Tomb of S. Pollio (near 31).
Earthquakes of 17 or 23	Slope houses (50, 51) and Tetragonos agora (61) rebuilt (possibly 28, 49).
14–37	Isis Temple (20) rededicated to Dicius Julius and Dea. Varius gymnasium-bath (16). Stoa of Damianus (7).
Before 27	Other repairs: Basilike stoa (21), Doric gate (19) to government center, stadium (104).
	Customs duty building near harbor to collect fish taxes.
Before 50	Marble Street (78) moved east and uphill;
ca. 50	Mazaeus and Mithridates gate connecting agora (61) and plaza. Monopteros for water E of this plaza.
1st c.	Medusa gate (66) W of Agora. Theater enlarged.
	Work to protect harbor: Dredging, constructing offshore islands, moles, and other harbor structures; diversion dams possibly for irrigation.
Late 1st/early 2nd c.	Harbor bath complex (92, 93, 94) and street from theater to harbor (83–87).
90–117	Serapeion (67) blocked way to Ortygia village. Embolos Street (36) paved.
First century C.E.	Four to six aqueducts built for fountains, baths.
92–93	Fountain/distribution chamber (17) at government center, fountain of C. L. Bassus.
1st c.	Fountains added to Memius monument and Tomb of S. Pollio.
Late 1st c.	Domitian Temple (30) was first Neokorate Temple.
100–4th c.	Imperial trade in marble.
Early 2nd c.	Latrine, houses at Embolos crossroads (43). Palace (76) above the theater (remodeled?).
102–114	Fountains: (15) along S road (110); of Trajan (on Kuretes St.) (36) fed by aqueduct from Tire, which also fed waterworks of Temple of Serapeion (67).

117–118	Marnas aqueduct, theater completed.
2nd c.	Theater gymnasium (79). Theater fountain house rebuilt, reservoir added on top.
119	Latrine building (43) related to Varius baths.
119–120 (117–123?)	Library of Celsus (55); Mazaeus and Mithridates Gate moved S, rebuilt at higher level as S gate of agora (61). Bouleuterion (22) enlarged.
130	Hadrianic construction: New Roman harbor with warehouses piers and gates. Street (83–87) to harbor. Shrine (41) S of Varius baths. Second Neokorate Temple (Olympeion) (98). Tridos Gate (52).
146–47	Dedication of Vedius gym (106) near stadium.
ca. 150	Earthquake. East gymnasium (12) built or repaired; Prytaneion (22), Slope Houses (50, 51).
ca. 2nd c.	Heroon (14). Parthian Monument (location unknown). Gymnasium and bath (16) E of Basilike stoa.
	Maximum size of city 280 hectares.
mid-2nd c.	Bouleuterion/Odeon widened.
169	Altar of Artemis at Tridos (52) enlarged. Gymnasium near Artemision built.
100 B.C.E.–200 C.E.	Delta growth 2 km.
By late 2nd c.	Marnas River captured by Little Meander River.
Unknown date	Paul's Grotto (71), a karst (?) cave used as early Christian chapel.
ca. 200	"Macellum" on hill opposite stadium entrance.
200–250	Corporation halls along street to stadium. Harbor Gates (87–88–89) rebuilt.
2nd–4th c.	Panayir quarries become shrines, tombs; stone shipped on Marnas River, at foot of Panayir.
Early 3rd c.	Theater plaza rebuilt with small fountain (72). Theater additions and analemma.
by 212	Tank added to fountain at (17).
211–217+	S stoa of Olympeion (95–97) rebuilt, with shrine (to Caracalla?).
262	Earthquake. Repairs to 11 major buildings: 50–51, 67, 7, 92–93–94, 41, 10, 66, 61, 55, 78, 75–104.
Early 4th c.	Agora repaired.
342–350	Repairs continue with imperial money: 92, 17, others.
4th c.	Tomb and fountain of S. Pollio given statues from unknown temple. Heroon (14) reused as "Tomb of Luke."
350–455	Frequent major earthquakes and fire.
350	Library of Celsus (55) ruined; converted to a fountain. Theater analemma (75) toppled into reservoir. North gate of Agora (61) rebuilt.
After 350	Government center (18–22) abandoned. Prytenaion (22) destroyed; statues of Artemis carefully buried.
	Kuretes St. (32–52) rebuilt.
quakes of 358–66	High waterline burst, sending water down the stair streets into the agora.
4th c.	New water arrangements: Heroon of Androklos (near 52) re-

	built as fountain; two fountains in oval atrium of Harbor Bath (92); fountain exedra (86) opposite Harbor Bath complex—all supplied by Marnas aqueduct (?).
	Late houses occupy former Xystoi area (94).
Mid-4th c. on	Imperial and other pagan shrines replaced by churches. Olympion became Christian center, its S porticoes remodeled as Byzantine buildings at 95–96–97.
	Altar of Artemis (52) in Tridos Plaza replaced by stoa with mosaic floor. Medusa Gate W of agora (66) rebuilt.
4th c.	Quarries activated S of Roman port and on Ideli Hill, W. of Bülbül Dag.
4th c. on	Malaria increased.
	New buildings in NW (96–103) built on new lands. Fill was river sediments and earthquake debris.
3rd quarter 4th c.	Plague kills one-third of urban population of the empire.
End of 4th c.	Tetragonous agora (61) rebuilt and West Gate (66) repaired and Shrine of Hadrian rebuilt by Theodosius I.
ca. 400	Retaining wall of the theater (75) repaired; theater reservoir out of use. Baths of Varius (41) and Shrine of Hadrian rebuilt.
	Destruction of the Olympion temple (98).
	Stadium addition (104) with early Christian church located in cemetery already needing repair.
400–450	Gate of Herakles (35) inserted in Kuretes St. Late antique house (102) on site of Hellenistic house.
4th–5th c.	Small church west of Basilike stoa (22).
5th c.	Shrine of Hadrian rededicated to Theodosius, new founder of city.
	Plateia (78) repaved.
Before 431	Quarry on E Panayir Dag (11) reused as Seven Sleepers cemetery with Church of St. Timothy (first bishop).
5th–6th c.	Church of St. John (152) built on Ayassoluk.
	Byzantine (or earlier) aqueduct (156) to serve Ayassoluk.
	Christian buildings (95–96–97) in former Olympion portico.
	Artemision (1) and Serapeion (67) become churches.
	Miscellaneous repairs and embellishments (83–87, 72, 84, 66).
	Fountain house (101) opposite stadium.
6th c.	E side of Tetragonos agora repaired with vaulting.
542	Plague kills one-third to one-half of the population from Wales to Central Asia (Thirgood 1981).
6th–12th c.	Ramparts (especially 77) rebuilt, some enclosing new areas.
	New Byzantine buildings: 81, 80, 85 (later a cistern for fountain opposite Harbor Bath).
	Palace above theater (76) rebuilt as banquet house.
after 650	Delta bypassed Roman port; new port at Panormos nearer coast.
	Tetragonous agora (61) no longer a population center.
Later Middle Ages	People lived around Ayassoluk, protected by the castle to N (151–154).

After 1000 C.E.	Sandy coastal barriers first noted; adjacent floodplain a swamp.
ca. 1327	Turkish conquest; Seljuk E of Ayassoluk became new focus of settlement with Isabey mosque and minaret (170).
ca 1400	Rampart enclosed Ayassoluk.
15th c. on	Ephesus landlocked; Turkish port at Alaman Gölü, NW of river; a paved way through swamps connected port with Seljuk.
18th and early 19th c.	Depopulation of area.
1890s	Formal excavation began under Austrians.
20th c.	Modern buildings of Seljuk set 2 m higher than medieval buildings.
	Quarries of Ideli Hill still in use; quarries in W of Pamucak Mountain near the sea, reactivated in modern times.
	Water table visible at the bottoms of some excavations (61, 98).
Today	Seljuk serves as market town for agricultural area.

Sources: Drews 1976; Kraft, Bruckner and Kayan 1996; Scherrer 1995, revised 2000; White 1998.

*Numbers after names of buildings refer to Ephesus map, Figure 5.14.

Miletus

During all periods	Occasional high seismic activity. Horst and graben movements, block structure between faults, rotation of adjacent coherent blocks. Fluctuations of sea level.
18,000 B.P.	Sea level 120 m lower than present day. Samos and Lade Islands connected to mainland.
6000–5500 B.P.	Sea at maximum; to Aydin (ancient Tralles), now 65 km inland.
from 5500 B.P.	Miletus a series of small islands.
4000–3000 B.P.	Wet period with cold winters fostered oak forests.
2nd half of 4th M	Miletus I: Late Chalcolithic. First settlements Miletus area.
3rd M B.C.E.	Miletus II: Early Bronze Age.
1st half of 2nd M B.C.E.	Minoan period at Miletus.
ca. 1925/1900–1759/29	Miletus III: Middle Bronze Age (MM IB–MMII)
ca. 1750/20–1490/50	Miletus IV: end of middle Bronze Age and earlier late Bronze Age (MM III/LM IA–LM IB/II),
1450–1300	Miletus V: Middle of the late Bronze Age (LH IIB–LH IIIA2)
13–11th c.	Miletus VI: End of the Bronze Age (LH IIIB–C),
1000	Bay of the sea extended E of Söke. West Market/Athena Temple area was small islands. Lion Harbor began to fill in.
8–7th c.	Miletus sent traders, miners, colonists to Black Sea.
650–550	Miletus had 75–90 colonies, first in the Black Sea area, then in Thrace.
6th c.	Residential quarters around Kalabak Hill; potteries utilized clay found 300 m south of city in Balat Formation.
500–300	Warmer period. Oak forests.
500–495 B.C.E.	City destroyed in Persian War.

5th c.	Population and site increased with enlarging delta.
412–05	Ports refurbished for Peloponnesian and Ionian wars.
Since 300	Heavy siltation began in Meander River valley.
3rd–1st c.	Settled area expanded. Six marble quarries at Bafa Sea.
3rd c.	Two gneiss quarries at Myus developed.
2nd c.	Myus engulfed by delta; residents moved to Miletus.
ca. 100	NE harbor already silting up.
60 and 23 B.C.E.	Two earthquakes.
by turn of era	Bafa Sea cut off from Latmian Gulf by alluvium; Lion Harbor half full of sediments.
1st c. C.E.	"Fountains of boiling hot water" in Meander River valley (Strabo).
17, 18, 49	Earthquakes.
150	Earthquake.
3rd quarter of 2nd c.	New construction: aqueducts, sewers, public latrines. Delta building passed peninsula at 1 mile per 250 years; filled in Lion Harbor at 0.2–0.5 meters per year.
262	Earthquake.
3rd–4th c.	Problems with water supply, silt deposits in harbors and swamps; malaria.
359 and 445	Earthquakes.
5–7th c.	Environment swampy-terrestrial. Partial sea walls around Miletus.
481–565	Large reservoir for the Stefania aqueduct, near modern Balat.
8th c. on	All harbors silted up.
By late 19th c.	Miletus 9 km from the sea; delta still prograding.
Last half of 20th c.	Drainage projects in valley.
July 16, 1955	Earthquake: relocation of village to Balat, 5 km south.
Late 20th c.	High water table; Lion Harbor quay submerged.

Sources: Schroder and Yalcin 1992; New excavations (1999–2000) of W.-D. Neimeier at the Athena Temple (personal communication, W. von Graeve; Tuttahs 1997.

Morgantina

Early Pleistocene	Faults running NW-SE.
2 M–10,000 B.P.	Serra Orlando ridge: sand, sandstone, clays, flysch, calcareous marls, diatomites, under limestones, marls, sandstones.
560 B.C.E.	Grazing of pigs in oak forests, hunting, forest products, wool.
ca. 450	Settlement moved to center of ridge: agora formed around Chithonian shrine and NE corner spring.
450–1st c. B.C.E.	Piped water system supplemented cisterns, wells. Drain system paralleled supply.
4th–3rd c.	Chithonian sanctuary of sandstone ashlar. Quarries at edges of agora.
317–276 or 250–220	Fountain at NE corner supplied by springs, rain.
3rd c.	Confined aquifer under city at maximum; many springs. East Hill artificially flattened for houses.
By ca. 200	Deposit of 10 m of alluvium in river valley triggered by clearing forest, farming on steep slopes, lack of terracing.

Before 70 B.C.E.	Pipeline (from Aidone spring?) built to supply agora fountain.
turn of era	Site deserted.
1693 C.E.	Earthquake at Aidone, Noto Antico, Syracuse; no known damage here.
1995	Water pressure along fractured rock caused landslides near South Gate.

Sources: Antonaccio 1997; Asheru 1980; Strabo V. 24.

Priene

1000 B.C.E.	Sea extended E past Söke site.
7th and 6th c.	Second location SSW of Söke; Naulochos farther SW(?).
late 5th c.	City and port are one (Tribute Lists).
before 350	Mausolus refounded Priene on final terrace site.
350–325	Streets, water supply, and drainage systems laid out.
330s	Port W of town.
250 on	Problems with Meander River.
1st c. B.C.E.	Priene 7.5 km from the coast.
26/25 B.C.E.	Earthquake.
68 C.E.	Earthquake.
ca. 100 C.E.	Renovation of core buildings.
238, 244, and 262	Earthquakes.
6th c.	Reservoir walls remortared.
12c	Site abandoned; monuments disintegrated.
1653	Earthquake of IX intensity; surface ruptured 70 km along river.
1891 and 1899	Earthquakes; the latter another IX; surface rupture 50 km.
1955	Earthquake (6.8) destroyed the town of Balat at Miletus; epicenter between Miletus and Priene.
Today	Priene 15 km from coast.

Sources: Altuna 1998; Badian 1996; Bottermann 1994; Bruckner 1986, 1991; Crowther 1996; Eisma 1978; Ilhan 1971; Radt 1993; Sherwin-White 1985.

Selinus

10–7th c. B.C.E.	Elamites used Gaggara spring on West Hill.
9th c.	River mouths as ports for Dorian Greeks.
beginning 7th c.	Greeks penetrated N, along rivers.
7–6th c.	Quarries on Manuzza for houses, foundations, roads.
6th c.	River walls; sewers; aqueduct from N springs; wells, cisterns.
570–56	Paved cart roads to move stone from quarries. Barone quarry to N; Cusa quarry to W. Two temples built on E Hill.
560 on	Metope stone from Portella di Misilbesi quarry; sculptures of Menfi stone from Belice V. quarry.
mid-6th c.	Massive destruction—earthquake? 550–490 ; many old buildings demolished.

ca. 550	First fortification of Manuzza; earliest river walls and sewers.
Last half of 6th c.	Temple G on E Hill.
6–5th c.	West Hill sanctuary with fountain, water conduit, rampart, etc.
510–480	Sides of Acropolis sheathed in stone against landslides.
ca. 500	Harbors formalized with stone edges, auxiliary buildings.
500–460	Temple E, monuments on Acropolis.
Early 5th c.	Agora on Manuzza especially for grain trade.
490–80	Three temples and altar built on acropolis.
by 450	Second wall around acropolis.
450–430	River drainage works; new river walls?
426 B.C.E.	Earthquake with probable damage.
409	Carthagenian invasion; destruction of Temple G.
408 on	Repair of acropolis temples and ramparts with N Gate and extensions.
4th century	Temple B completed?
200 B.C.E. on	Harbors silted up; still out of use today.
17 C.E.	Earthquake in eastern Sicily and Calabria, affected unpopulated Selinus(?).
361–65	Severe earthquake in Belice valley.
6th c.	Earthquake (?) caused great damage.
8th and 9th c.	Earthquake series—the worst in 852—near Castelvetrano, NW of Selinus.
9–11th c.	Arabs utilized quarries.
12th c.	Earthquake ruined one, damaged two temples.
13th–19th c.	Aragonese rulers utilized quarries.
Early 20th c.	Gales revealed harbor walls.
1968	Major earthquake in Belice valley.

Sources: de la Geniere and Theodorescu 1980–81.

Syracuse

Mid–8th c. B.C.E.	Rivers facilitate Greek traders' exploration inland.
648	First quarries.
6th c.	Potteries NW of the Great Harbor in blue clay district.
6–5th c.	Mole connected Ortygia with mainland. Wells, cisterns in Santa Lucia area.
5th c.	Arethusa's spring made into fountain. Galermi aqueduct built.
402	Quarries at Cave di Santa Panagia and Scala Greca used to build rampart and Euryethos fortress.
1st half 1st c.	Catacomb use in Santa Lucia area.
1st c. C.E.	Amphitheater built in old quarry.
41–54	Improved water supply, new Ortygia bath.
3rd quarter 1st c.	Earthquake affected potteries.
2nd half of 2nd c.	Port facilities repaired.
220 C.E. on	Catacombs of Santa Lucia, Santa Maria di Gesu, Vigna Cassia.
330+	Catacomb of San Giovanni.
Post-Roman	Submergence of Ortygia's edges.
1140	Violent earthquake closed springs in Great Harbor.

1169	IX–X earthquake.
Late Middle Ages	Isthmus from Ortygia to mainland at N partly above water; used as cemetery.
1542	Great earthquake destroyed old isthmus mole.
After 1542	Erosion and deposition along north shore changed shoreline.
1543–56	Charles V built fortress at new isthmus; street pattern from Ortygia to mainland changed to NW.
1693	7.7–9 magnitude earthquake, with tsunami, was most destructive: Roof of Latomie quarry fell in; cracks in other quarries and catacombs.
1728	Cathedral (former Athena Temple) given baroque facade.
1757	Great earthquake; remodeling of Ortygia palace.
1798	Napoleon's ships watered from Arethusa fountain, still flowing amply.
18–19th c.	Ancient aqueduct of Paradiso latomie still flowing.
1846	Great earthquake.
About 1915	Plains drained for agriculture.

Sources: Aureli et al. 1994, 1998; Collin-Bouffier 1987; Drögemüller 1967; Ercoli and Specioli 1988b; Farbicus 1932; Gargallo 1970; Mirisola 1996; Villard 1994.

Appendix B: Glossary of Technical Terms

Definitions are from *The Concise Oxford Dictionary of Earth Sciences* (1991), *The Random House Dictionary of the English Language*, second edition (1987), and *The New Shorter Oxford English Dictionary on Historical Principles* (2 vol.) (1993).

Achaea: the Roman province of which Argos was a part and Corinth was the capital.
adobe: sun-dried brick, made of clay frequently mixed with straw.
adyton: Inner shrine of a Greek temple.
aerosol: colloidal substance of small particles suspended in air, either natural (i.e., from volcanoes) or man-made (dust or chemicals).
aggradation: raising the bed of a river.
agora: central public area for a Greek city.
alveolization: having surface cavities.
alluvial, alluviation: applied to the environments, action, and products of rivers. Alluvial deposits (alluvium) are clastic, detrital materials transported by a stream and deposited on a floodplain.
analemma: retaining wall.
anhydrite: an evaporative mineral used for cements.
anhydrous mode: with all water removed, particularly the water of crystallization.
anta (anta, pl.):   a rectangular pillar projecting from the side of a door or the corner of a building. When it projects from a wall, it is called a pilaster.
apatite: a phosphate mineral found in igneous and metamorphic rock, and in fossil bones; used as fertilizer and to make phosphoric acid and other chemicals.
aqueduct: a long-distance waterline, sometimes underground, sometimes at ground surface, and sometimes raised on a wall or bridge.
aquiclude: a layer of stone or clay that confines the flow of water. See aquifer.
aquifer: a body of permeable rock capable of storing water in significant quantities; the water moves through the rock. A *confined aquifer* is sealed above and below by impermeable material. An *unconfined aquifer* has as its top the water table.
archaeology: the study of the material remains of the human past by specially trained

persons who excavate and who use the data they find to make historical judgments.

archaic: a Greek period from the seventh to early fifth century B.C.E.

architectonic: highly organized in manner and structure; like architecture.

arenaceous: sandy.

argillaceous: clayey.

Artemision: temple dedicated to Artemis, especially the one at Ephesas.

aseismic: free from earthquakes.

ashlar: standard blocks of masonry.

a.s.l.:   above sea level.

Attic foot: a Dorian measurement equal to 29.46 cm.

augite: the most common pyroxene, found in igneous and metamorphic rocks.

baldachin: literally a miniature temple, or more commonly any permanent ornamental canopy.

B.C.E.:   before the common era.

beach rock: beach sand deposited in the intertidal zone, cemented by precipitation of needle-like crystals of aragonite in the pores between grains, solidifying in as little as 10 years.

bed: the smallest distinguishable unit of classification of stratified rock.

bedding plane: a well-defined planar surface that separates beds in sedimentary rock. Each bed accumulated during a separate deposition period.

biotite: a common mineral, dark and lustrous.

bioturbation: the disturbance of sedimentary deposits by living organisms.

bipedales: two-foot bricks, of the Roman era.

birefringence: the ability of minerals to split planes of polarized light into two rays.

bothros: hole or pit especially for drink offerings poured out for gods of the netherworld.

bouleuterion: council chamber.

B.P.:   before the present.

brecia: sedimentary rock with angular clasts.

Bronze Age: use of bronze tools began in Greece before 3000 B.C.E., and at other dates elsewhere. The Bronze Age in the Mediterranean is thought to extend from the late third into the late second millennium.

calcaire plaque: see sinter.

calcarenite: (a) calcareous sediment with a high percentage of quartz clasts in a calcareous matrix; (b) a clastic limestone in which both clasts and matrix are calcareous.

calcareous: deep-sea, fine-grained deposit of small organisms containing more than 30% calcium carbonate.

carbon 14 (C^{14}):   radiocarbon dating, a method for dating organic material that is applicable to about the last 70,000 years. It relies on the assumed constancy over time of atmospheric ratios of C^{14} and its known rate of decay, of which half is lost in a period (the "half-life") of every 5730 plus or minus 30 years.

catacomb: an underground cemetery, especially one consisting of tunnels and rooms with recesses dug out for coffins and tombs.

catchment: area from which a surface stream or groundwater system derives its water.

cavea: seats of a Greek theater, carved out of a hill and arranged along an arc of a circle.

C.E.:   common era (the past 2000 years).

cella: the enclosed central structure of a classical temple.

chamotte: crushed, fired clay, also called grog.

chlorite: a mica-like mineral.

chora: the area around a city, its hinterland.

chryselephantine: made of gold and ivory; pertains to statues of the fifth century B.C.E.

cistern: a water holder constructed or dug below ground surface.

classical: (a) Greek culture of the fifth and fourth century B.C.E.; (b) characteristic of Greek and Roman antiquity; (c) conforming to Greek and Roman models in literature and art; (d) of the finest quality in its class. In this work, we restrict the term to definition (a).

clast: particle of broken-down rock

clay: a natural earthy material that is plastic when wet, consisting essentially of hydrated silicates of aluminum; used for making bricks, tiles, pottery, etc.

colonnade: a row of regularly spaced columns, usually supporting an entablature and one side of a roof.

conglomerate: coarse-grained rock with rounded clasts greater than 2 mm in size; it may be produced by a variety of processes.

Corinthian: an order of architecture employing acanthus leaf capitals, developed in the late fifth century and hence later than Ionic and Doric.

cummingtonite: a mineral sometimes found in igneous rocks.

detric: derived from the mechanical breakdown of rock by weathering and erosion.

diagenetic: changed under low temperature and low pressure.

diatomites: sediments made of the silaceous cell walls of diatoms, laid down in lakes or deep seas.

dip: angle of inclination of a plane.

dip-slip fault: a fault with relative displacement parallel to the dip of the fault plane.

discontinuity: (a) a break in sedimentation forming prismatic blocks of stone; (b) a boundary or layer at depth, marked by a significant change in the speed of transmission of seismic waves, indicating change in type of rock. In this book, we usually mean definition (a). Discontinuities may be subvertical (normal to the stratigraphy and orthogonal to themselves).

doline (swallow-hole, sinkhole): steep-sided, enclosed depression in a limestone region, usually at a site of joint density that focuses drainage vertically. Dolines enlarge by solution and collapse. The floor of a doline may be part of a horizontal or quasi-horizontal cave system.

Doric: an order of architecture developed in mainland Greece and its western colonies.

Doric foot: 32.6 cm, about 12.5 inches.

drain: a general term for a pipe or conduit (natural or man-made) by which liquid empties. See sewer.

drawdown: the lowering of a water surface as the result of deliberate extraction of groundwater.

ekklesiasterion: assembly of the citizens; sometimes, the place of such meetings.

encaustic: a stucco with pigmented wax burned to fix it in place, to protect and improve the surface or architectural details such as volutes.

entablature: the part of a classical building between the supporting columns and the roof.

eustatic: worldwide change (in sea level).

exotic: introduced from elsewhere, as rocks carried by a glacier.

failure: the process by which a body looses cohesion under stress, and divides into two or more parts, usually by a brittle fracture.
  failure stress: the stress that causes a failure.

fault: approximately plane surface of a fracture in rock, caused by brittle failure, and along which observable relative displacement has occurred between adjacent blocks.

> normal fault: a dip-slip fault of more than 50° in which the upper hanging wall moves downward relative to the footwall. Normal faults cause horsts and grabens.
>
> strike-slip fault: a fault with horizontal displacement parallel to the strike of a vertical or subvertical fault plane.
>
> transform fault: a strike-slip fault occurring in the sea at the boundaries of lithosphere plates, in which the movement of the crustal blocks is reversed (transformed) in comparison with a similar strike-slip fault on land. These faults usually occur at right angles to the ridge itself and indicate the direction of spreading.

flysch: thick, redeposited clastic materials.

fossils: remains of once-living organisms, dating from before the last Ice Age. Types of fossils associated with particular layers were the first means of correlating dates and strata in different regions. The order and names of the strata were determined in the nineteenth century. Quantitative methods now make the dating much more precise.

fountain: a structure for discharging jets of water, often with architectural embellishment and sculpture. A fountain may be the facade of a spring or the outlet of an aqueduct.

friability: easily crumbled, said of soil or rock.

fuzzy set: see set.

gateway city: one commanding connections between its tributary area and the outside world, heavily committed to transportation and wholesaling (Burghardt 1971).

geology: (a) the science that deals with the dynamics and physical history of the earth, the rocks of which it is composed, and its physical, chemical, and biological changes; (b) the geologic features and processes occurring in a given region on the earth or a planet.

Geometric: a style of vase painting in Greece between the tenth and eighth centuries B.C.E. that utilized abstract rectilinear and curvilinear shapes; hence, the period as a whole.

geomorphology: study of the landforms on the earth's surface and the processes that formed them. Geomorphology explains the origin and development of landforms, the relation of landforms to underlying geologic structure, the history of geologic change as recorded in landforms, and the relation of local tectonics—both events and processes—to local landforms.

graben: a downthrown, linear, crustal block bordered lengthways by normal faults. Regional grabens may be called rifts.

gradient: rate of ascent or descent; an inclined surface.

grog: fired and crushed pottery, also called chamotte.

gypsum: the first mineral to precipitate from evaporating sea water, because it is very insoluble. It can also occur from the reaction of sulfuric acid on limestone in volcanic areas. Selenite, satin spar, and alabaster are varieties.

ha: see hectare.

half graben: a similar feature to a graben, but bordered only on one side by faults; these are associated with tilt-block tectonics.

head: the height of water above a certain datum point, considered together with the pressure head (related to atmospheric pressure) and the velocity of the water. Water flows from points of greater hydraulic head to lesser, along the hydraulic gradient.

hectare (ha): 10,000 square meters or 2.471 acres.

Helladic: Bronze Age culture of mainland Greece, from 2900 to 1100 B.C.E.
Hellenistic: Greek culture from the late third through the first century B.C.E.
Holocene era: 10,000 B.P.–present.
horst: upthrown block between two steep-angled fault blocks.
hydraulic head: the energy possessed by a unit weight of water at a particular point, measured by the level of the water above a certain datum level, and combining elevation head, (atmospheric) pressure head, and velocity head.
hydrogeology: geology of water within the earth's crust, including its physics, chemistry, and environmental relationships.
hydrography: measurement, description, and mapping of the surface waters of the earth.
hydrostatic: equilibrium and pressure of liquids.
Imperial Roman: the period of the greatest extent of the Roman Empire, late first century B.C.E. to early fifth century C.E.
incubation: sleeping in a temple in hopes of experiencing a vision and a cure.
insulae: city blocks (Roman).
interstitial: one mineral filling in the spaces in another mineral.
Ionia, Ionic: The west coast of Anatolia (Asia Minor), settled by Greeks from the tenth century B.C.E. on; also, the style of architecture developed there, especially the volute capital.
Iron Age: a historical period after the Stone and Bronze Ages, in the Mediterranean area beginning approximately 1200 B.C.E. and lasting some 2000 years.
isostatic compensation: introduction, into a model, of flexural adjustment of the lithosphere, topographic corrections, or presence of low-density roots [of mountains] to account for gravity anomalies.
isotropic: with physical properties the same when viewed from any direction; said of physical or optical characteristics.
joint: a discrete brittle fracture in a rock, evincing little movement parallel to the plane of fracture but slight movement normal (perpendicular) to it.
Jurassic era: approximately 213–144 Ma.
ka: thousand.
karren: solutional features in karst terrane, at the surface of an outcrop of hard linestone, specifically runnels (shallow troughs) a few millimeters deep and fissures extending several meters into the linestone.
karst: a region underlain by limestone or other carbonate rock (named for such a region in the Dinaric Alps of Yugoslavia) and characterized by landforms—shafts and caverns—created by dissolution of the stone by carbonization.
   karst aquifers: characterized by large void spaces (caves), high hydraulic conductivity, flat water tables, and extensive networks of solution channels, often with turbulent flow.
lacustrine: related to a lake, especially of sediments laid down in lakes.
lens: a body of rock thick in the middle and thinner toward the edges, similar in shape to a biconvex lens; specifically, a lens-shaped water body, like the freshwater suspended above the brackish water in the rock of Humei Hill at Miletus.
limestone: a sedimentary rock mainly of calcite or dolomite, often organic, chemical or detrital in origin.
liniation: parallel lines formed by alignment of minerals or by tectonic movement.
lithography: description of stones (English, not American, usage). See petrology.
lithosphere: the upper layer of the solid earth, comprises oceans, continents, and the brittle part (tectonic plates) of the uppermost mantle.
lithotype: category of stone.

liturgy system: public service combined with financial support of necessary public functions, such as baths, entertainment, parades. A feature of late Hellenistic and Roman social organization.

load: the mass of rock overlying a geologic structure, or the total amount of material carried by a river.

l/s: liters per second.

Ma: million years [ago].

m.a.s.l.:   meters above sea level.

macrofossils: fossils large enough to be seen and studied without the use of a microscope.

Maenads: bacchantes; frenzied or raving women, especially during festivals of Dionysos (ancient Greek).

maquis (French):   wild, bushy land.

marble: metamorphosed limestone, hard enough to take a polish.

marl: calcareous clay.

Mesostructure: the pattern of cracks, joints, strata, and local faults that make up the discountinuities of an area of study; they range in size from centimeters to a hundred meters or so.

Mesozoic (Cretaceous) era: 248–213 million years ago.

metamorphic rock: minerals recrystallized by pressure, temperature, or volatile content.

microfracturing: small, local cracks, no more than tens of centimeters long and 0.1 mm wide.

Miocene era: 24.6–5.1 Ma.

morphological setting or system: a theoretical construct about the relationships of the physical properties of a natural system, such as angles, porosity, rock type, and moisture content.

multivariate: statistical collection, classification, and interpretation of numerical facts or data, with more than one variable, using mathematical theory to impose order and regularity on disparate elements; [abstract] analysis separating entities into constituent parts.

muscovite: a rock-forming mineral, often found in clastic sediments.

nappe: a horizontally thrust rock layer.

   phreatic nappe: groundwater in horizontally thrust rock.

Neogene era: ca. 24.6–2 Ma, just before the Quaternary, subdivided into Miocene and Pliocene epochs.

normal: being at right angles to a line.

Nymphaeum: fountain, frequently ornamented with statues of guardian water nymphs.

oekist: founder of a settlement.

olivine: rock-forming mineral, usually olive green, occurring in igneous rock with pyroxene; alters to serpentine during weathering.

omphalos: from the Greek for "navel"; used for a navel-shaped stone said to mark the center of the world.

ooze-galleries: passageways arranged to collect trickling underground water and deliver it as needed. See *qanat*.

opisthodomos: western treasury room of a Greek temple.

opus sectile: a kind of mosaic with pieces cut for each location; regular mosaic is made of standard-size tiles.

oracle: an utterance, often obscure, given by a priest or priestess at a shrine, as the answer by a god to a human inquiry.

orthostates: large stone blocks revetting the lower wall of a Greek cella.

overpressure: pressure developed in rapidly deposited sediments that have been sealed so liquids cannot escape.

Paleozoic (Permian) era: nearly 600 million years ago.

palestra: a Greek gymnasium.

paragenesis: a mineral assemblage all formed at the same time.

pentellic: a white marble from hills at Athens, used especially in the fifth century B.C.E.

perennial river: a stream that flows all year long.

pericline: folded, as rock.

peripteras: having columns all around, as a Greek temple.

permeable: permitting fluids (e.g., water) to flow through.

petrology: the study of rocks in general, including their occurrence, field relations, structure, origins and history, mineralogy, and textures. American usage for "lithography."

Phaiax: Greek engineer of the fifth century B.C.E., called Feace or Faiace in Italian.

phreatic nappe: see nappe.

piezometric surface: hypothetical surface defined by the level to which water in a borehole rises.

piston: a disk or cylindrical part tightly fitting and moving within a cylinder to compress or move a fluid in the cylinder (air or water) or to transform energy by a fluid entering or expanding within the cylinder.

pithoi: very large clay vessels.

plate: a segment of the lithosphere with little volcanic or seismic activity, but with almost continuous belts of earthquakes and usually volcanic activity along the boundaries. Seven large plates are distinguished: African, Antarctic, Eurasian, Indo-Australian or Indian, North American, Pacific, and South American. Smaller plates are the Arabian, Caribbean, Cocos, Nasca, and Philippine, whereas some microplates are the Groda, Hellenic, and Juan de Fuca plates.

Pleistocene era: approximately 2 Ma to 10,000 B.P.

Pliocene era: last 5.1–2 Ma of the Neogene part of the Tertiary era.

polis: an independent Greek city.

polity: a community organized as a state. Greeks of the archaic and classical periods lived in separate polities (city-states) united by one culture.

polje: a large, flat depression with underground drainage through a karst network.

polygonal: many sided; having three or more (usually) straight sides. In masonry, used to distinguish ashlar from less regular shapes of stone.

poros: sandy limestone of the Pleistocene era; an archaeological term, loosely equivalent to calcarenite.

porosity: the ratio of the volume of the pores, as a rock stratum, to the total volume of the mass.

portico: the porch of a classical building, or a free-standing building consisting of a roof supported by columns or piers. See colonnade.

pozzo (Italian):   well shaft. A line of pozzi often reveal a qanat.

primary sources: see sources.

process: near-surface mechanism that creates landforms.

progradation: outward building of a sedimentary deposit, such as the seaward advance of a delta or an alluvial fan, by deposit of sediments.

pyroxenes: a mineral group that includes jadite.

qanat (Persian; also called *foggara* in Arabic):   an underground aqueduct of the Middle East, characterized by a line of airholes on the ground surface, spaced about 35 m apart.

quartz: a crystalline rock-forming mineral.

quartz arenite: a sandstone having more than 95% quartz grains.

R: according to the Richter scale for measuring earthquake intensity.

recharge: the downward movement of water from the soil to the water table.

reelaboration: (a) production of a natural substance by natural agencies; (b) conversion by chemical processes.

rejuvenation: the marked increase in rate of erosion when a land mass is relatively elevated; streams respond by incision.

Republican: Roman Latin culture up to the late first century B.C.E.

reservoir: a natural or artificial place where water is collected and stored, especially water for supplying a community, irrigating land, furnishing power, etc. Reservoirs are generally much larger than cisterns and may be completely subsurface, partially sunk in the ground, or protruding aboveground.

resurgence: the reappearance of surface water that had been diverted underground in a karst region.

ria: drowned river valley in an area of high relief.

rift, rift valley: a split between two bodies that were once joined, especially an elongated trough of regional extent bounded by two or more faults.

sanctuary: a scared or holy place, with or without a cult building.

sandstone (arenite):   sedimentary rock formed from lithified sand bound together with mud and cement; its ingredients are quartz, feldspar, mica, and general rock particles.

scree: like a talus, but found inland in deserts or glaciated areas.

secondary sources: see sources.

sediment, sedimentary: material derived from rock, living organisms, or chemical precipitation, and deposited at or near the earth's surface.

seismic: pertaining to earthquakes.

set: a number, group, or combination of things of similar nature, design, or function. fuzzy set: a set defined so as to allow for imprecise criteria of membership.

sewer: an artificial conduit, usually underground, for carrying off wastewater in an urbanized area.

shear: fracture along a plane from forces acting parallel to the plane.

sherds: fragments of ceramics. Many can be dated within 25 years by material and form.

silt: soil of fine texture.

sinter: calcareous or siliceous material deposited at springs, in pipes, or in limestone caves, from the German; in French, *calcaire plaque.*

socle: base for a column or pedestal.

soprintendenza: office of the superintendent of an Italian archaeological region.

sources: primary sources are the writings of participants or contemporary observers, state archives, and other contemporary documents; secondary sources are later histories or commentaries. In this sense, coins and archaeological findings in general are primary sources, but the explication of them is secondary.

spoilia: materials reused from a destroyed building, especially decorative features such as reliefs and architectural members like capitals.

spring: a flow of water aboveground where the water table intersects the ground surface. A trickle or ooze of water is a seep. Reappearance of water in a karst region is a resurgence. Water that rises under pressure from a permeable stratum overlain by impermeable rock is an artesian well.

stoa (portico):   A linear Greco-Roman building with short end walls, a long back wall, and a row of widely spaced columns along the front, having multiple uses and locations in ancient cities.

Stone Age (divided in to Old/Paleolithic, Mesolithic, and New/Neolithic Ages):   Beginning at least 2,500,000 B.P. with the use of stone tools and continuing in some parts of the world until the present.

stratigraphy: the study of stratified rocks in terms of time and space, correlating rocks from different localities by comparing fossils, rock units, and time intervals; thus, the relative spatial and temporal arrangement of rock strata. The assembly of materials found together in one layer, at one place, is highly informative to the trained eye of the archaeologist or geologist.

stratum: rock in a layer or bed. The term does not imply thickness or extent (unlike "bed").

stream terrace: see terrace.

stress: a measure of the intensity of a force acting upon a body, as a function of its area.
   comprehensive stress: changes the volume of a body.
   shear stress: changes the shape of the body.

strike: the compass direction of a horizontal line on an inclined plane.

structure: (a) the position of a bed or stratum as indicated by the dip and strike. (b) the coarser composition of a rock, as contrasted with its texture.

stylobate: foundation for a colonnade, whether free-standing or part of a temple.

subsidence: (a) falling to the bottom, precipitating; (b) a depression of the earth's crust that allows sediment to accumulate.

subvertical: nearly vertical in orientation, normal (perpendicular) to the stratigraphy.

succession: sequence, especially (ecology) the progressive replacement of one community by another until a climax community is established.

syncline: basin or trough-shaped fold whose upper components are younger than those below.

synoecism: union of several towns or villages into or under one capital city (Greek).

talus: piles of rock fragments at the foot of a cliff or slope.

tectonic: pertaining to deformation of the earth's crust.

temenos: a sacred enclosure or precinct.

temple: a building or place set aside for worship of a deity. Originally having the same meaning as sanctuary, it came to mean specifically the building in the holy place.

tepe: Turkish word for hill.

terminus ante quem: the first limiting point, chronologically; opposite of *terminus ad quem*, the last limiting point.

terrace: a nearly flat portion of a landscape terminated by a steep edge.
   stream or river terraces: terraces produced by the actions of rivers.

terrane: a fault-bounded area of region with a different stratigraphy, structural style, and geologic history from those of adjacent areas (particularly a North American usage, distinguishing geological terrane from general tracts of land).

Tertiary era: 65–2 Ma.

tetragonos: four-sided.

tholos: a circular (Greek or Roman) building.

tilt-block tectonics: a tectonic variation in which crustal blocks are rotated along normal faults that tend to become nearly horizontal at depth; important at the passive margin of a rifted continental mass where the blocks are of thick sedimentary sequences (e.g., Ionia).

topology: (a) the topography of an area and how this reflects its history; (b) the branch of mathematics that deals with the properties of figures and surfaces that are independent of their size and shape, unchanged by deformation, that is, the way parts are interrelated or arranged.

transgression: an advance of the sea to cover land, due to rise of sea level.

travertine: a calcium carbonate rock deposited at hot springs

tribute lists: a payment made periodically by one state or ruler to another as a sign of submission to ensure peace and protection.

thrust: a low-angle reverse fault in which the upper part overhangs the lower.

tsunami: Japanese term for tidal wave, now widely adopted.

tufa: tuff is defined in the *Oxford English Dictionary* as "stratified cellular rock" and can be carbonate. In the *Concise Oxford Dictionary of Earth Sciences,* also produced in England, tuff is "compacted volcanic ash," rare but not impossible for the eastern Mediterranean, but *tufa* is a sedimentary rock made from calcium carbonate. The word *tuff* is used loosely by archaeologists to refer to both kinds of rock.

tuff: see tufa.

unconformity: a large break in the geological record, observable as the overlying of rocks by others that would not normally succeed them directly, and having a different direction or plane of stratification.

vocabulary: a collection of signs or symbols constituting a system of nonverbal communication, especially a group of forms characteristic of a style of architecture.

water table: the upper surface of groundwater or the level below which an unconfined aquifer is permanently saturated with water.

wattle-and-daub: a building technique using upright posts interwoven with twigs or branches and plastered with a mixture of clay and straw.

well: a hole drilled, bored, or dug into the earth to obtain water. Wells are usually vertical, and their sides are not waterproof.

xyst (xysto, xystos):   in ancient Greek and Roman architecture, a covered portico, frequently surrounding a courtyard.

zircon: a mineral concentrated in river sands.

Notes

Chapter 1

1. "We" is used to reflect team consensus on a point, "I" when the opinion or activity of Crouch are indicated. "Personal communication" indicates a letter to or interview with Crouch.

Chapter 3

Scientific contributor to Agrigento section is geologist L. Ercoli. Personal knowledge of engineer G. Trasatti, water manager E. Trasatti, and tunnel expert C. Micceché was crucial.

Chief scientific contributor to the section on Morgantina is geologist G. Bruno, with help from geologists P. Atzori, A. Cannata, P. Mazzoleni, S Nicosia, A. Pezzino, E. Renna, F. Schilirò, and engineer A. Sciortino. Geologist S. Judson taught us to see Morgantina's ridge.

Scientific contributor to the section on Selinus was L. Ercoli with field assistance by P. Cipolla. A. Aureli provided maps.

Scientific contributors to the section on Syracuse are A. Aureli and M. de Meio, hydrogeology and water management history; L. Ercoli and G. Speciale, geology; E. Bongiovanni, general knowledge of the site; R. Maugeri and R. Ruggeri, catacombs as aqueducts.

1. Dates of geological, political, and constructional events have been combined in the Chronological Tables of Appendix A.

2. Akragas was the Greek name. The Romans called it Agrigento. During the Middle Ages, the settlement on the western summit was called Girgenti. The name Agrigento was adopted again in the twentieth century.

3. In the Italian and French archaeological and geological literature, both natural springs and outfalls of *passages* are termed "sources" which is confusing for English speakers.

4. Along the road from Palermo to Agrigento, past Comitini, a byroad leads to Macalube, whence extraordinary eruptions of methane gas were described by Solinus in the third century C.E. These still occur today, "a strange and terrifying spectacle with tall tongues

of fire," coming from tiny conical volcanoes 1.5 to 3 feet tall filled with salty bubbling mud (*Blue Guide: Sicily*).

5. Bird's eye view by Chatelet (1785), fig. 4 in Cotecchia 1996; see also Cotecchia's fig. 19, Structural Model of the Agrigento Area.

6. When Arnone (1948) studied the galleries, his assistant was the father of Calagero Micceché.

7. Speleologist R. Ruggieri has participated in a new investigation of these passages. See Archeoclub d'Italia 1996.

8. See Morgantina section for an analogous weakening process that dewatered the site, contributing to its abandonment.

9. Tombs here have a mix of local, Greek, and Siculo-geometric pottery. Cups were buried because wine drinking was a high-status activity, indicating the emergence of aristocracy. Lyons (1991) and Antonaccio (1997) surmise a mixed culture.

10. Mineralogical analysis was performed by the nonoriented powders method, using a Siemens X-ray diffractometer (XRD) and Ni-filtered CuKa radiation. The analytical constraints were kV=40, mA=20, input window 1/2, and output window = 0.2°. Major chemical elements were determined by X-ray fluorescence (XRF); loss on ignition by heating at 990°C; and FeO and MgO contents by titration with $KMnO_4$ and by atomic absorption, respectively. Porosity of the terra-cotta was measured with a Carlo Erba 2000 mercury porosimeter.

11. Included minerals are biotite, sometimes chloritized, and muscovite, as well as zircons and apatite. In three samples, there are pale green augitic pyroxenes, detrital in character, often highly fractured, and largely reelaborated during the movement of the sediment.

12. "Cult was used as a means of marking the socio-political layout of state territory, defining its extent not only culturally (splitting 'wild' from 'civilized' space), but also in political terms, since boundaries separate states and are thus prime point of definition of state membership. If, therefore, we advance as a general rule the idea that sanctuaries reflected contemporary perceptions of areas crucial to the orientation of settlement on a regional level, it becomes clear that they provide the earliest evidence for large-scale regional perceptions" (Morgan 1990: 7). This claim to dominate was an important reason for scattering the temples widely over the site: "Sanctuaries were founded together with the city itself, as if to affirm that the establishment of a new community is unthinkable without also defining the territory. . . . Sanctuaries on the outskirts of their territory . . . display that the act of occupying, possessing, and exploiting such lands was sanctified by the gods" (Greco 1996; compare with Puglies-Carratelli 1989).

13. Only on West Hill have the dedications survived. For the rest of the site, we will refer to the temples by their traditional letters.

14. Zeus Meilichios was now invoked by the Punic name of Chryszor or the Syrian name of Baal-Shamin (Tusa 1979–82). Worshippers set up new stelae of Zeus as Baal-Hammon and Kore as Tanit (Le Dinahet 1984), both Middle Eastern names. Temples A and O on the acropolis became Punic sanctuaries with added mosaic floors incorporating white outlines of "angels," solar disks, staffs, and bull's head wreaths (or bucrania) Mertens 1989). The Syrian names Baal-Shamin and Baal-Hammon suggest traders from Palmyra or Syrian slaves or both.

15. A comparative study of the best-known quarries at Selinus, Corinth, and Ephesus would be illuminating.

16. Sixteenth through eighteenth-century Spanish settlements in the New World set aside land for later colonists (Crouch 1983).

17. Aristotle (*Politics* 330a&b) knew that malaria and other diseases were related to

"marsh vapors" as did Hippocrates (Book 7), but they seemed unaware of the dangers of unsanitary living conditions and poor nutrition, though they knew of contagion from sick persons (Villard 1993, 37–44). Seasonal epidemics, centered in low marshy areas, broke out during the Athenian siege of 413 and later sieges, especially after a very hot summer. Thucydides (VI.90, VI.103, VII.14.3, VII.25.1–2) deemed the Cyane (and the area it watered) a "scourge" during the Athenian-Syracusan war in which the large number of deaths from disease (malaria?) broke the will of the Athenian army.

18. Note that the name Epipolai referred to different sections of the high plateau at different times in the long history of the city (Drögemüller 1969): Such mobility of nomenclature has also made it difficult to be certain of the locations of various suburbs of the city.

19. The speleological group of R. Maugeri and friends report (1992) an irregular tunnel under the amphitheater, which they explored and mapped. Its form is strikingly like a karst passage. Map in Crouch collection.

20. W. Vetters points out (personal communication) that the change in relation of Ortygia's edges to the sea is likely to be from tectonic uplift of the center of the island. This would-be horstlike behavior is similar to the uplift of the horst from which flow the Pisma and Pismotta artesian springs in the plain. An uplift in Calabria (the toe of the Italian boot) in 1994 was reported by Miyauchi, Dai Pra, and Sylos Labini (W. Vetters, personal communication); this was a continuance of the process that left nine sets of sea terraces at +200 to −8 m from the present level in the Mediterranean area. In some places, the shore has lowered 10 m in the last 2500 years, but the Mediterranean average is −1 to −3 m (Ercoli and Speciale 1988).

21. Thanks to R. Ruggieri for showing this to me.

22. Although the island was named after the legendary birthplace of the goddess Artemis (see Ephesus section), it was the nymph Arethusa who captured the imaginations of the ancient Greeks. Fleeing a lecherous river god, she swam under the waters of the Mediterranean from Greece or Egypt (in different versions of the legend), reappearing at the western shore of the island. Her symbolic portrait head appears on ancient coins of both Syracuse and Morgantina (Drögemüller 1969; Aureli et al. 1997; Antonaccio 1997).

23. The book of Deuteronomy (2.13), of the mid-second millennium, and the book of Jeremiah (38.6), written in the last quarter of the seventh century B.C.E., refer to cisterns cut in rock. Pausanias, Aristotle, Plato, and Hippocrates ranked rainwater as healthiest, followed by spring water, then cistern water, because the latter became stagnant if held too long. Hellmann (1990) noted that hotter, drier areas south of Athens (like Sicily) had many cisterns, which can often be distinguished from wells by the presence of hydraulic cement, because wells were by definition not waterproof. Her study also noted the use of metal grids to prevent things from falling into cisterns and wells.

24. The martyrdom of Saint Lucy in 304 under Diocletian established her catacomb as primary in status (see Agnello's map in Coarelli and Torelli 1984: 263). After Constantine's edict of toleration in the 330s, the catacomb of San Giovanni developed around the crypt of Sta. Marziano, second bishop of Syracuse, martyred under Gallienus and Valerian in the 250s. His tomb was the seed from which the sanctuary of San Giovanni developed after 313, with the tomb and catacomb lying beneath the church.

Chapter 4

Chief scientific contributor to the section on Argos is P. G. Marinos with the aid of S. Plessas. P. Gaki-Papanastasiou and J. Knauss also gave personal assistance.

Principal scientific contributor to this section on Corinth is P. G. Marinos with the help of S. Plesas and N. Mourtzas, all engineering geologists. D. Romano's maps for the Corinth Excavations project have been indispensable.

Chief scientific contributors to the section on Delphi are engineering geologist P. G. Marinos and his doctoral student E. Kolaiti.

1. An early investigation of Argos area geology was that of A. Philippson (1892) who noticed that because of the particular combination of rainfall paths, stone types (limestone and conglomerates), and the tilt of the land, the Western Peloponnesos was better watered than many other parts of Greece. Philippson began the investigation of groundwater flow, being particularly interested in dolines, swamps, the many small springs and few large ones such as Kephalari, and especially springs and swamps related to inland dolines or found along the coast. He commented on the underground circulation network, cautioning that the visible springs of today may not be the same as the ancient ones of the same names, owing to earthquakes and other ground changes that could alter the outflow. He described rivers of three categories: perennial, periodic, and those dry except after torrential rain. Near Argos he pointed out the spring-fed Inachos River, but named the source as Kephalari, whereas the name of the spring and village is actually Kephalovryso. Some 70 years later, J. Dercourt (1962) published an important study under the aegis of the Greek geological service. He named three rivers of the plain, the Inachos from the north and west, the Xerias (modern Charadros) from the west, and the torrent Tsipiana (Megalo Rema or Manessi north of Nauplia?), each with additional tributaries. His studies of the lithography of the mountains published not only general descriptions of the extent and depth of the layers of Pindos and Tripoli limestones, and the appearances of flysch, dolomite, and sandstone, but also preliminary investigations of the structural and tectonic history of the area.

The geohydrology was studied by contributors to the fifth international symposium on underground water tracing, which met in Athens in 1986 and issued a useful proceedings the same year, edited by A. Morfis and H. Zojer. Our other major source for the geology has been the work of E. Zangger (formerly known as Finke) (with Niemi 1988; with Van Andel 1990a, Ph.D. in 1993).

2. An early scientific study of springs south of Argos was conducted by N. J. Papakis (1966) on the Kiveri/Argos spring. Papakis recommended building a dam in the sea to capture the fresh water to be pumped northward for irrigation. When I visited the area in 1985, the dam was still in place, but the pumping of irrigation water had been abandoned as unsatisfactory (Morfis and Zojer, 1986: photograph of the spring and dam).

3. As early as the Mycenaean era, there were efforts at Tiryns, on the eastern edge of the plain, to cope with torrential streams that brought debris as well as water to the low area around the city. The people built an upstream dam to divert the river to the south of both hills now called Prophet Elias, east and southeast of their city, after a catastrophe about 1200 B.C.E. that buried a stretch of the coast under several meters of debris (Knauss, 15 A and B, 1996, Fig. 13 p. 111; Runnels 1995). See also T. K. Smith (1994).

4. The ancient stories recount that the Argolid was once covered by the sea—Poseidon's realm—which then receded. Piérart (1988) thinks that the contrast in resources and life patterns between the northern plain—dry because of no irrigation—and the southern watery area was worked out in the development of two sets of myths and two conflicting dynasties during the second millennium: one dynasty centered at Argos and one at Mycenae; their border lay south of Mycenae (Piérart 1992, Appendices). Poseidon, having lost the dominance of the plain to Hera, showed his anger by controlling the rain so that the rivers of the Argolid, all but the Erasinos at Lerna, were dry in summer. Once the plain was recognized as Hera's, the Heraion at its northeast edge became a settlement, a vast necropolis, and shrine, as early as the eleventh century (Hall 1995). People from all the

cities of the Argive Plain came there to worship, display their riches, and compete at games (Hägg 1987). To the south, however, the lagoons and Lake Lerna marked the transition from Hera's dominion to Poseidon's.

5. The geological terraces we have been discussing are the large-scale models for terraces constructed by humans for agricultural purposes; the latter began to be built in the eastern Mediterranean between 1200 and 900 B.C.E. (Wertime 1983: 446).

6. Although D. Romano calls this pattern *per strigas*, the term has a different meaning in Greek city planning. In the Greek city of the archaic period, *per strigas* (by furrows or stripes) does not mean a grid; rather, it refers to one long axial street with cross streets meeting at right angles but without small cross streets parallel to the main axis (Crouch 1994, chap. 5).

7. The ancient Tibetan medical treatise, Gyu-zhi, states that "the best therapeutic environment is said to be a place of great natural beauty." Thanks to G. Hirsch for bringing this to my attention (Hirsch 2000).

8. Minoan and Mycenaean material, but not necessarily a settlement, dates from 1600 B.C.E. on (Higgins and Higgins 1996). The settlements stood on flysch or alluvium. The economy was based on cattle rearing, with transhumance to high valleys in the summer and to the lower plains in the winter (Müller 1992). Mycenaean tombs preceded Roman ones to the east and west of the site.

9. Archaic Greek society understood political crisis as evidence of divine wrath, like famine and earthquake. In cases recorded in inscriptions, people came to Delphi to ask the Pythia for guidance in solving pressing social issues. By the third quarter of the eighth century, the three major issues were founding new cities, locating sites for new cities, and altering the political arrangements within city-states. In the eighth and seventh centuries, Delphi gave directions for seeking a new site or approval to lay city foundations. According to Malkin (1989) and Morgan (1990), the oracle of Delphi supported new social orders, comprehensive reforms, and some tyrants.

10. Modern skeptics tend to discount such stories as myths, but during my several visits to the site we were admonished to speak softly near the fountain of Kastalia lest loud voices set off landslides. This was especially necessary before the rock was consolidated in the 1990s with tie rods and cement.

11. Unfortunately I have lost the source of this quotation.

Chapter 5

Principal scientific contributors to the case study on Miletus are T. Güngör, Y. Birsoy, and H. Brückner, geologists; Ü. Özis, A. Alkan, G. Tuttahs, engineers; and J. Kraft and I. Kayhan, geographers.

Scientific contributors for the section on Priene are T. Güngör, and Y. Birsoy, geologists, and Ü. Özis, A. Alkan, and C. Ortloff, engineers.

Scientific contributors to the section on Ephesus are W. Vetters and T. Güngör, geologists; civil engineers Ü. Özis with A. Alkan, O. Baykan, and A. Atalay; and fluids engineer C. Ortloff. Shoreline locations have been determined by H. Brückner, I. Kayan, and J. Kraft.

1. Similarly, a paleomagnetic survey along the coast of Turkey as far south as Izmir revealed the rotation of adjacent coherent blocks about vertical axes and in opposite directions since twenty million years ago. The local rotation patterns can be very complex, especially in the Karaburun Peninsula west of Izmir, which rotated 44° downward, and at Izmir, 37° upward. At Izmir the rotation is dated 11.3 million years ago (Kissel et al. 1987).

The brittle upper part of the lithosphere possibly does not follow the lower ductile part's motions in the subduction process, or the lower ductile motions may be more complex than has been recognized. Measured rotation in extension zones without commensurate strike-slip faulting, and/or possible large rotation (72°) resulted from successive episodes of faulting.

2. The Iron Age began during 1500–1000 B.C.E. in Anatolia, but flourished from 1200 onward, as is known from Hittite texts (Yalçin 1993).

3. The environmental costs of metallurgy in the Milesian colonies of the south shore of the Black Sea are likely to have been comparable to the effects of the silver mines at Lavrion in Attica where the smelting required an estimated 1 M tons of charcoal, made from 2 M acres of coppice (small trees cut down periodically for fuel), to produce an estimated 1.5M tons of lead, an estimated 0.0035 M tons of silver, and 1.5M tons of slag as byproduct (Hughes 1994); metallurgic slags as indexes of ancient industry are well known in Anatolia (Wertime 1983: 451). Other mines cost two to eight times as much in fuel, even when the use of regrowth trees was factored in (Hopkins 1978b: 56): one ton of Spanish silver required ten tons of trees and produced 400 tons of lead, and the gold mines near Leon moved more than 70 million tons of earth (placer mining). Therefore, in Wertime's opinion, the change from copper and bronze to iron was mainly a matter of fuel efficiency, since with iron the results were many times more metal per unit of fuel.

4. Sinope, the first Greek colony and emporium in the Black Sea (early eighth century), gave its name to both a kind of iron and a pigment made from iron oxide. The Greeks' switch from bronze to iron, which they called *sideros*, and steel (carbonized iron), which they called *chalybs*, is recorded in those names, taken from peoples in this ore-producing district of northeast Anatolia where tradition says smelting methods were invented (Drews 1976, citing Xenophon IV. vii.44–5). This nomenclature can be dated from the Geometric era (Yalçin 1993). The main exporter of iron—and most distant colony of the northeast coast—was Trapesus (today: Trabezon), a well-established emporium by about 756, according to Xenophon (IV.vii.44–5). Trapesus and other colonies of this Black Sea coast exploited no fewer than 53 ancient iron deposits within an area extending 75 km along the coast and 15 km inland. There were also rich silver mines in the hinterland of Trapesus, 14 of the 71 still producing today in Turkey. Strabo (XI 14.9, xiii 3.19, xii 3.300) discusses these mines.

5. Some iron samples from Miletus are now located in the Bochum (Germany) Museum of Chemistry, Minerals, and Metalography. It would be interesting to test Minoan and Mycenaean metal work for isotopic differences or similarities with metals from the two Milesian areas. Iron remained important at Miletus, where it was used as money in fourth century B.C.E. (Yalçin 1993). Continuing extraction of iron ore from this area is revealed by news of the sinking of 2 ore ships in Dec. 2001 near Suiope (*Los Angeles Times*, Dec. 10, 2001.)

6. The lower basin, with which we are concerned, was 3710 km², and between 1937 and 1946 the discharge per unit of surface was 5.95 m³/yr. The maximum discharge in the Aegean basin was 9.35 ³/yr and the maximum for Turkey was 128.50 m³/yr from the Manavgat River near Antalya on the south coast (Karatekin 1953). In the 1950s, schemes for harnessing the Great Meander River included a generating potential of 75,000 kilowatts; irrigable land of 150,000 hectares; already irrigated land of 22,000 hectares; reservoir projected and approved 1000 M m³; reservoir under construction 0.45000M m³—figures that suggest the ancient potential when the land was more densely forested.

7. The residents of Myus moved to Miletus in the second century, taking with them most of the hewn stone from their city; only the foundations of a temple (later incorporated into the base of a Byzantine fortress) have been found to mark the old city.

8. Many kilns in the mountainous areas around Miletus have been discovered by Peschlow Bindokat. If a kiln were located near sources of clay and wood for fuel, only the finished ceramics would require transport. Ceramics were much less bulky than raw clay and logs, and therefore less expensive to transport. Inscriptions from Delos indicate that the sanctuary needed over five tons of firewood annually at a cost of more than 200 talents (Linders, 1988: 268), suggesting the savings generated by not having to transport the wood for firing ceramics.

9. If each ship had 200 men, that would be 16,000 Milesians, suggesting a population of 64,000 (Calhooon 1970: 427) including the hinterland as well as the city. Since the district between Kalabak Hill and the rampart could have had a population of 20,000–24,000 at this time (von Graeve 1999 and earlier), with perhaps that many again on Humei Hill and four or five times as many in the villages of the hinterland, these numbers of ships and men are reasonable. However, experts do not agree on estimates of ancient populations (Crouch 1972).

10. Study of the archaeology of sewers and latrines in Roman Italy by A. O. Koloski-Ostrow (forthcoming) indicates that although Roman cities provided the convenience of public latrines, their knowledge of hygiene was poor. Latrines were dark, which cut down on flies, but difficult to clean.

11. During this interval of residence at their port, the citizens of Priene were invited to Panhellenic games under the name of Naulochos (Botermann 1994), and they belonged to the Ionian confederation (according to Herodotus) and to the Athenian League. However, they paid only a modest assessment of one talent, an amount that indicates the small size of the town (Landucci Gattinoni 1992; von Berchen 1979). For a while, Priene and Naulochos had the same coin images, which has caused some confusion between them (Rumschied 1994), but I think this can be explained by the fact that for a while they were one and the same.

12. Ever since I learned of Alexander the Great's formation of a great army strong enough to defeat the Persian army in battle and to take over the Persian empire, and confident enough to turn their backs on a defeated enemy and go off to India, I have wondered not only *how* they did it but, even more, *why* they were impelled to begin. I have speculated that it could be the decades-long, drought-induced famine that motivated Alexander and his armies to risk so much in conquering the Persian empire and exploring all the way to India (cf. Camp 1982). Because of famine, ancient people had to risk more and more of their accumulated resources, taking on wage labor, selling livestock, mortgaging land, splitting the household, selling the land, and finally migrating permanently to a safer and richer area (Gallant 1991: 141, Fig. 5.1). At such a time, the dangers of belonging to a conquering army were less than those of staying at home to starve. It seems to me that a quarter-century or a three-decade-long drought could be enough to make the survivors ready to risk all on a hazardous venture. The expansion of Greek culture into the Middle East was necessary for the Macedonians. This elaborate scheme suggests a leader or an army that does not expect to return to Macedonia in a few months. The brain and manpower drain continued through most of the third century, although Alexander died in 323, as the drought was intensifying.

13. Athenaeus, a fifth-century B.C.E. historian, reported that Ephesus was founded in response to a Delphic foundation oracle (*Delpnosophistae* 8.361) in about 900 B.C.E.

14. Architects Chersiphron of Knossos and Theodoros of Samos, who began the temple in 565–40 B.C.E., wrote a book (which has since been lost), about the building and devised methods for moving the enormous stones needed for column drums and other elements of the Temple (Placzek 1982, vol. 1: 413).

15. In the 1950s, schemes for development of the river noted that there were 65,000

hectares of drainable marsh and 7000 hectares of area to be protected from flooding (Karatekin 1953).

16. A map with three of the Ephesus quarries and the Teos quarry (but not so designated) is published as figure 1 in Moens et al. 1988a, and a detail showing Teos and Ephesus (Efes) by Herz on page 89 of the same book.

17. Published in Engelmann et al. 1980.

18. We are grateful to excavator Prof. I. H. Ataç for permission to inspect and photograph this reservoir.

19. It is useful to compare harbor management at Ephesus with that along the coast of Israel (Raban 1988). Bronze Age coastal changes in Israel resemble those at Ephesus and Miletus during the classical period. Protected sites were soon full of silt unless they were flushed by artificial channels. Cutting artificial outlets pointing away from the wind kept estuaries free of sand. The new channel served as both a drain and an extra boat entrance to the harbor and an alternate path for winter storms carrying heavy debris. Wave-carried sand could be used as a building material, as discussed by Oleson (1985) for the harbor at Sebastos. The undeniable relation of the ports to the evolving geology of the site at Ephesus has been an interactive process between nature and human action.

20. Kraft remarked on the challenge of excavation in the harbor because the bottom was sticky and stinking black mud from residential latrines (personal communication).

21. My study of Palmyra, Syria (Crouch 1969), indicates that the big shift in trade routes came not in the second century but in the third quarter of the third century c.e., when the Sassanian conquest of Mesopotamia blocked overland routes to India, shifting trade to sea routes that led to Egyptian ports.

Chapter 7

1. In South America, the use of paleoclimate data to reconstruct models of civilizations is much more advanced than in Old World studies (Ortloff and Kolata, 1993)

Chapter 8

1. See Corn (1996) for an interesting discussion of how historians use objects.

2. The books of Peter F. Smith, architect and psychologist, who studies urban design (1979 ff.) may be consulted for illuminating discussion of the interaction between external arrangements and internal satisfactions.

Bibliography

Journal Abbreviations

AA	*Archäologischer Anzeiger*
AAAG	*Annals of the American Association of Geographers*
AAL/R	*Atti D. R. Accademi dei. naz. Lincei/Rendiconti*
AJA	*American Journal of Archaeology*
ANNPISA	*Annuario della Scuola Normale di Pisa*
ANNSCATENE	*Annuario della Scuola Archeologica di Atene e delle Mission Oriente*
ANRW	*Aufstieg und Niedergang der römischen Welt*
ARCHCL	*Archeologia Classica*
ASS	*Archivo Storia Siciliano*
BAR	*British Archaeological Reports*
BCH	*Bulletin de Correspondance Hellénique*
BSA	*Annual of the British School at Athens*
Bull. Geol. Soc. Greece	*Bulletin of the Geophysical Society of Greece*
Bull. Geol. Soc. Turkey	*Bulletin of the Geological Society of Turkey*
Earth Plan. Sci. Lett	*Earth and Planetary Science Letter*
EOS	*EOS Transactions of the American Geophysical Union*
FDELPHES	*Fouilles de Delphes*
FiE	*Forschungen in Ephesos*
IASH Bull.	*Bulletin of the International Assn. of Scientific Hydrology*
Ist. Mitt	*Istanbuler Mitteilungen*
IvE	*Inschriften von Ephesos*
JDAI Istanbul	*Jahrbuch of the German Archaeological Institute, Istanbul*
JFA	*Journal of Field Archaeology*
JGEOPHYS	*Journal of Geophysical Research*
JHS	*Journal of Hellenic Studies*
JÖAI	*Jahresheften des Österreichisches Archäologisches Institut, Vienna*
JRS	*Journal of Roman Studies*
JSAH	*Journal of the Society of Architectural Historians*

MÉLROME   *Mélanges d'Archéologie et d'Histoire de l'École Francaise*
                               de Rom
Miner. Petrogr. Acta   *Mineral Petrograph Acta*
NAC   *Numismatic e Antichita Clasica*
NFg   Neue Folge
NOTSC   *Notizie degli Scavi di Antichità*
REG   *Revue des Études Grecques.*
RGPGD   *Revue de Geographie Physique et de Geologie Dynamique*
Z. Deutsch. Geol. Ges.   *Zeitschrift der Deutsches Geologische Geschichte*
ZPE (Bonn)   *Zeitschrift für Papyrologie und Epigraphik*

Ancient Sources

Aelian, C. A. (170–235 C.E.), *Various Histories.* IX.421

Ammianus Marcellinus (330–395 C.E.). *The Chronicles of Events.* Ed. and trans. J. Rolfe. London, 1935.

Aristotle (384–322 B.C.E.). *Politics,* 330a&b; 1267b22–1268a14 (Miletus); 1327a *Meteorologia,* 1.14 (Lerna); Pseudo-Aristotle, *De Mundo* (listed below); 1330a (town siting); 1330b (malaria).

Arrianus (died ca. 180 C.E.). *Anabasis Alexandri (History of Alexander the Great).* 24.5 (Priene).

Athenaeus (5th century B.C.E.) historian, cited in Strabo.

Chronicle of Paros, a historic epigraphic text: Ol.CXXXV/IG; XII.v.444.

Cicero (106–43 B.C.E.). *II Verres* I2.75, II.3.136, II.4,43 and 75; 11.4; IV(Second Pleading).53, N.S. 118 (all on Syracuse) (pp. 477–78 in the R. Garnett translation, 1900)

Corpus Inscriptionis Latinae (CIL). x, 1624 (Ephesus earthquake)

Corpus Instriptionum Graecarum (CIG). 1843 ff.

Corpus Scriptorum Historiae Byzantinae. I. Belker, ed. 838–39. Includes: Kedronos, G. *Synopsis Historion.* Vol. 24.

Diodorus [Siculus] (fl. 60–30 B.C.E.). *History.* IV.23, 4; V.2–4; X.26 (Argos); XI.25 and 78 (Morgantina); XII.10 and 82.3 (Selinus River); XIII.63.3 and XIV.1.1 (Selinus defenses); XIII.84.1 (lake at Agrigento); XIII.6.3.3 and XIV.1.1 (Solinus defenses); XVI.17.5 (Selinus abandoned by Syracuse in 405); XIII.43–44 (Selinus-Segestsa struggles); XIV.7.1, 70.5, 71.1, 421; XIV.42 (Syracuse); XV.49.5 (dolines); XVI.17.5 (Selinus defenses); XX.56.3 and 79.5, XXI.82.3 (Selinus-Segestsa struggles); XXII.10.2 (Selinus attacked); XXIV.1.1.

Dioscorides (Dioscurides) (40–90 C.E.). *De Materia Medica.* I.73.1 (Temple of Asklepios at Agrigento).

Frontinus (35–103 C.E.). *Water Supply of the City of Rome.*

Glycas, M. (12th c. C.E.). *Michaelis Glycae Annales, Corpus Scriptorum Historiae Byzantinae.* vol. 16, 1838: 473.

Gyu-Zhi, an ancient Tibetal medical treatise, quoted in *Tibetan Medicine and Psychiatry,* cited by Hirsch 2000.

Herodotus (484?–430/420 or 490–425 B.C.E.), *History.* I,17 (Lydians)); I,142 (grain at Agrigento); 1,146 (Greeks at Miletus marrying Carians); VI,7,8,28,76 (dolines); VII,153 (Rhodian foundation of Gela and Akragas).

Hesiod (fl ca. 800 B.C.E.). *Works and Days.*

Hieronymus, see Jerome.

Hippocrates (fl. ca. 460 B.C.E.). *Airs, Waters, Places.* Bk. 7 (malaria).

Historia Augusta (Augustan History). (4th c. C.E.). Trans. D. Magie. Harvard University Press. 1960–61.

Homer (9–8th or 8th–7th c. B.C.E.), *Iliad; Odyssey* 1.1884 (Miletus).

Inscriptions of Ephesos (I.E.) 1062 (Mt. Pion and water); 1441 (Ephesus walls).

Jeremiah (7th–6th c. B.C.E.), 2.13 and 38.6 (cisterns).

Jerome (Hieronymus). (347–419/20 C.E.). *Vita Sancti Hilarionis* and *Commentiorum in Isaiam.* In *Patrologia Graeco-Latina*, Vols. 23 and 24.

Justinian (3rd century C.E.). *Novellae.* XXIV.viii.9.

Kedronos, G. (alt.: Cedrenus, Cedrinus) (fl. ca. 1100). *Synopsis Historion.* I. Bekker, ed. Vol. 24 of *Corpus Scriptoriumm Historiae Byzantinae*, 1838–39.

Libanius (314–393 C.E.). *Libanni Opera.* R. Föster, ed. Vol. 2. Leipzig: Or. 118.291.

Livy, T. (59/64 B.C.E.–17 C.E.). *History.* XXV,26.8 (Syracuse siege) XXVI,40 (Agrigento).

Orosius (fl. 414 C.E.). *Histories against the Pagans.* V,6,2 (Mt. Etna).

Palladius (fl. 250 C.E.). *De Re Rustica (On Farming).* 12 vol.

Patrologia Graeco-Latina. Includes Jerome (4th c.), Marianus (709) and Sartori. J.-P. Migne, editor and publisher. Paris, 1856–75. 93 volumes.

Pausanias (143–176 C.E.), *Description of Greece.* P. Levi. translator. I.21.9 (Lerna); II.15–16 and 18–19 (Peloponnesos); II.15.5 and II.24.6 and 25.4–5 (Peloponnesos); II.17.4–5, II.20.6 and II.25.8 (Argos Heraion); II.37.1 and II.38.2 (Lerna); IV.35, 4.4.3 (Syracuse); III,7,5–6, IV.4.3, 23.3 and IV.34.9–10 (Argos); IV.35.2 (Peloponnesos); V.7.3 (Syracuse foundation oracle); VI.2.4f (Artemis at Ephesus); VII.2.5 (Miletus); VII.2.8. (Ephesus); VII.24.6 (dolines); VIII.24.5. (deforestation); VIII.72 (Peloponnesos); X.4.1 (urban elements); X.5.9, X.9.1, and X.10.5 (Delphi); X.37.5 7 (Lerna).

Pindar (518/522–after 446/438 B.C.E.). *Odes.* [C.M. Bowra, trans. London: Penguin Books, 1969]. O1.xiii,61 (Peirene Fountain at Corinth). Scholia XIII,32, on Acrocorinth.

Plato (428/27–348/47 B.C.E.), *Letters* II,313a–b; III,319a; VII,347a; VIII,349c.; *Laws (Leges et Epinomis).* IV.704d; *Critias*, III.D–E (deforestation).

Pliny the Elder (23/24–79 C.E.), *Natural History* II,93; IV.xii or xiii.66 (earthquakes, V.29 (Ephesus); V.31 (grain at Agrigento); V.112 (Miletus); XIII.29 (deforestation); XXXI.86 (Agrigento); XXXV 179,184 (bitumen found near Agrigento); XXXVI.1–.159–60 (kilns).

Plutarch (46–119+ C.E.), *Lives: Lysimachus.* 3 vols. K. Ziegler, ed. Teubner XXX (1914–35).

Polybius (200–118 B.C.E.). *Histories.* I.v.5, V.lxxxviii–lxxxix (Temple of Zeus at Agrigento).

Procopius (ca. 500–after 562 C.E.). *Wars; Buildings; Secret History.*

Pseudo-Aristotle. *De Mundo.* Adapted and translated (from Greek to Latin) by Apuleius (124–179 C.E.). Includes a section on earthquakes.

Seneca, L. A. (4 B.C.E.-65 C.E.). *Naturales Quaestiones* and *Epistulae Moral.*

Servius (fl. 4th century C.E.). *As. Aeneus* (commentary on Virgil). VII.84 (fountains).

Solinus, C. Julius. (3rd c. C.E.). *Itinerarium Antonini Augusti.* V.18–19.

Solon (630–560 B.C.E.). Code of Laws for Athens.

Strabo (64/63 B.C.E.–21/24 C.E.). *Geography.* I.3.16 (Ephesus); III.5.12 (Selinus); V.2.5 (deforestation); VI.2.4 (Morgantina); VI.2.7 (Agrigento); VI.2.9 (dolines); VI.5.22 (foundation of Syracuse); VIII.6.8 (dolines); VIII.6.11 (Argos); VIII.6.15 (Argos vs. Tiryns); VIII.6.22 (Corinth settlers to Syracuse); VIII.24.5 (Lerna); XI.14.9; XII.3.300; XII.8.15–17 (boiling fountains at Miletus); XIII.3.19; XIV.1.3 (foundation of Miletus); XIV.1.6 (Miletus); XIV.1.21 (Ephesus); XIV.2.6 (Lerna); XIV.2.9 (Miletus); XVI.1.24 (dams at Ephesus); XIV.6.54 (to come); XXII–XXIV (Priene). On earthquakes: I.iii.16, IX.421, XII.viii.17–18, XIV.ii.5. Reference to *Dephosophistae* VIII.3.61.

Tacitus, P. or G. C. (ca. 55/56-ca. 120 C.E.). *Histories.*

Theophrastus (390 or 372–287/86 B.C.E.). *Inquiry into Plants (De Historia Plantarum)*. V.1, X.6 (Sicilian grain); III.ii.5.6; IV.1.2–5; V.2–7; VII.3; VIII.4.4–4.6 (grain from Agrigento); XVII.3.6.13.

Thucydides (ca. 460–ca.400 B.C.E.). *History of the Peolponnesian War*. II.93, II.viii.4 (earthquakes); VI.90, VI.103, VII.14.3, VII.25.1–2 (Syracuse).

Virgil/Vergil (70 B.C.E.-19 C.E.). *Georgics* I.103 (grain at Agrigento); II.520.

Vitruvius, M. P. (first century B.C.E.). *The Ten Books on Architecture*. VII.2 (stucco), 6 (marble); VIII.1.6–7 (water); X (machines); M. H. Morgan, trans. and ed. New York: Dover Publications, 1960.

Xenophon (431–before 350 B.C.E.). *Hellenica*. IV.vii.44–45

Modern Sources

Abaddie-Reynal, C. 1998. Un example de régression du phénomène urbain: Argos aux IVe– VIIe siècles ap. J.-C. In Pariente, A., and G. Touchais, eds. *Argos et l'Argolide: Topographie et urbanisme*. École Française d'Athènes. Paris: De Boccard Édition. 397–402.

Abrams, E. 1994. *How the Maya Built Their World: Energetics and Ancient Architecture*. Austin: University of Texas Press.

Abrams, P., and E. W. Wrigley, eds. 1978. *Towns in Societies: Essays in Economic History and Historical Sociology*. Cambridge: Cambridge University Press.

Adam, J.-P. 1989. Consequences du seisme de l'an 62 à Pompei. In Helly, B. and A. Pollino, eds. *Tremblement de terre: Histoire et archeologie*. IV. *Rencontres internationales d'archeologie et d'histoire d'Antibes*, 2–4 Nov. 1983. Vallbonne. 165–68

Adamesteanu, D. 1956. Saraceno ed il problema della penetrazione Rodio-Cretese nell Sicilia Meridionale. *ARCHCL* 8: 121–96, 26 plates, and map.

Agnello, S. L. 1954. Recenti explorazioni nelle catacombe syracusane di S. Lucia. I. *RAC* 30: 7–60.

———. 1955. Recenti explorazioni nelle catacombe syracusane di S. Lucia. II. *RAC* 31: 7–50.

Aksu, A. E., T. Konuk, A. Ulug, M. Duman, and D. J. W. Piper. 1987. *Quaternary Tectonic and Sedimentary History of Eastern Aegean Shelf Sea Area*. Chamber of Turkish Geological Engineers, Ankara, Turkey.

Akurgal, E. 1995. *Anadolu Uygarliklan (Anatolian Civilization)*. 5th ed. Istanbul: Net Yayinlan. Cited in Altunel, 1998.

Alaimo, R., and P. Ferla, 1981. *Guidebook for the Excursion in Sicily*. AIPEA. 7th International Clay Conference, Sept. 6–12, Bologna and Pavia, Italy. Palermo: International Assn. for the Study of Clays.

Albritton, C. C. Jr. 1980. *The Abyss of Time*. San Francisco: Freeman, Cooper. Reprinted 1986, Los Angeles: J. P. Tarcher, Inc.

Alcock, S. 1993. *Graecia Capta: The Landscape of Roman Greece*. Cambridge: Cambridge University Press.

Alkanoglu, E. 1978. Geologiisch-petrographische und geochemische Untersuchungen am Südostrand des Menderes Massivs in Westanatolien/Türkey. Publ. Diss., Univ. Bochum, 1–166, 1 Anhang, 1 Geol. Karte, sahlr. Abb. Bochum.

Allan, W. 1970. Ecology, Techniques, and Settlement Patterns. In Ucko, P. J., R. Tringham, and G. W. Dinbleby, eds. *Man, Settlement and Urbanism: Proceedings of the Research Seminar in Archaeology and Related Subjects*. Cambridge, Mass.: Schenkman Publishing 211–25.

Allen, H. 1970. Excavations at Morgantina (Serra Orlando), 1967–69: Preliminary Report X. *AJA* 74 : 359–83 and plates 911–98.

————. 1972–73. Per una definizione della facies preistorica di Morgantina: l'età del ferro. *Kokalos* 18–19: 147 ff.

————. 1976–77. The Effect of Population Movements and Diffusion on Iron Age Morgantina. *Kokalos* 22–23: 479 ff.

Almeida-Taxeria, M. E., ed. 1990. *Proceedings of the Course on Natural Hazards and Engineering Geology: Preservation and Control of Landslides and Other Mass Movements.* Luxembourg: Commission of the European Communities.

Altenhöfer, E. 1974. Die Theaterhalle von Milet, *Mélanges Mansel (Mansel'e Armagan).* Ankara: Türk Tariih Kuurumu Basimevi: 607 ff.

Altunel, E. 1998. Evidence for Damaging Historical Earthquakes at Priene, Western Turkey. *J. of Earth Sciences* 7: 25–35.

Alzinger, W. 1970. Ephesos: Wasserversorgung, Nymphaeum, Brunnenhäuser, München, Druckenmüller. In *Paulys Realencyclopädie der classischen Altertumswissenschaft.* Suppl. 12: 1604–08. ·

————. 1972. *Die Ruinen vom Ephesos.* Berlin-Vienna: Verlag A. F. Koska.

Amadori, M., et al. 1992a. Integrated Interpretation of Geophysical Survey Results. In *Bollettino di Geofisica Teorica ed Applicata* 34.134–35 (June–September):: 215–17.

Amadori, M., N. Feroci, and I. Versino. 1992b. Geological Outline of the Selinunte Archaeological Park. In *Bollettino di Geofisica Teorica ed Applicata* 34.134–35 (June–September): 87–99

Amandry, P. 1952. Observations sur les monuments de l'Héraion d'Argos. *Hesperia* 21.

————. 1977. La Fountain Castalie. *BCH* Suppl. 4: 195–227

————. 1978. Notes de topographie et d'architecture delphiques. 7: La Fontaine Castalie (compléments). *BCH* 102.1: 221–41.

————. 1988. À propos de monuments de Delphes: Questions de chronologie (1): *BCH* 112: 591–610.

————. 1993. L'opisthodome du Temple d'Apollon: Notes de topographie et d'architecture delphiques. *BCH* 117: 263–83.

Ambraseys, N. N. 1970. *Early Earthquakes in the Near and Middle East,* pt. 11: *Historical Earthquakes before 17 a.d.* Reports No. SC/1473/69 and SC/21299/70, Paris: UNESCO. 70.

————. 1971. Value of Historical Records of Earthquakes. *Nature* 232 (Aug. 6): 375–380.

————. 1973. Earth Science in Archaeology and History. *Antiquity* 47: 229–31

————. 1975. Studies in Historical Seismicity and Tectonics. *Geodynamics of Today.* Royal Society of London: 9–16.

————. 1978. Historical Seismicity and Tectonics. In W. C. Brice, ed. *The Environmental History of the Near and Middle East Since the Last Ice Age.* London: Academic Press. 185–210.

————. 1994a. Material for the Investigation of the Seismicity of Libya. *Libyan Studies* 25: 7–22.

————. 1994b. Material for the Investigation of the Seismicity of Central Greece. *Materials of the CEC Project "Review of Historical Seismicity in Europe"* Vol. 2, 1994.

————. 1996. *Seismicity of the East Mediterranean and Middle East.* I: *Engineering Seismology and Earthquake Engineering.* Research Report No. 96–3 (June). Civil Engineering Dept., Imperial College, London.

Ambraseys, N. N., and C. F. Finkel. 1992. The Seismicity of the Eastern Mediterranean Region during theTurn of the Eighteenth Century. *Ist. Mitt.* 42: 148–62. 323–43.

————. 1995. *The Seismicity of Turkey and Adjacent Areas 1500–1800.* Istanbul: Eren.

Ambraseys, N. N., and J. A. Jackson. 1997. Seismicity and Strain in the Gulf of Corinth (Greece) since 1694. *J. of Earthquake Engineering* 1.33: 433–74.

Ambraseys, N. N., and C. Melville. 1988. In Lee, W. H. K., ed. *Historical Seismograms and Earthquakes of the World*. New York: Academic Press. 181–200.

Ambraseys, N. N., and P. Pantelopoulos. 1989. The Fokis (Greece) Earthquake of 1 August 1870. *European Earthquake Engineering* 1: 10–18.

Ambraseys, N. N., and D. P. White. 1996. *Seismicity of the East Mediterranean and Middle East?I*. ESEE Research Report No. 96–3, June. London: Civil Engineering Dept., Imperial College.

Amore, C., G. Gandolfi, E. Giuffrida, L. Pagganelli, and A. Zanni. 1979. A caratteristiche morfologiche, tessiturali e composizionali del litorale del Golfo di Cania. *Miner. Petrogr. Acta* 23: 47–75.

Andolfo, U. 1967. Considerations on Economic Posibilities and Development of Turkish Marbles. *Bull. Min. Res. Expl. Inst.*, Turkey 68 (April): 125–35.

Andreau, J. 1973. Histoire des séismes et histoire économique: le tremblement de terre de Pompéi (2 ap. J.-C.). *Annales; économies, sociétés, civilisations* 28: 369–95

Andrews, G. F. 1975. *Maya Cities*. Norman: University of Oklahoma Press.

Angeliakis, A. N., and A. S. Issar, eds. 1995. *Diachronic Climatic Impact on Water Resources*. NATO ASI Series. Berlin: Springer Verlag.

Antonaccio, C. 1997. Urbanism at Archaic Morgantina. *Acta Hyperborea* 7: 167–89.

Appleton, J. 1986. Some Thoughts on the Geology of the Picturesque. *Garden History* July-Sept.: 270–91.

———. 1986. The Role of the Arts in Landscape Research. In Penning-Rousell, E. C., and D. Lowenthal, eds. *Landscape Meanings and Values*. London: Allen and Unwin. 26–47.

Archeoclub d'Italia, Sede di Agrigento. May 1996. *Gli ipogei di Agrigento: valorizzazione e fruizione*. Atti del Convegno Studi, Agrigento 16, with new map of the *ipogei* of Girgenti (historical center).

Armandry, P., J.-M. Luce, and D. Rousset.1990. Autour de Delphes. In Delphes: Races cultes et jeux: 100 ans de fouilles et de decouvertes. *Les Dossiers d'Archaeologie* 15 (July–August): 26–31.

Armijo, R., H. Lyon-Caen, and D. Papanastassiou. 1991. A Possible Normal-Fault Rupture for the 464 b.c. Sparta Earthquake. *Nature* 351: 137–39.

Armijo, R., B. Meyer, G. C. P. King, A. Rigo, and D. Papanastassiou. 1996. Quaternary Evolution of the Corinth Rift and Its Implications for the Late Cenozoic Evolution of the Aegean. *Geophysical J. International* 126: 11–53.

Arnone, L. 1948. *Gli ipogei dell'Agrigentino*. Reprinted Agrigento: Edizioni Ente Provinciale per il Turismo, 1991, with additional essays.

Aschermann, H. 1973. Man's Impact on the Several Regions with Mediterranean Climates. In di Castri, F., and H. A Mooney., eds. *Mediterranean Type Ecosystems: Origin and Structure*. New York: Springer-Verlag. 363–71.

Asgari, N. 1989. Zwei Werkstücke für Konstantinopel aus denprokonnessischen Steinbrüchen. *Ist. Mitt.* 39: 49–63.

Ashby, T. 1935. *The Aqueducts of Ancient Rome*. Oxford.

Asheri, D. 1980. La colonizzazione greca. In Gabba, E., and G. Vallet, eds. *La Sicilia antica. VI: Indigeni, Fenici-Punici e Greci*. Naples: Società Editrice Storia di Napoli del Mezzogiorno Continale e della Sicilia. 89–142.

———. 1988. A propos des sanctuaires extraurbains en Sicile et Grande-Grèce: Théories et témoignages. In Mactoux, M.-M., and E. Geny, eds. *Mélanges Pierre Lévéque*. Annales Littéraire de l'Université de Besancon 367, Vol. 79. Paris: Centre de Recherches d'Histoire Ancienne. 1–15.

Assessorato Regionale Beni Culturali Ambientali di Agrigento. ca. 1996. *La valle dei templi:*

Tra iconografia e storia. Published at Agrigento. For papers by R. Mirisola, E. Bongiovanni, and G. Piccione, see individual citations. Also contains essays by E. De Miro, G. Schirò, L. Mascoli and G. Vallet, G. Fiorentini, G. Giarrizzo, and L. Trizzino.

Associazione Geotecnica Italiana. 1991. *The Contribution of Geotechnical Engineering to the Preservation of Italian Historical Sites*. Florence, 1991. Includes: Introduction, 13–36; Humanities and Geotechnics, 37–50; and case studies, including three by Ercoli and colleagues.

Atalay, E. 1976–77. Antiker Marmorsteinbruch bei Ephesos. *JÖAI* 51: 59–60.

Attas, M., J. M. Fossey, and L. Yaffe. 1987. An Archaeometric Study of Early Bronze Age Pottery Production and Exchange in Argolis and Korinthia (Corinthia), Greece. *JFA* 14.1 (spring): 88–90.

Atzori, P., C. D'Amioco, and A. Pezzino. 1974. Realizione geo-petrografica preliminare sul cristallino della acatena peloritana (Sicilia). *Rivista Minneraria Siciliana* 148–150: 1–8.

Atzori, P., G. Bruno, D. P. Crouch, P. Mazzolini, and A. Pezzino. 1995. Hydrogeological, Petrographical and Geochemical Characterization of Morgantina (Sicily, Italy) Hydraulic Network. *Proc. of First International Congress: Science and Technology for the Safeguard of Cultural Heritage in the Mediterranean Basin*. Nov. 27–Dec. 2, Catania-Syracuse, Italy. 407–15.

Aupert, P. 1977. Un édifice dorique archaïque à l'emplacement du stade. *BCH* Suppl. 4: 229–245.

———. 1979. *FDELPHES*. II: *Topographie et architecture: Le stade*. Paris: De Boccard Édition: passim, especially 19–31, 45, 81–82, 173.

———. 1981. Argos II: Patirimoine et urbanisme. *BCH* 105: 893–99.

———. 1982. Argos: Thermes A. *BCH*: 637–39.

———. 1985. Un sérapieion argien? *CRAI*: 151–175.

———. 1988. Argos: Temple et Thermes A. *BCH* 112: 710–15.

———. 1989. Argos: l'Aqueduc nord. *BCH* 13. II: Études chroniques et rapports. 772–31.

———. 1991. Les thermes comme lieus de culte. In *Les Thermes romains: Actes de la table ronde organizée par l'École Française de Rome*.

———. 1994. L'eau curative à Argos. In Ginouvés, R., A.-M. Guimier-Sorbets, J. Jouanna, and L. Villard, eds. *L'eau, la santé et la maladie dans le monde grec*. Paris: Actes de Colloque à Paris, 25–27 Nov. 1992. 193–200.

Aureli, A. 1983. Problems arising from surexploitation of coastal aquifers in Sicily. *GA & I* 157–68.

Aureli, A., and M. De Maio. 1994. Fresh Water Supplying Syracuse at the Time of the Greeks and Romans. Paper for Geology and Urbanization conference in Athens. Manuscript in the files of D. P. Crouch.

Aureli, A., D. P. Crouch, L. Ercoli, and M. De Maio. 1997. Agrigento and Syracuse: Geology and the Human Environment. In Marinos, P. G., and G. C. Koukis, eds. *The Engineering Geology of Ancient Works, Monuments and Historical Sites: Proceedings of the Symposium of the International Association of Engineering Geology*. Athens. Rotterdam: A. A. Balkema. 305–64.

Aurenhammer, M. 1996. Sculptures of Gods and Heroes from Ephesus. In Koester, H., ed. *Ephesos: Metropolis of Asia*. Valley Forge, Pa.: Harvard Theological Studies, Trinity Press International. 251–80.

Autino, P. 1987. I terremoti nella Grecia classica. *Mem. Inst. Lombardo*, Milan. 388 no. f: 355–446.

Badian, E. 1966. Alexander the Great and the Greeks of Asia. In *Ancient Society and Institutions*. Oxford: Basil Blackwell. 37–69.

Baird, D. W. 1971. Review of Mediterranean Alpine Tectonics. In Campbell, A. S., ed. *Geology and History of Turkey.* Petroleum Exploration Society of Libya, 13th Annual Field Conference. 139–58.

Baker, D. G. 1981. Botanical and Chemical Evidence of Climatic Change. In Rotberg, R. I., and T. K. Rabb, eds. *Climate and History: Studies in Interdiscliplinary History.* Princeton, N.J.: Princeton University Press. 233–40.

Baker, M. N., and R. E. Horton. 1936. Historical Development of Ideas Regarding the Origin of Springs and Ground-water. *EOS: Transactions of the American Geophysical Union* 395–400.

Ballance, M. H., and O. Brogan. 1971. Roman Marble: A Link between Asia Minor and Libya. In Campbell, A. S., ed. *Geology and History of Turkey.* Petroleum Exploration Society of Libya, 13th Annual Field Conference. 33–36.

Bammer, A. 1961–63. Zur Topographie und städtbaulichen Entwicklung von Ephesos. *JÖAI* 46: 136–57.

———. 1964–65. Zum jüngeren Artemision von Ephesos: Die Terrasse des jüngeren Artemision und eustatische Schwankungen des Meeresspiegels. *JÖAI* 47: 126–31.

———. 1968–71. Ein Rundfries mit Bukranien und Girlanden. *JÖAI* 49: 33–37.

———. 1972a. Architektur. Sonderdruck der *JÖAI* 50, Beiblatt: 381–406.

———. 1972b. *Die Architektur des jüngeren Artemision.* Wiesbaden: F. Steiner. Includes Die geographische Lage des Heiligtums, 3 ff.

———. 1974. Recent Excavations at the Altar of Artemis in Ephesus. *Archaeology* 27: 202–5.

———. 1983. Architecture et société en Asie Mineure au IVᵉ siècle. *Architecture et Société.* Rome: École Français de Rome. 271–300.

———. 1986–87. Ephesos in der Bronzeit. *JÖAI* 57: 1–47.

———. 1988. *Ephesos: Stadt an Fluß und Meer.* Graz: Akademische Druck- u. Verlagsanstalt.

———. 1995. Natur und Ruine in Ephesos. *JÖAI*: 98–115.

Bammer, A., R. Fleischer, and D. Knibbe. 1974. *IvE* I.23 (1.1a), with text repeated in *Führer durch das archäologische Museum in Selcuk-Ephesos.* 98–100.

Banaka-Dimaki, A., A. Panayotopoulou, and A. Oikonomou-Laniado. 1998. Argos à l'époque romaine et paléochrétienne: Synthèse des donnés archéologiques. In Pariente, A., and G. Touchais, eds. *Argos et l'Argolide: Topographie et urbanisme.* École Française d'Athènes. Paris: De Boccard Édition. 327–36.

Barajaru-Gléni, K. 1998. Contributions à l'étude de la topographie de la ville antique d'Argos. In Pariente, A., and G. Touchais, eds. 1998. *Argos et l'Argolide: Topographie et urbanisme.* École Française d'Athènes. Paris: De Boccard Édition. 271–90, with maps.

Barajaru-Gléni, K., and A. Pariente. 1998. Argos du VIIᵉ au IIᵉ siècle av. J.-C. In Pariente, A., and G. Touchais, eds. *Argos et l'Argolide: Topographie et urbanisme.* École Française d'Athènes. Paris: De Boccard Édition. 165–78.

Barbano, M. S., and M. Cosentino. 1981. Il terremoto siciliano dell'11 genaio 1693. *Rend. Soc. Geol. It. 4.*

Barone, V., and S. Elia, 1979. *Selinunte: Vicende storiche-illustrazione dei monumenti.* Palermo: S. F. Flaccovio Edizioni.

Barra Bagnasco, M. 1996. Housing and Workshop Construction in the City. In Pugliese-Carratelli, G., ed. *The Western Greeks.* Venice: Bompiani. 353–436.

Bartlett, W. H. 1853. *Pictures from Sicily.* London: A. Hall, Virtue.

Basile, B., G. De Stefano, and G. Lena. 1988. Landings, Ports, Coastal Settlements and Coastlines in SE Sicily from Prehistory to Late Antiquity. In Raban, A., ed. *Archaeology of Coastal Changes.* BAR Int. Series 404.2: 15–33.

Beck, B. F., and W. L. Wilson, eds. 1987. *Karst Hydrogeology: Engineering and Environmental Applications*. Rotterdam: A. A. Balkema.

Bell, Malcolm. 1976. Il tempio greco in Sicilia: Architettura e culti. *Atti della 1a riunione cientifica della Scuola di Perfezionamento in Archeologia Classica dell'Universita di Catania. Siracusa, 24–27 Nov. 1976*: 140–47 and plates 41 and 42.

———. 1980. Stylobate and Roof in the Olympieion at Akragas. *AJA* 84: 359–72 and plate 47.

———. 1986–87. La fontana elleniistica di Morgantina. *Quad. Messina.* (Instituto di Archeologia) 2: 111 ff.

———. 1988. Excavations at Morgantina, 1980–88: Preliminary Report. *AJA* 92: 313–42.

———. 2000. The Agora of a Hellenistic Royal City. Lecture at the National Gallery of Art, Washington, D.C. Apr. 27.

Bell, Martin, and J. Boardman, eds. 1992. *Past and Present Soil Erosion: Archaeological and Geographical Perspectives*. Oxbow Monograph 22

Belloni, S., B. Martinis, and G. Orombelli. 1972. Karst of Italy. In Herak, M., and V. T. Stingfield, eds. *Karst: Important Karst Regions of the Northern Hemisphere*. Elsevier: Amsterdam. 85–128.

Benndorf, O. 1906. Zur Ostkunde und Stadtgeschichte von demmselben (Ephesos). *FiE* 1: 9–110.

Bennett, J. W. 1976. *The Ecological Transition: Cultural Anthropology and Human Development*. New York: Pergamon Press.

Bérard, Jacques. 1941a. *La colonisation grecque de l'Italie Méridoniale et de la Sicile dans l'antiquité*. Paris: De Boccard Édition.

Bérard, Jean. 1941b. *Bibliographie topographique des principales cités grecques de l'Italie Meridionale et de la Sicile dans l'antiquité*. Paris: De Boccard Édition.

Berchen, D. von. 1979. Alexandre et la restauration de Priene. *Museum Helveticum* 27: 198–205

Berger, S. 1991. Great and Small Poleis in Sicily: Syracuse and Leontinoi. *Historia* 40: 129–42.

Bessac, J. C. 1988. Problems of Identification and Interpretation of Tool Marks on Ancient Marbles and Decorative Stones. In Herz, N., and M. Waelkens, eds. *Classical Marble: Geochemistry, Technology, Trade*. NATO ASI Series E: Applied Sciences, Vol. 153. Dordrecht: Kluwer Academic Publishers; 41–53, with many illustrations.

Betson, R. P. 1977. The Hydrology of Karst Urban Areas. In Dilamarter, R. R., and S. C. Csallany, eds. *Hydrologic Problems in Karst Regions: Proceedings of the International Symposium, April 1976*. Bowling Green, Ky.: W. Kentucky University Press. 162–75.

Beven, K. 1987. Towards a New Paradigm in Hydrology. In Rodda, J. C., and N. C, Matalas, eds. *Water for the Future: Hydrology in Perspective*. Publication No. 164. Wallingford, Oxfordshire, U.K.: IAHS. 393–403.

Biers, J. C. 1985. *Corinth. XVII. The Great Bath on the Lechaion Road*. Princeton, N.J.: American School of Classical Studies.

Biers, W. R. 1978. Water from Stymphalos? *Hesperia* 47: 171–84, plates 45–49.

Bilde, P. G., I. Nielsen, and M. Nielsen, eds. 1993. *Aspects of Hellenism in Italy: Towards a Cultural Unity?* Danish Studies in Classical Archaeology, Acta Hyperborea 5, Museum Tusculanum P. University of Copenhagen.

Billot, M.-Fr. 1990. Note de typologie et d'histoire. *Hesperia* 59: 95–140.

———. 1992. Apollon Pythéen et l'Argolide archaique: Histoire et Mythes. *Arxaiognosia* (Athens) 6.1–2 (1989–90): 96–98.

Bintliff, J. L. 1975. Sediments and Settlement in Southern Greece. In Davidson, D. A., and

M. L. Shackley, eds. *Geoarchaeology: Earth Science and the Past*. Boulder, Colo.: Westview Press. 267–75.

———. 1977. *Natural Environment and Human Settlement in Prehistoric Greece*. Suppl. Series 28 i and ii. Oxford: BAR

———. 1992. Erosion in the Mediterranaean Lands: A Reconsideration of Pattern, Process, and Methodology. In Bell, Martin, and J. Boardman, eds. *Past and Present Soil Erosion: Archaeological and Geographical Perspectives*. Oxbow Monograph 22. 125–31.

Birks, H. H., ed. 1988. *The Cultural Landscape*. Cambridge: Cambridge University Press.

Birmingham, J. 1990. Epilogue. The Archaeology of Colonization: Observations from *Terra Australis*. In Descœudres, J.-P., ed. *Greek Colonists and Native Populations*. Oxford: Clarendon Press. 635–48.

Birot, P. 1959. Geomorphologie de la region de Delphes. *BCH* 83: 258–74.

Birsoy, Y. K. ca. 1992. Radial Flow towards a Karstic Spring. Manuscript in D. P. Crouch collection.

Bivon, L. 1990–91. Cataclismmi e calamita naturali: Loro incidenza nella vita socio-economica e politica della Sicilia tardo-antica—la documentazione epigrafica. *Kokalos* 36–37: 331- 339.

Blalock, H. M. Jr. 1964. *Causal Inferences in Nonexperimental Research*. New York: W. W. Norton. Reprinted 1972.

Blegen, C. 1920. Corinth in Prehistoric Times. *AJA* 24: 1–13.

Blegen, C., O. Broneer, and R. Stillwell. 1930. *Corinth III*. Part I. *Acrocorinth: Upper Peirene*. Princeton, N.J.: American School Classical Studies.

Blue Guide: Sicily. 1975. London: Ernest Benn.

Blue Guide: Turkey. 1983. Paris: Hachette.

Blumenthal, E. 1963. *Die altgriechische Siedlungskolinisation im Mittelmeerraum unter besonderer Berücksichtigung der südküstre Kleinasiens*. Geographische Studien, 10. Universität Tübingen

Blumenthal, M. 1941. Géologie des montagnes de la Transversale d'Eskipazar et leurs sources minérales (Vilâyet de Çankiri). Ankara, Turkey: Maden Tetkik ve Arama Enstitusu Mecmuasi 335035 and 571–93 and 4 pls.

Bogli, A. 1980. *Karst Hydrology and Physical Speleology*. Berlin: Springer-Verlag.

Bommelaer, J.-F. 1983. La construction du temple classique de Delphes. *BCH* 107: 191–216.

———. 1990. Les portiques des Delphes. *Annual of the British School at Athens* 85: 33–51.

———. 1991. *Guide de Delphes*. Paris: École Française d'Athènes and De Boccard Édition.

———. 1992. Le monument argien des "Sept contre Thèbes." *BCH* suppl. 22: 265–304.

Bommelaer, J.-F., and J. Des Courtils. 1994. *La salle hypiostyle d'Argos*. Etudes peloponnesiennes 10. Athens: French School.

Bommelaer, J.-F., P. Amandry, G. Roux, F. Croissant, D. Laroche, and J. Marcadé. 1990. Les sanctuaires et les temples. In Delphes: Races cultes et jeux: 100 ans de fouilles et de decouvertes. *Les Dossiers d'Archaeologie* 15 (July–August): 38–57.

Bonacasà Carra, R. M., 1987. *Agrigento Paleocristiana*. Palermo: Assessorato Regionale Beni Culturali Ambientali.

Bongiovanni, E. ca. 1996. Aspetto idrologica. *La valle dei templi: Tra iconografia e storia*. 12–13.

Bonnet, P. 1920. Plan and Construction [of Priene]. *Revue Archeologique* 11: 366–71.

Bookidis, N., and J. E. Fisher. 1972. The Sanctuary of Demeter and Kore on Acrocorinth. Preliminary Report for 1969–70. *Hesperia* 41: 283–331.

———. 1974. The Sanctuary of Demeter and Kore on Acrocorinth. Preliminary Report for 1971–73. *Hesperia* 43: 267–307.

Bookidis, N., and R. S. Stroud. 1987. *Demeter and Persephone in Ancient Corinth*. Princeton, N.J.: American School of Classical Studies.

Boschi, E., G. Ferrari, and E. Guidoboni. 1993. PERSEUS Project: A GIS (Geographic Information System) for Historical Seismic Data, Organisation and First Results. Presented at the 23rd European Seismological Commission, Prague, Sept. 7–12, 1992.

Boschi, E., E. Guidoboni, and D. Mariotti, with collaboration of Dr. Messina of the State Archive, Syracuse, Italy. 1998. I terremoti dell'area siracusana e i loro effetti in Ortygia. Part of the data-base PERSEUS. 15–36.

Bosi, C., R. Cavallo, and V. Francaviglia. 1973. Aspetti geologici e geologico-tecnici del terremoto della Valle del Belice del 1968. 2 maps of the "Valle del Belice: Zona maggiormente colpita dal terremoto del 1968." *Mem. Soc. Geo. It.* 12.

Boswell, P. G. H. 1936. Problems of the Borderlands of Archaoelogy and Geology in Britain. *Prehistorical Society Proceedings*, N.S. 2: 149–160.

Botermann, H. 1994. Wer baute des neue Priene? Zur Interpretation der Inschriften von Priene Nr. 1 und 156. *Hermes* 122: 162–187.

Bottema, S., G. Entijes-Nieborg, and W. van Zeist, eds. 1990. *Man's Role in the Shaping of the Eastern Mediterranean Landscape*. Rotterdam: A. A. Balkema.

Bous, B., P.-Y. Péchoux, and J.-J. Dufaure. 1984. Nature, répartition et histoire du risque seismique dans l'aire méditerranéenne. In *Homage à Pierre Birot: La mobilité des paysages méditerranéans*. Toulouse. 1984. *Revue Géographique des Pyrénées et du Sud-ouest*. Travaux II: chap.9, 277–301.

Bousquet, B. 1977. Etudes delphiques. *FDELPHES* 3.5: 91–105; *BCH* Supp 4.

Bousquet, B., J.-J. Dufare, and P.-Y. Péchoux. 1983. Effects of Earthquake; Comment repérer les effects seismiques dans les paysages méditerranéens. In Helly, B., and A. Pollino, eds. *Tremblement de terre: Histore et archeologie. 4: Rencontres internationales d'archeologie et d'historie d'Antibes, 2–4 Nov. 1983*. 23–37 and 39–62.

Bousquet, J. 1993. Le tholos de Delphes et les mathematiques pre-Euclidiennes. *BCH* 115.1: 285–313.

Bradford, J. 1957. *Ancient Landscapes*. G. Bell & Sons: London, 1957; reprinted Bath: C. Chivers, 1974.

Brein, F. 1976–77. Zur ephesischen Topographie. *JÖAI* 51: 65–76.

Brice, W. C. 1978. The Desiccation of Anatolia. In Brice, W. C., ed. *The Environmental History of The Near and Middle East since the Last Ice Age*. London: Academic Press: 141–47.

Brill, R. H., and J. M. Wampler. 1967. Isostope Studies of Ancient Lead. *AJA* 71: 63–77.

Brinker, W. 1990. Wasserspeicherung in Zisternen: Ein Beitrag zur Frage der Wasserverstörung früher Städte. Dissertation, Technical University of Braunschweig, Germany.

Brinkmann, R. 1971. The Geology of Western Anatolia. In Campbell, A. S., ed. *Geology and History of Turkey*. Petroleum Exploration Society of Libya, 13th Annual Field Conference: 171–209.

———. 1976. *Geology of Turkey*. Amsterdam: Elsevier.

Brizzolari, E., L. Orlando, S. Piro, and L. Versino. 1992. Ground Probing Radar Survey in Selinus Archaeological Park. *Bollettino di Geofisica Teorica ed Applicata* 34.134–35 (June–September): 181–19.

Broneer, O. 1954. *Corinth. I, Pt IV: The South Stoa and Its Roman Successors*. Princeton, N.J.: American School of Classical Studies.

Brückner, H. 1986. Man's Impact on the Evolution of the Physical Environment in the Mediterranean Region in Historical Times. *GEOJ* 13.1: 7–17.

———. 1990. Changes in the Mediterranean Ecosystem during Antiquity: The Geomorphological Approach as Seen in Two Examples. In Bottema, S., G. Entijes-Nieborg, and W. van Zeist, eds. *Man's Role in the Shaping of the Eastern Mediterranean Land-scape*. Rotterdam: A. A. Balkema. 127–37.

———. 1992. Human-Induced Erosion Processes in Mediterranean Countries: Evidences from Archaeology, Pedology and Geology. *Geoöko Plus* (Bensheim): 97–110.

———. 1993. Preliminary report on the work of J. Kraft and I. Kayan, and of H. Brückner. *JÖAI*: 25–26.

———. 1994. Geomorphologie und Paläo-environment der Milesia. *JDAI Istanbul*: 329–30.

———. 1995a. Geomorphologie und Paläo-environment der Milesia. *AA*: 329–33.

———. 1995b. Geoarchäoloigische Forschungen. *JÖAI*: 24–25.

———. 1996. Geoarchäologie an der türkischen Ägäisküste. *Geographische Rundschau (GR)*. 48/50.10: 568–74.

———. 1997a. Coastal Changes in Western Turkey: Rapid Delta Progradation in Historical Times. *Bulletin d'Institut Oceanographique* (Monaco), Special No. 18: 63–74.

———. 1997b. Geoarchäoloigische Forschungen in der Westtürkei: Das Beispiel Ephesos. In T. Breuer, ed. *Geoarchäoloigische Forschungen im Mittelmeerraum und in der Neuen Welt*. Passauer Schriften zur Geographie 15.

———. 1998. Coastal Research and Geoarchaeology in the Mediterranean Region. In D. H. Kelletat, ed. *German Geographical Coastal Research: The Last Decade*. Tübingen: Institute for Scientific Cooperation: 235–57, esp. "Deltas of Küçük Menderes and Buyük Menderes," 246–53.

Brückner, H., and W. W. Jungmann, with A. Look, I. Kayan, and J. Kraft. 1995. Geologische Sondierungen. *JÖAI*: 20–21.

Bruno, G., and S. Nicosia. 1994. Caratteri geologici e idrogeologici dell'area archeologica di Morgantina (Sicilia Centrale). *Atti del 2° semin. intern: "Il sistema uomo-ambiente tra passato e presente," 3–6 giugno, Ravello (Italia)*.

Bruno, G., and E. Renna. 1998. La rete idrica di Morgantina: Tentativo di definizione del livello dell'acqua allomtermp delle comdotte in terracotta. *Atti del convegno "Cura aquarum in Sicila," May 1998, Syracuse, Sicily*. Leiden: Sichting Babeseh, 2000. 69–78.

Bruno, G., and A. Sciortino 1994. Geology and Water of Morgantina. Manuscript in D. P. Crouch collection.

Bruno, G., and D. Sgobba. 1993. Indizi neotettonici nell'area delle Murge centrali, desunti da misure statistico-strutturali eseguite su stalagmiti. *Itinerari Speleologici* 7 (Dec.): 11–22.

Bruno, G., A. Cannnata, I. Gonzalez Diez, and F. Schilirò. 1992. Mineralological and Geotechnical Characteristics of the Apulian and Sicilain Bentonite Clays. *Miner Petrogr. Acta* 35-A: 331–43.

Brunt, P. A. 1966. Athenian Settlements Abroad in the Fifth Century b.c. *Ancient Society and Institutions: Studies Presented to Victor Ehrenberg on His 75th Birthday*. Oxford: Basil Blackwell. 71–92.

———. 1971. *Italian Manpower, 225 b.c.–a.d. 13*. Oxford: Clarendon Press.

Bryson, R. A., and C. Padoch, 1981. On the Climates of Hiistory. In Rotberg, R. I., and T. K. Rabb, eds. *Climate and History: Studies in Interdiscliplinary History*. Princeton, N.J.: Princeton University Press. 3–17.

Buchanan, R. H., E. Jones, and D. McCourt, eds. 1971. *Man and His Habitat*. New York: Barnes and Noble.

Burdon, D. J. 1964. *Karst Groundwater Investigations: Greece*. Rome: FAO.

Burdon, D. J., and A. Dounas. 1964(?). Hydrochemistry of the Parnassos-Ghiona Aquifers and Problems of Sea-Water Contamination in Greece. No. 57 of IASH, *Groundwater in Arid Zones*: A10.

Burdon, D. J., and N. Papakis. After 1961. Methods of Investigating the Groundwater Resources of the Parnassos-Ghionna Limestones. No. 57 of IASH, *Groundwater in Arid Zones*: 143–59.

———. 1963. *Handbook of Karst Hydrogeology with Special Reference to the Carbonate Aquifers of the Mediterranean Region*. Athens: FAO/Institute for Geology and Subsurface Research.

Burés Vilasseca, L. 1998. Les estructures hidràliques a la ciutat antiga: l'exemple d'Empúrries. Barcelona: Biblioteca de Catalunya (Ph.D. dissertation).

———. 2000a. Empúries: A City without an Aqueduct. In Jansen, G. C. M., ed. *Cura aquarum in Sicilia*. Annual Papers on Classical Archaeology, suppl. 6. Leiden: Stichting Babesch. 65–72.

———. 2000b. *Tarraco Scipionum Opus*. A Roman-Spanish City Built on Karst. In Jansen, G. C. M., ed. *Cura aquarum in Sicilia*. Annual Papers on Classical Archaeology, suppl. 6. Leiden: Stichting Babesch. 79–84.

Burford, A. 1960. Heavy Transport in Classical Antiquity. *Economic History Review*, Second. Series, 13.1: 1–18.

Burghardt, A. F. 1971. A Hypothesis about Gateway Cities. AAAG 61: 269–85.

Burns, A. 1974. Ancient Greek Water Supply and City Planning: A Study of Syracuse and Acragas. *Technology and Culture* 15: 389–412.

———. 1976. Hippodamus and the Planned City. *Historia* 21: 414–28.

Butti, K., and J. Perlin. 1980. *A Golden Thread: 2500 Years of Solar Architecture and Technology*. Palo Alto, Calif.: Cheshire Books (Van Nostrand Reinhold Co., New York).

Buttrey, T. H., K. T. Erim, T. D. Groves, and R. R. Holloway. 1989. *Morgantina Studies*, II: *The Coins*. Princeton, N.J.: Princeton University Press.

Butzer, K. W. 1965. *Environment and Archaeology: An Introduction to Pleistocene Geography*. Chicago: Aldine.

———. 1980. Holocene Alluvial Sequences: Problems of Dating and Correlation. In Cullingford, R. A., D. A. Davidson, and J. Lewin, eds. *Timescales in Geomorphology*. Chichester: J. Wiley and Sons. 132–42.

———. 1982. *Archaeology as Human Ecology: Method and Theory for a Contextual Approach*. Cambridge: Cambridge University Press.

Cahill, N. 1992. Koukkyra Melaina and the Distribution of Land in Greek Colonies. Paper given at the 94th Annual Meeting of the American Archaeological Society; abstract in AJA 97 (1993): 345–46.

Calderone, S., M. Caltabiano, E. De Miro, D. Deorsola, and L. Campagna. 1990. *Gli edifici publici civili di Agrigento antica*. Agrigento: Museo Archeologico Regionale di Agrigento and Soprintendenza Beni Culturali ed Ambientali di Agrigento.

Calhoon, J. B. 1970. Space and the Strategy of Life. *Ekistics* 29.175: 427 ff.

Camp, J. M. 1979a. Water Supply of Ancient Athens from 3000–86 b.c. Ph.D. dissertation, Princeton University.

———. 1979b. A Drought in the Late 8th Century b.c. *Hesperia* 48: 397–411.

———. 1982. Drought and Famine in the Fourth Century b.c. In Studies in Athenian Architecture, Sculpture, and Topography Presented to Homer A. Thompson. *Hesperia* Suppl. 20: 9–17.

Campbell, A. S., ed. 1971. *Geology and History of Turkey*. Petroleum Exploration Society of Libya, 13th Annual Field Conference.

Capel, J., F. Huertas, and J. Linares. 1985. High Temperature Reactions and Use of Bronze Age Pottery from La Mancha, Central Spain. *Miner Petrogr. Acta* 29-A: 563–75.

Carapezza, M., R. Aliamo, S. Calderone, and Marco Leone. 1983. I materiali e l'ambiente delle sculture di Selinunte. In V. Tusa, ed. *Le sculture di selinunte*. Edizioni Sellerio: 32–41.

Carpenter, R., and A. W. Parsons. 1936. *Corinth III. 2: The Defenses of Acrocorinth and the Lower Town.* Cambridge, Mass.: Harvard University Press for American School of Classical Studies.

Carrozzo, M. T., P. Colella, M. di Filippo, Y. Quarta, and B. Torro. 1992. Detailed Gravity Prospecting and Geological-Structural Setting of the Selinunte Archaeological Site. *Bollettino di Geofisica Teorica ed Applicata* 34.134–35 (June–September): 101–8.

Carta Geologica d'Italia. 1971. Foglio 636, Agrigento at 1:10,000.

Carter, J. C. 1990. Metapontum: Land, Wealth, and Population. In Descœudres, J.-P., ed. *Greek Colonists and Native Populations.* Oxford: Clarendon Press. 405–42.

Castagnoli, F. 1956. *Ippodamo di mileto e l'urbanisitica a pianta ortogonale.* Rome. Issued as *Orthogonal Town Planning in Antiquity.* Cambridge, Mass.: 1971.

Cavallari, C. 1887. *Die Stadt Syrakus in Alterthum* (German translation of Cavallari-Holm, *Topographie archaologie di Sircusa*). Strasburg: J. H. Editiones Heitz (Heitz-Münden).

Cavallari, C., and S. Cavallari. 1872. *Bull. Comm. Ant. e B. Arti in Sicilia.* V. Tab. 1, map of Selinus (Fig. 4.3).

Cavallari, F. S. 1882. *Sulla topografia di Talune città greche in Sicilia e dei loro monumnti.* Palermo.

————. 1951. Scoperte nelle die nipopve arteroe stradi. *NOTSC* 23: 261–34.

Cavallari, F. S., and A. Holm, with C. Cavallari. 1883. *Topografiia archeologica di Siracusa.* Palermo: Tipografia del Giornale "Lo Statuto."

Çeçen, K. 1996. *The Longest Roman Water Supply Line.* Istanbul: Türkiye Sinai Kalkimma Bankasi.

Changes of Climate. 1963. Proceeding of the Rome Symposium. 1961, organized by UNESCO and the World Meteorological Organization. Paris: UNESCO.

Changnon, S. A. Jr. 1973. Inadvertent Weather and Precipitation Modification by Urbanization. *J. of Irrigation and Drainage Development:* 7–41.

Charneux, P. 1957. Rome et la Confédération Achaienne. *BCH* 81: 197–202 (re the Argos calendar).

Chatzidakis, P., G. Katsikis, and. M. Vartis-Matarangas. 1997. Delos Sacred Greece: Characterization of the Building Stones, Their Origin and Decay Factors. In Marinos, P. G., G. C. Koukis, G. C. Tsiambaos, and G. C. Stourjaras, eds. *Engineering Geology and the Environment: Proceedings of the Symposium of the International Association of Engineering Geology, Athens.* Rotterdam: A. A. Balkema. 3089–94.

Childs, W. A. P. 1993. Herodotos, Archaic Chronology, and the Temple of Apollo at Delphi. *Ist. Mitt.* 43: 399–441.

Cirrincione, R. 1992. Relazioni strutturali e composizionali fra Cummingronite, Orneblenda ed Attinodo coesistenti in Porfiriti K-calc-alcarine della Sicilia nord-orientale. *Miner Petrogr. Acta* 35: 125–38.

————. 1995. *Petrologia delle Porfiriti K-calc-alcaline del conglomerato tortoniano. Sicilia centrale e nor orientale.* Tesi di Dottorato di Ricerca in Petrologia delle Associasioni Magmatiche Viciclo, Univ. di Catania.

Ciurcina, P. n.d. (ca. 1985). *Syracuse: Historical Town.* Syracuse: Edizioni Esclusiva Italia Sebastiano.

Club Alpino, Italia. Sezione di Napoli. 1987. *La cavita artificiali: Aspecti storico-morfologici e loro utilizzo.* Naples.

Coarelli, F., and M. Torelli. 1984. *Guide archeologiche.* Vol. 13: *Sicilia.* Rome: Laterza. Includes: Selinunte, 73–103; Siracusa e il suo territorio, 222–82.

Cobet, J. 1997. Die sind die Stadt: Zur Stadtbefestigung des antiken Milet, in Milet 1995–94. *AA* 2: 249–84.

Cole, K. C. Nov. 16, 1999. Time, Space Obsolete in New View of Universe. *Los Angeles Times.*

———. Nov. 17, 1999. How Faith in the Fringe Paid Off for One Scientist. *Los Angeles Times*.

———. Nov. 18, 1999. Unseen Dimensions Hold Theory Aloft. *Los Angeles Times*.

Cole, S. G. 1988. The Uses of Water in Greek Sanctuaries. In Hägg, N., N. Marinator, and C. Nordquist, eds. *Early Greek Cult Practice*. Stockholm: Skrifter Utgivna av Svennsska Inst.i Athen, 4°, 38: 161–65.

Collin-Bouffier, S. 1987. L'alimentation en eau de la colonie grecque de Syracuse: Reflexions sur la cité et sur son territoire. *Mél. Rome Antiquité*, 99.2: 661–91 and 2 large maps.

———. 1994. Marais et Paludisme èn Occident Grec. *BCH* Suppl.: 321–36.

———. 2000. Quelles fonctions pour la *kolymbethra* d'Agrigente? In Jansen, G. C. M., ed. *Cura aquarum in Sicilia*. Annual Papers on Classical Archaeology, suppl. 6. Leiden: Stichting Babesch. 37–43.

Collingwood, R. G. 1939. *Autobiography*. Oxford: Oxford University Press.

———. 1946/1967. *The Idea of History*. Oxford: Clarendon Press.

La colonisation grecque en Mediterranée occidentale. Actes de la rencontre scientifique en hommage à Georges Vallet. Rome-Naples, 15–18 Nov. 1985. Paris: De Boccard Édition, 1999.

Columba, G. M. 1893. *Gli studi geografici nel i secolo dell'impero Romano*. Turin: Clausen.

———. 1906. *I porti della Sicilia*. Rome: Officina Poligrafica Italiana. Reprint 1991 Palermo: Accademia Nazionale di Sciencae, Lettere e Arti.

Commissione Grappelli. 1968. *La geotecnica negli interventi sugli Antichi centri Abitati sugli edifici monumenti*. Associazione Geotecnica Italiana. Vol. 1: 109–124. Includes papers by A. Croce, E. De Miro, B. Fanelli, R. Jappelli, V. Liguori, P. Morandi, N. Nociulla, E. Pace, A. Pellegrina, and P. Rossi Doria. See esp. Croce et al. 93–103 and 110–23.

Constantinidis, C. V., J. Christodoulias, and A. I. Sofianos. 1988. Weathering Processes Leading to Rockfalls at Delphi Archaeological Site. In Marinos, P. G., and G. C. Koukis, eds. *The Engineering Geology of Ancient Works, Monuments and Historical Sites: Proceedings of the Symposium of the International Association of Engineering Geology, Athens*. Rotterdam: A. A. Balkema: 201–5.

Corn, J. J. 1996. Object Lessons/Object Myths? What Historians of Technology Learn from Things. In W. D. Kingery, ed. *Learning from Things*. Washington, D.C.: Smithsonian Institution Press. 35–54.

Cornell, T., and J. Matthews. 1982. *Atlas of the Roman World*. New York: Phaidon-Oxford.

Cosomopoulos, M. B. 1998. Le Bronze Ancien 2 en Argolide: Habitat, urbanisation, populaton. In Pariente, A., and G. Touchais, eds. *Argos et l'Argolide: Topographie et urbanisme*. École Française d'Athènes. Paris: De Boccard Édition. 41–56.

Cotecchia, V. 1996. Geotechnical Degradation of the Archaeological Site of Agrigento. In *Geotechnical Engineering for the Preservation of Monuments and Historic Sites. A. Croce Memorial Symposium. Napoli, Oct. 1996*. Rotterdam: A. A. Balkema: 3–9 with 3 fold-out maps and other illustrations.

Coulton, J. J. 1977. *Ancient Greek Architects at Work: Problems of Structure and Design*. Ithaca, N.Y.: Cornell University Press.

———. 1987. Roman Aqueducts in Asia Minor. In Macready, S., and F. H. Thompson. *Roman Architecture in the Greek World*. Occasional Papers (N.S.) 10. London: Society of Antiquaries of London. 72–84.

Courbìn, P. 1998. Le Temple d'Apollo Lycien à Argos: Quelques suggestions. In Pariente, A., and G. Touchais, eds. *Argos et l'Argolide: Topographie et urbanisme*. École Française d'Athènes. Paris: De Boccard Édition. 61–69.

Courby, F. 1925. La Terrasse du Temple. *FDELPHES* 3.2: 176–81 reporting work of 1912–13.

Croce, A. 1985. Old Monuments and Cities: Research and Preservation. *Geotechnical Engineering in Italy*. Associazione Geotecnica Italiana: 374–81.

Croce, A., E. De Miro, G. B. Fenelli, R. Jappelli, V. Ligouri, R. Morandi, N. Nocilla, E. Pace, A. Pellegrino, and P. Rossi Doria. 1980. La città di Agrigento e la valle dei templi. Questioni di stabilità del territorio e di conservazione dei monumenti. In *Atti XIV Convegno Nazionale di Geotecnica (28–31 Ottobre)*. Florence: Associazione Geotecnica Italiana. 91–93 and 110–23.

Crouch, D. P. 1969. Palmyra. Ph.D. dissertation, University of California at Los Angeles.

———. 1972. A Note on the Population and Area of Palmyra. *Mélanges de l'Université Saint-Joseph*. 47.241–50.

———. 1984. The Hellenistic Water System of Morgantina, Sicily: Contributions to the History of Urbanization. *AJA* 88: 353–65, plates 46–47.

———. 1985. *History of Architecture: Stonehenge to Skyscrapers*. New York: McGraw-Hill, chaps. 5 and 6.

———. 1988. Planning Water Management for an Ancient Greek City. International Hydrology Program. UNESCO. Int. Symposium 24–29 April 1988. *Hydrological Processes and Water Management in Urban Areas*. Duisburg, Germany.

———. 1989. Water Management at Agrigento and Morgantina. In *Wassernutzung und Wasserbau im antiken Italien*. Mitteilungen of the Leichtweiss-Institut für Wasserbau, Technical University of Braunschweig, Germany. 103:155–174.

———. 1990. The Agora at Morgantina: Design and Drains. Presented at the session on Architecture and the Design of Urban Spaces, at the College Art Assn. annual meeting, New York.

———. 1993. *Water Management in Ancient Greek Cities*. New York: Oxford University Press.

———. 1995. Avoiding Water Shortages: Some Ancient Greek Solutions. In Angeliakis, A. N., and A. S. Issar, eds. *Diachronic Climatic Impact on Water Resources*. NATO ASI Series. Berlin: Springer-Verlag: 129–60.

———. 1996 Priene's Streets and Water Supply. In De Haan, N., and G. C. M. Jansen, eds. *Cura aquarum in Campania*. Leiden: Stichting Babesch. 137–43.

———. 1997. Ancient Cities as Four-Dimensional Models. In Marinos, P. G., G. C. Koukis, G. C. Tsiambaos, and G. C. Stourjaras, eds. *Engineering Geology and Environment: Proceedings of the Symposium of the International Association of Engineering Geology, Athens*. Rotterdam: A. A. Balkema. 3095–99.

Crouch, D. P., and J. J. Johnson. 2001. *Traditions in Architecture: Africa, America, Asia, and Oceania*. New York: Oxford University Press.

Crouch, D. P., D. J. Garr, and A. I. Mundigo. 1983. *Spanish City Planning in North America*. Cambridge, Mass.: MIT Press.

Crouch, D. P., T. Güngör, and A. Alkan. 1997 . Priene and Pompeii: Cities in Geologically Active Settings. In Marinos, P. G., G. C. Koukis, G. C. Tsiambaos, and G. C. Stourjaras, eds. *Engineering Geology and the Environment: Proceedings of the Symposium of the International Association of Engineering Geology, Athens*. Rotterdam: A. A. Balkema. 3107–11.

Crowley, T. J., and G. R. North. 1991. *Paleoclimatology*. New York: Oxford University Press.

Crowther, C. V. 1996. I Priene 8 and the History of Priene in the Early Hellenistic Period. *Chiron* 26: 195–238.

Cullingford, R. A., D. A. Davidson, and J. Lewin, eds. 1980. *Timescales in Geomorphology*. Chichester: J. Wiley and Sons.

Cultrera, G. 1938. Rovine di un antico stabilimento idraulico in contrada Zappalà. *Notizie degli Scavi* 261–301, Tab. 1.

————. 1942. *Il santuario Rupestre Presso S. Biagio in Agrigento*. Palermo: Atti della Reale Accademia di Scienze Lettere e Arti, Serie IV, Vol. 3, part II, Fasc. 4, 21: 609–27.

Cuomo di Caprio, N. 1992. *Morgantina studies III: Fornaci e officine da vasaio tardo-ellenistiche*. Princeton, N.J.: Princeton University Press. Esp. App. 3, Indagini techni-oche e analisi di microscopia ottica, 126–80.

Daina, A., T. Macaluso, S. Monteleone, G. Pipitone, S. Vernuccio, V. Agnesi, and U. D'Angelo. 1974. *Studio della Franosita del territorio di Agrigento*. No. 36: 5–14, with maps. Palermo: Istituto di Geologia.

Dall'Aglio, M. 1970. Geochemistry of Stream and Ground Waters from Western Sicily: The Changes in Spring Water Chemism after the 1968 Earthquake. *Atti Convegno Internazionale Sulle Acque Sotterranee*: [First] International Symposium on Ground-Water, Palermo. Palermo: E.S.A. 226–239.

Dall'Aglio, M., and C. Tedesco. 1968. Studio geochimico de idrologico di sorgenti della Sicilia. *Rivista Mineraria Siciliana* 19, parts 112–14 (July–Dec.): 171–210 with map of Agrigento, 208.

Dalli Cardillo, A., and N. Sciangula. 1960s. *I sotterranei di Agrigento: Documentazione fotografica sul labirinto e su altri ipogei dell'Agrigentino*. Agrigento: Azienda Autonoma Soggiorno e Turismo; reprint 1992.

D'Angelo, U., and S. Vernuccio. 1994. Note illlustrative della carta geological Marsala (F degrees 617 scala 1: 50.000). *Bollettino della Societa Geologica Italiana* 113.1: 55–67.

D'Angelo, U., S. Vernuccio, and G. Ruggieri. 1971. *Carta neotettonica del foglio 271 and 636, Agrigento*. Carta Geologica d'Italia, at 1:10,000.

Darcque, P. 1998. Argos et la plaine argiène à l'époque mycénienne. In Pariente, A. and G. Touchais, eds. *Argos et l'Argolide: Topographie et urbanisme*. École Française d'Athènes. Paris: De Boccard Édition. 103–15. With map.

Darwin, C. B. 1956. The Time Scale in Human Affairs. In Thames, W., ed. *Man's Role in Changing the Face of the Earth*. Chicago: University of Chicago Press.

Davidson, D. A. 1970. Terrain Adjustment and Prehistoric Communities. In Ucko, P. J., R. Tringham, and G. W. Dinbleby, eds. *Man, Settlement and Urbanism: Proceedings of the Research Seminar in Archaeology and Related Subjects*. Cambridge, Mass.: Schenkman Publishing. 17–22.

————. 1980. Erosion in Greece during the First and Second Millennia b.c. In Cullingford, R. A., D. A. Davidson, and J. Lewin, eds. *Timescales in Geomorphology*. Chichester: J. Wiley and Sons. 143–59.

————. 1985. Geomorphology and Archaeology. In Rapp, G., and J. A. Gifford, eds. *Archaeological Geology*. New Haven, Conn.: Yale University Press. 25–55.

Davidson, D. A., and M. L. Shackley, eds. 1975 *Geoarchaeology: Earth Science and the Past*. Boulder, Colo.: Westview Press.

Davis, O. K. 1992. Rapid Climatic Change in Coastal Southern California Inferred from Pollen Analysis of San Juaquin Marsh. *Quaternary Research* 37: 89–100.

Davis, O. K., J. Jirikowic, and R. N. Kalin. 1992. Radiocarbon Record of Solar Variability and Holocene Climatic Change in Coastal Southern California. In K. T. Redmond, ed. *Proceedings of the Eighth Annual Pacific Climate (Paclim) Workshop, March 10–13, 1991*. California Dept. of Water Resources, Interagency Ecological Studies Program Technical Report 31.

De Angelis, F. 1994. The Foundation of Selinus: Overpopulation or Opportunities. In Tsetskhladze, G. R., and F. De Angelis, eds. *The Archaeology of Greek Colonisation*. Oxford Committee for Archaeology, Monograph 40. 87–110.

Debazac, E. F. and G. Mavrommatis. 1971. Les grandes divisions écologiques de la végétation forestière en Grèce continentale. *Bulletin de la Societe Botanique de France* 118: 429–52.

De Caro, S. (n.d.). La citta sannitica: Urbanistica e architettura. *Pompeii*: 23–46.

———. 1992. Lo svilluppo urbanistico di Pompeii. *Atti e Memorie della Societa Magna Grecia*, 3rd Series 1: 67–90, plates 3–8.

De Haan, N., and G. C. M. Jansen, eds. 1996. *Cura aquarum in Campania.* Annual Papers on Classical Archaeology. Suppl. 4. Leiden: Stichting Babesch.

de Jesus, P. S. 1980. *The Development of Prehistoric Mining and Metallurgy in Anatolia.* Oxford: BAR International Series 74.

de la Genière, J. 1978. La colonisation grecque en Italie Meridoniale et en Sicile et l'acculturation des non-grecs. *Revue Archéologique* 2: 257–76 and plates.

———. 1979. À propos de deux sondages extra-muros de l'acropole de Sélinonte. *Philias Charin: Misc. in onore di E. Manni.* Vol. 4 of 6 vols. Rome: 1295–99, plates 1 and 2.

———. 1982. Selinonte: Recherches sur la topographie urbaine 1975–1981. ANNPISA Ser. 3. 12: 469–79.

de la Genière, J., and D. Theodorescu. 1979. Ricerche topografiche nell'area di Selinunte. AAL.R. s. 8, 34: 1–12 and 385–95.

———. 1980–81. Contribution à l'histoire urbanistique de Sèlinonte. *Kokalos* 26–27: 973–96.

Delphes: Races cultes et jeux: 100 ans de fouilles et de decouvertes. *Les Dossiers d'Archaeologie* 15 (July–August) 1990.

Demand, N. H. 1990. *Urban Relocation in Archaic and Classical Greece.* Norman: University of Oklahoma Press.

De Miro, E. 1957. Il quartiere ellenistico-romano di Agrigento. AAL.R. *Classe di Scienze Fisiche*: 135–40 and 4 pages of plates.

———. 1985. La casa greca in Sicilia: Testimonianze nella Sicilia centrale dal VI al III sec. a. C. *Philias Charin: Misc. in onore di E. Manni.* Vol. 6 of 6 vols. Rome: 708–37.

———. 1988. Architettura civile in Agrigento ellenistica-romana e rapporti con Anatolia. *Quaderni del'Instituto de Archeologia dell'Universita' di Messina* 3: 63–72

———. 1990. *Le necropoli greche di Agrigento dai recenti scavi.* Agrigento: Museo Arceologico Regionale. Catalog for an exhibition June 16–Sept. 30, 1990. Abridged from his article in *Le necropoli di Agrigento.* Rome: L'Erma di Bretschneider, 1988.

Demirtasli. 1978. Vousciam and Early Alpine Events in the Taurus Belt. Presented at the *First Geological Conference of the Middle East.* Ankara, Turkey.

Deorsola, D., D. Gulli, R. Panvini, and F. Valoruzzi. 1998. *Veder greco: Le necropoli di Agrigento.* (catalog). Rome and Agrigento: Museo Archeologico Regionale.

de Polignac, F. 1984. *La naissance de la cité grecque: Cultes, espace et société VIIIᵉ–VIIᵉ siècles avant J.-C.* Paris: La Découverte. Also published as: *Cults, Territory, and the Origins of the Greek City-State.* Trans. J. Lloyd. Chicago: University of Chicago Press.

———. 1998. Cité et terra à l'époque géometrique: Une modèle argien? In Pariente, A., and G. Touchais, eds. *Argos et l'Argolide: Topographie et urbanisme.* École Française d'Athènes. Paris: De Boccard Édition. 145–62.

Dercourt, J. 1962. *Etude geologique de la region comprise entre les plaines d'Argos et de Tripolis.* Geological and Geophysical Research, Vol. 8, No. 3. Athens: Institute for Geology and Subsurface Research.

———. 1977. Equisse géologique du Nord du Péloponèse. In Bintliff, J. L. *Natural Environment and Human Settlement in Prehistoric Greece.* Suppl. Series 28 i and ii: Oxford: BAR. 415–26.

Déroche, V., and P. Pétridis. 1993. Agora romaine et Villa Sud-est. *BCH* 117: 611–44.

Descœudres, J.-P., ed. 1990. *Greek Colonists and Native Populations.* Oxford: Clarendon Press.

des Courtils, J. 1992. L'architecture et historie d'Argos dans la première motié du Vᵉ siècle avant J.-C. *BCH* Suppl.: 241–51.

de Vries, J. 1981. Measuring the Impact of Climate on History: The Search for Appropriate Methodologies. In Rotberg, R. I., and T. K. Rabb, eds. *Climate and History: Studies in Interdiscliplinary History.* Princeton, N.J.: Princeton University Press. 19–50.

de Waele, J. A. 1971. *Akragas graeca: Die historische Topographie des griechischen Akragas auf Sizilien. I: Historischer Teil.* Rome: Archeologische Studiën van het Nederlands Historisch Instituut te Rome, Deel III: 3–97.

———. 1985. La popolazione di Akragas antica. In *Philias Charin: Misc. in onore di E. Manni.* Vol. 3 of 6 vols. Rome: 749–60.

de Waele, J. A., and M. G. Amadasi. 1980. Agrigento: Gli scavi sulla Rupe Atenea (1970–75). *NOTSC* 34: 395–452.

de Wit, H. E. 1988. New Data Used to Test Paestum's Environmental History. In Marinos, P. G., and G. C. Koukis, eds. *The Engineering Geology of Ancient Works, Monuments and Historical Sites: Proceedings of the Symposium of the International Association of Engineering Geology, Athens.* Rotterdam: A. A. Balkema. 2113–16.

de Wit, H. E., P. van der Gaag, and J. Sevink. 1988. Geological Answers to Archaeological Questions: An Example. In Marinos, P. G., and G. C. Koukis, eds. *The Engineering Geology of Ancient Works, Monuments and Historical Sites: Proceedings of the Symposium of the International Association of Engineering Geology, Athens.* Rotterdam: A. A. Balkema. 1613–22.

di Castri, F., and H. A. Mooney, eds. 1973. *Mediterranean Type Ecosystems: Origin and Structure.* New York: Springer-Verlag.

Dickinson, W. R. 1971. Plate Tectonics in Geologic History. *Science* 74.4005 (Oct. 8): 107–13.

Dietrich, B. C. 1992. Divine Madness and the Conflict at Delphi. *Kernos* 5: 41–58.

Dietz, S., and I. Papachristodoulou, eds. 1988. *Archaeology in the Dodecanese.* Copenhagen: National Museum of Denmark, Dept. of Near Eastern and Classical Antiquities.

Dilamarter, R. R., and S. C. Csallany, eds. 1977. *Hydrologic Problems in Karst Regions: Proceedings of the International Symposium, April 1976.* Bowling Green, Ky.: W. Kentucky University Press.

Dincauze, D. F. 1987. Strategies for Paleoenvironmental Reconstruction in Archaeology. *Advances in Archaeological Method and Theory* 11: 255–319.

Dirkin and Lister 1983.

Dirkin, M. K., and C. J. Lister, 1983. The Rods of Diogenis: An Ancient Marble Quarry in Eastern Crete. *BSA*78: 69–96.

Di Vita, A. 1953. L'elemento punico a Selinunte nel IV e nel III sec. a.C. *ARCHCL* 5: 39–47.

———. 1956a. La penetrazionne siracusana nella Sicilia sud-orientale alla luce delle più recenti scoperte archeologiche. *Kokalos* 2.2: 3–31 and 177–210.

———. 1956b. Per l'architettura d l'urbanistica greca d'età arcaica: La stoa del *Temenos* del Tempio C e lo sviluppo programmato di Selinunte. *Palladio* N.S. 17: 3–31.

———. 1967. L'acropoli di Selinunte nelle sue varie fasi e l'origine dell'urbanistica "Ippodamea" in Occidente. *Palladio* N.S. 17: 33–60.

———. 1979. Contributi per una storia urbanistica di Selinunte. *Philias Charin: Misc. in onore di E. Manni.* Vol. 3 of 6 vols. Rome: 803–29, 4 figs. and 1 plate.

———. 1984a. Selinunte fra il 650 ed il 409: Un modello urbanistico coloniale. *ANNSCATENE* 62 (N.S. 46): 1–53 and 2 plates

————. 1984b. La fortificazioni di Selinunte classica. ANNSCATENE 62 (N.S. 46): 469–79.

————, ed. 1985. *Sikanie: Storia e civiltà della Sicilia greca.* Antica Madre Series. Milan: Garzanti.

————. 1990. Town Planning in the Greek Colonies of Sicily from the Time of Their Foundations to the Punic Wars. In Descœudres, J.-P., ed. *Greek Colonists and Native Populations.* Oxford: Clarendon Press. 343–63.

————. 1996. Urban Planning in Ancient Sicily. In Pugliese-Carratelli, G., ed. *The Western Greeks.* Venice: Bompiani. 263–308, 397. Includes Selinus plan by D. Mertens.

Dolci, E. 1988. Marmora Lunensia: Quarrying Technology and Archaeological Use. In Herz, N., and M. Waelkens, eds. *Classical Marble: Geochemistry, Technology, Trade.* NATO ASI Series E: Applied Sciences, Vol. 153. Dordrecht: Kluwer Academic. 77–84.

Dominguez, A. J. 1989. *La colonizacion griega en Sicilia: Griegos, Indigenoas y Punicos en la Sicilia arcaica: Interacción y aculturacion.* Oxford: BAR International Series 549 (ii).

Doukellis, P., J.-J. Duraure, and É. Fouache. 1995. Conteste géomorphologique et historique de l'acqueduc de Nicopolis. BCH 119: 211–33.

Douzouugli-Zachou, A. 1998. Le plaine argiène. In Pariente, A., and G. Touchais, eds. *Argos et l'Argolide: Topographie et urbanisme.* École Française d'Athènes. Paris: De Boccard Édition. 32–38.

Doxiadis, C. A. 1962. *Architectural Space in Ancient Greece.* Cambridge, Mass.: MIT Press.

————. 1972. *The Method for the Study of the Ancient Greek Cities.* 5 vols. Athens: Ekistics Center

Drake, T., and W. M. Jordan, eds. 1985. *Geologists and Ideas: A History of American Geology.* Geological Society of America.

Drews, R. 1976. The Earliest Greek Settlements on the Black Sea. JHS 96: 18–31.

Dreybrodt, W. 1990. Karstification: A Model. J. of Geology Sept.: 639–55.

Driesch, A. von den, and Dr. Peters [sic]. 1992. Siedlungsabfall versus Opferreste: Essgewohnheiten im archaischen Milet. Ist.Mitt. 42: 117 ff.

Drögemüller, H.-P. 1969. *Syrakus: Zur Topographie und Geschichte einer griechischen Stadt.* Heidelberg: C. Winter Universitätsverlag.

Drorvinis, V. K. 1998, Téménonion, Le Port Antique d'Argos. In Pariente, A., and G. Touchais, eds. *Argos et l'Argolide: Topographie et urbanisme.* École Française d'Athènes. Paris: De Boccard Édition. 291–313.

Ducat, J. 1965. Les dates de fondation des colonies grecques de Sicile. *Bulletin de la Faculté des Letres de Strasbourg.* 43: 413–18.

Dufaure J. 1977. Neotectonique et morphogenese dans une peninsule mediteraneene. RGPGD 18: 27–58.

Dufaure, J. J., and A. Zainanis. 1980. Styles néotectoniques et étagements de niveaux marins sur un segment d'arc insulaire, le Péloponnèse. *Proc. Conf. Niveaux Marins et Tectonique Quaternaire dans l'Aire Méditerranéenne.* Paris: CNRS. 77–107.

Dunbabin, T. H. 1948. *The Western Greeks.* Oxford: Oxford University Press.

Duncan-Jones, R. P. 1963. City Popuation in Roman Africa. JRS 52: 85–90.

Dürr, S., R. Altherr, J. Keller, M. Okrusch, and E. Seidel. 1977. The Median Aegean Crystalline Belt: Stratigraphy, Structure, Metamorphism and Magmatism. In Closs H., et al., eds. *Mediterranean Orogens.* Stuttgart: Schweizerbart.

Eck, W. et al., eds. 1992. ZPE (Bonn). 90.

Eddy, J. A. 1981. Climate and the Role of the Sun. In Rotberg, R. I., and T. K. Rabb, eds. *Climate and History: Studies in Intercliplinary History.* Princeton, N.J.: Princeton University Press. 145–67.

Eder, W. 1994. Entwurf eines denkmalpflegerischen und touristischen Gesamtkonzepts für die Stadtruine Milet. *JDAI Istanbul* 275–81.

Edgerton, S. Y. Jr. 1974. Florentine Interest in Ptolemaic Cartography as Background for Renaissance Painting, Architecture, and the Discovery of America. *JSAH* 33.4 (Dec.): 275–92.

Edlund-Berry, I. E. M. 1989–90. The Central Sanctuary at Morgantina (Sicily): Problems of Interpretation and Chronology. *Scienze dell' Antichita.* 3–4: 327–338.

Ehrenberg, V. 1937. When Did the Polis Rise? *JHS* 57: 147–59.

Eisma, D. 1978. Stream Deposition and Erosion. In Brice, W. C., ed. *The Environmental History of the Near and Middle East since the Last Ice Age.* London: Academic Press. 67–81.

Elderkin, G. W. 1910. The Fountain of Glauke. *AJA* 19–50.

Emlyn-Jones, C. J. 1980. *The Ionians and Hellenism: A Study of the Cultural Achievment of the Early Greeek Inhabitants of Asia Minor.* London: Routledge and Kegan Paul.

Engelmann, H. 1991. Beiträge sur Ephesischen Topographie. *ZPE* (Bonn) 89: 275–95.

———. 1993. Celsusbibliothek und Auditorium in Ephesos. *JÖAI* 105–109.

Engelmann, H., D. Knibbe, and R. Merkelbach, eds. 1980. *Inschriften griechischer Städte Kleinasiens.* 14: *Die Inschristen von Ephesos*, Part IV: 44 ff. The Trophime poem is no. 1062.

Engels, D. 1990. *Roman Corinth: An Alternative Model for the Classical City.* Chicago: University of Chicago Press.

Ercoli, L. 1989. Considerazioni preliminari per una ricerca geologica sull'evoluzione dell'Acropoli di Selinunte. Internal report for the Dept. of Structural Engineering and Geotechnics, University of Palermo. 137–52.

———. 1991. Agrigento. *The Contributon of Geotechnical Engineering to the Preservation of Italian Historical Sites.* Florence: X ECSMFE (no pagination).

———. 1994. Gli ipogei dell'antica Akragas in rapporto all'assetto geostrutturale della Formazione di Agrigento. Presented at the 8th Congresso del Conservazine Nazionale Geologico, Rome.

———. 1997. Quarries and Roads of Selinus. Manuscript in D. P. Crouch collection.

———. 2000. Ancient systems of hydraulic supply in Sicily and surrounding small islands. In Jansen, G. C. M., ed. *Cura aquarum in Sicilia.* Annual Papers on Classical Archaeology. Leiden, Stichting Babesch. Suppl. 6: 57–65,

Ercoli, L., and D. P. Crouch. 1998. Mesostructure of Rocky Masses and Microform Karst Surface in Ancient Systems of Hydraulic Supply. In Marinos, P. G., G. C. Koukis, G. C. Tsiambaos, and G. C. Stourjaras, eds. *Engineering Geology and the Environment: Proceedings of the Symposium of the International Association of Engineering Geology, Athens.* Rotterdam: A. A. Balkema: 3143–49.

Ercoli, L., and A. Musso, 1991. Agrigernto. In *The Contribution of Geotechnical Engineering to the Preservation of Italian Historical Sites.* Florence: Associazione Geotecnica Italiana. 113–21.

Ercoli, L., and G. Speciale. 1988a. Rock Weathering and Failure Processes in the 'Latomia del Paradiso' (Syracuse, Italy). In Marinos, P. C., and G. C. Koukis, eds. *The Engineering Geology of Works, Monuments and Historical Sites: Proceedings of the Symposium of the International Association of Engineering Geology, Athens.* Rotterdam: A. A. Balkema: 771–778.

———. 1988b. Environment and History of the Latomiae of Syracuse (Sicily): Some Key Questions in the Preservation Problems. A. G. I. (expanded version of 1988a).

————. 1991. Siracusa. In *The Contribution of Geotechnical Engineering to the Preservation of Itlaian Historical Sites*. Florence: Associazione Geotecnica Italiana. 127–28

Ercoli, L., and C. Valore. 1990. A Proposed Study of the Long Term Deformation Processes of Natural Slopes and Man-made Structures of Selinus. Internal report prepared for the conference, *Reading Historic Sites through Geotechnical Evidence*. Florence: Associazionne Geotecnica Italiana.

Ercoli, L., V. Tusa, and C. Valore. 1991. Selinunte. In *The Contribution of Geotechnical Engineering to the Preservation of Italian Historical Sites*. Florence: Associazione Geotecnica Italiana. 125

Erguvanli, K. 1972. The Physico-Mechanical Properties of Turkish Marbles and Proposals for Their Classiification. *Bull. Geol. Soc. Turkey* 15.2: 153–70.

Erim, K. 1958. Morgantina. *AJA* 62:79–90.

Erim, K., et al. 1991. *Morgantina Studies. 2: The Coins*. Princeton, N.J.: Princeton University Press.

Erinc, S. 1978. Changes in the Physical Environment in Turkey since the End of the Last Glacial. In Brice, W. C., ed. *The Environmental History of the Near and Middle East since the Last Ice Age*. London: Academic Press. 87–110.

Errington, R. M. 1988. Aspects of Roman Acculturation in the East under the Republic. In *Alte Geschichte und Wissenschaftsgeschichte. Festschrift für Karl Christ zum 65 Geburtstag*. Darmstadt. 140–57.

Eschebach, H. 1977. Die innerstädtische Gebrauchswasserversorgung dargestellt am Beispiel Pompejis. In Boucher, J. P., ed. *Journées d'études sur les aqueducs romains*. Paris: 18–132.

————. 1979a. Die Gebrauchswasserverssorgung des antiken Pompeji. *Antike Welt* 10.2: 3–24.

————. 1979b. Pompeii: La distribution des eaux dans une grande ville romain. *Les Dossiers d'Archéologie* 38: 74–81.

Evans, G. 1971. The Recent Sedimentation of Turkey and the Adjacent Mediterranean and Black Seas: A Review. In Campbell, A. S., ed. *Geology and History of Turkey*. Petroleum Exploration Society of Libya, 13th Annual Field Conference: 385–406.

Evans, H. B. 1994. *Water Distribution in Ancient Rome: The Evidence of Frontinus*. Ann Arbor: University of Michigan Press.

Evans, R. K., ed. 1983. *Symposium (1978): Deforestation, Erosion, and Ecology: Contributions 1–2. JFA* 10.

Fabbri, B. 1992. About the Clay Used in the Majolica Altar of S. Cristina Church in Bolsena (Central Italy). *Miner Petrogr. Acta* 35-A: 161–70.

Fabricius, K. 1932. Das antike Syrakus: Eine historisch-archäologische Untersuchung. *KLIO* 28 Neue Folge. 15: 1–30.

Fahlbusch, H. 1982. *Vergleich antiker griechischer und römischer Wasserversorgungsanlagen*. Mitteilungen of the Leichtweiss-Institut für Wasserbau, 73. Braunschweig: Technischen Universität Braunschweig.

Fallico, A. M. 1971. Siracusa: Saggi di scavo nell'area della Villa Maria. *NOTSC*: 581–639.

Fant, J. C. 1988a. *Papers on the Roman Marble Quarrying and Trade*. Oxford: BAR International Series.

————. 1988b. The Roman Emperor in the Marble Business: Capitalists, Middle-men or Philanthropists? In Herz, N., and M. Waelkens, eds. *Classical Marble: Geochemistry, Technology, Trade*. NATO ASI Series E: Applied Sciences, Vol. 153. Dordrecht: Kluwer Academic Publishers: 147–58.

Fazello, F. Thommaso, 1558. *De Rebus Siculis Decades Duae, Nunc Primim in Lecem*

Editae. Panormi: Appud Ionnem Matthaeum Maidam, et Franciscum Carraram. Republished in Catania, 1749.

Fekete, K. 1977. The Development and Management of Karst Water Resources in the City of Pecs and Its Vicinity, Hungary. In Dilamarter, R. R., and S. C. Csallany, eds. *Hydrologic Problems in Karst Regions: Proceedings of the International Symposium, April 1976*. Bowling Green, Ky.: W. Kentucky University Press. 279–85.

Ferla, P., R. Alaimo, G. Falsone, and F. Spatafora. 1984. Studio Petrografico delle Macine di Età Arcaica e Classica da Monte Castellazzo di Poggioeale. *Sicilia Archeologica*. 56/Anno XVII: 1–30.

Finley, M. I. 1971. Archaeology and History. *Daedalus* (winter): 168–86.

———. 1973. *Problemes de la Terra en Grece Ancienne*. Paris-La Haye: Mouton.

———. 1975. *The Use and Abuse of History*. London: Chatto and Windus.

———. 1976. Colonies: An Attempt at a Topology. *Transactions of the Royal Historical Society*. 26: 167–88.

———. 1977. The Ancient City: From Fustel de Coulanges to Max Weber and Beyond. *Comparative Studies in Society and History* 19: 305–27.

———. 1985. *Ancient History: Evidence and Models*. London: Chatto and Windus.

Fiorentini, G. 1993–94. Attività della Soprintendenza ai Beni Culturali e Ambientali di Agrigento. *Kokalos* (conference in 1988): 491 ff.

Fischer, D. H. 1981. Climate and History: Priorities for Research. In Rotberg, R. I., and T. K. Rabb, eds. *Climate and History: Studies in Interdiscliplinary History*. Princeton, N.J.: Princeton University Press. 241–50.

Fleming, N. C. 1968. Holocene Earth Movements and Eustatic Sea Level Change in the Peloponnese. *Nature*. 217 (March 16): 1031–32.

———. 1969. *Archaeological Evidence for Eustatiic Change of Sea Level and Earth Movements in the Western Mediterranean during the Last 2000 Years*. Special Paper 109. Geological Society of America. 1–125.

———. 1972. Eustatic and Tectonic Factors in the Relative Vertical Displacement of the Aegean Coast. In Stanley, D. J., ed. *The Mediterranean Sea*. Stroudsburg, Pa.: Dowden, Hutchinson, and Ross: 189–201.

———. 1992. Predictions of Relative Coastal Sea-Level Change in the Mediterranean Based on Archaeological, Historical, and Tide-Gauge Data. In Jeftic, L., J. D. Milliman, and G. Sestini, eds. *Climate Change and the Mediterranean*. London: Edward Arnold. 247–81.

Fleming, N. C., and C. O. Webb. 1986. Tectonic and Eustatic Cosastal Changes During the Last 10,000 Years Derived from Archaeological Data. Z. *Geomorph*. N.F. Suppl. 62: 1–29.

Fletcher, C. H. III, H. J. Knebel, and J. C. Kraft. 1990. Holocene Evolution of an Estuarine Coast and Tidal Wetlands. *Geological Society of America Bulletin* 102: 283–97.

Foglio della Carta d'Italia dello I. G. M., 267, Aragona, a map of ancient and modern (as of 1937) water lines to Agrigento. In A. Aureli collection.

Foley, A. 1998, Ethnicity and the Topography of Burial Practices in the Geometric Period. In Pariente, A., and G. Touchais, eds. *Argos et l'Argolide: Topographie et urbanisme*. École Française d'Athènes. Paris: De Boccard Édition. 137–44.

Folk, R. L., and S. Valastro Jr., 1985. A Successful Technique for the Radiocarbon Dating of Lime Mortor. In Rapp, G., and J. A. Gifford, eds. *Archaeological Geology*. New Haven, Conn.: Yale University Press. 303–11.

Fontenrose, J. 1978. *The Delphic Oracle, Its Responses and Operations with a Catalogue of Responses*. Berkeley: University of California Press.

Forchheimer, P. 1923. Wasserleitungen. *FiE* 3: 224–55.

Forrest, W. G. 1957. Colonization and the Rise of Delphi. *Historia* 6: 160–75.

Foss, C. 1979. *Ephesus after Antiquity:A Late Antique, Byzantine, and Turkish City*. Cambridge: Cambridge University Press.

Fossel, E., and G. Langmann. 1972. Nymphaeum des C. *Laekanius Bassus*. In Grabungen in Ephesos von 1960–1969. *JÖAI* 50: 301–10.

Foucart, M. P. 1864. Memoire sur les ruines et l'histoire de Delphes. *Archives des Missions Scientifiques*. 2nd series, 5.2: 1–105.

Fougères. G., and J. Hulot. 1910. *Sélinonte*. Paris. Libraire Générale de l'Architecture et des Arts Décoratifs (Schmid).

Foxhall, L., and H. A. Forbes. 1982. *Sitometreia:* The Role of Grain as a Staple Food in Classical Antiquity. *Chiron* 12: 41–90.

Frasca, M. 1983. Una nuova capanna "Sicula" a Siracusa, in Ortigia: Tipologia dei materiali. *MÉLROME* 95.2: 565–98.

Frederiksson, R. 1999. From Death to Life: The Cemetery of Fusco and the Reconstruction of Early Colonial Society. In Tsetskhladze, G. R, ed. *Ancient Greeks West and East*. Leiden: E. J. Brill. 229–65.

Freedberg, D. 1997. The Failure of Pictures: From Description to Diagram in the Circle of Galileo. Lecture given at the Center for Advanced Study in the Visual Arts, National Gallery of Art. December.

Freyberg, V. 1973. Geologie des Isthmus von Korinth. *Erlanger Geologische Abhandlungen* 95: 1–183. With English summary.

Friesen, S. 1996. The Cult of the Roman Emperors in Ephesos: Temple Wardens, City Titles, and the Interpretation of the Revelation of John. In Koester, H., ed. *Ephesos: Metropolis of Asia*. Valley Forge, Pa.: Harvard Theological Studies, Trinity Press International. 229–50.

Friesinger, H., and F. Krinzinger, eds. 1999. *100 Jahre österreichische Forschungen in Ephesos. Akten des Symposiums Wien 1995*. Archäologische Forschungen 1 = DenksdhrWien 260.

Frontinus, S. J. First century C.E. *The Stratagems and Aqueducts of Rome*. C. E. Bennet, ed. Cambridge, Mass.: Loeb Classical Library, 1925. Also published as: Frontinus, S. J. *The Two Books on the Water Supply of the City of Rome*. C. Hershchel, ed. Cambridge, Mass.: New Eng. Water Works Association., 1973.

Furon, R. 1952–53. Introduction à la geologie et hydrogeologie de la Turquie. *Memoires du Museum National d'Histoire Naturelle*. Serie C. Sciences de la Terre. Vol. 3. Paris: 1–99, 102.

Gabba, E. 1981. True History and False History in Classical Antiquity. *JRS* 71: 50–62.

Gabba, E., and G. Vallet, eds. 1980. *La Sicilia antica. 6: Indigeni, Fenici-Punici e Greci*. Naples: Società Editrice Storia di Napoli del Mezzogiorno Continale e della Sicilia. Includes: Asheri, D. (1980) La colonizzazione greca, 89–142.

Gage, M. 1978. The Tempo of Geomorphological Change. *J. of Geology* 78: 619–25.

Gaki-Papanastasiou, P. 1994. Geomorphological Evolution of the Argive Plain (Greece). Ph.D. dissertation, University of Athens.

Galanakis, D., A. Mettos, D. Mataragas, and M. Triantafyllis. 1997. Landslide of Malakasa Area, Geological and Tectonic Regime. In Marinos, P. G., and G. C. Koukis, eds. *Engineering Geology and the Environment: Proceedings of the Symposium of the International Association of Engineering Geology, Athens*. Rotterdam: A. A. Balkema. 665–76.

Galanopoulos, A. G 1955. Historic Seismicity of Greece. *Geological Annals of Greece*. Laboratory for Geology and Paleonotology of the Athens University: 83–121. In Greek but with appendix 1 in English: Chronological list of "Seismic History of Central Greece."

Gale, N. H., Z. A. Stos-Gale, and J. L. Davis. 1984. The Provenance of Lead Used at Ayia Irini, Keos. *Hesperia* 3.4: 369–406.

Gallant, T. W. 1991. *Risk and Survival in Ancient Greece*. Stanford, Calif.: Stanford University Press.

Galloway, P. ca. 1992. *The Archaeology of the Ethnohistorical Text*. Manuscript in D. P. Crouch collection.

Garbrecht, G. 1987. Der Sadd-el-kafara, die älteste Talsperre der Welt. *Historische Talsperren*. Stuttgart: Verlag K. Wittwer. 97–110.

Gargallo, P. 1970. The Ports of Ancient Syracuse. *Archaeology* 23.4 (Oct.): 312–17.

Garnsey, P. 1983. Grain for Rome. In Garnsey, P., K. Hopkins, and C. R. Whitaker, eds. *Trade in the Ancient Economy*. Berkeley: University of California Press. 118–130.

———. 1988. *Famine and Food Supply in the Greco-Roman World*. Cambridge: Cambridge University Press.

Garnsey, P., and I. Morris. 1989. Risk and the *Polis*: The Evolution of Institutionalised Responses to Food Supply Problems in the Ancient Greek State. In Halstead, P., and J. O'Shea, eds. *Bad Year Economics*. Cambridge: Cambridge University Press. 86–105.

Garnsey, P., and R. Saller. 1987. *The Roman Empire: Economy, Society and Culture*. London: Duckworth; Berkeley: University of California Press. Esp. chap. 2, Government without Bureaucracy, 20–40.

Gasperini, P., F. Bernardini, G. Valensise, and E. Boschi. Feb. 1999. Defining Seismogenic Sources from Historical Earthquake Felt Reports. *Bulletin of the Seismological Society of America* (Berkeley, Ca.) 89: 94–110.

Gates, M.-H. 1994–1997. Reports on Archaeology in Turkey. AJA vols. 98–101.

Gatier, P.-L. Tremblement du sol et frissons des hommes: Trois séismes en Orient sous Anbastase. In Helly, B., and A. Pollino, eds. *Tremblement de terre: Histore et archeologie. 4: Rencontres internationales d'archeologie et d'historie d'Antibes, 2–4 nov. 1983*. Vallbonne. 87–94.

Gay, P. 1974. *Style in History*. New York: McGraw Hill.

Gell, W. 1810. *The Itinerary of Greece with a Commentary on Pausanias and Strabo*. London: T. Payne.

Gentili, G. V. 1951. Siracusa: Scoperte nelle Due Nuove Arterie Stradali. *NOTSC* 23: 261–334. See also his 1956 and 1961 reports.

———. 1991. Il grande Tempio Ionico di Siracusa: I dati topografici e gli elementi architettonici Raccolti find al 1960. *Historia* 40: 61–84.

Georgoudi, S. 1974. Quelques problèmes de la Transhumance dans la Grèce ancienne. *REG* 87: 115–85.

Germann, K. 1981. Lagerstätteneigenschaften und Herkunftstypische Merkmalsmuster von Marmoren am Südwestrand des Menderes-massive (Südwestanatolien). *JDAI Istanbul* 96: 215–35.

Germann, K., and A. Peschlow-Bindokat. 1981. Steinbrüche von Milet und Herakleia am Latmos. *JDAI Istanbul* 96: 157–214.

Gerschenkron, A. 1968. *Continuity in History and Other Essays*. Cambridge, Mass.: Belknap Press of Harvard University Press.

Giannotti, G. P, L. Lombardi, and G. Sidoti. 1972. Schema idrogeologico della sicilia occidentale. *Rivista Mineraria Siciliana*. 133–35: 3–26.

Giesheimer, M. 1989. Gènese et développement de la Catacombe St.-Jean à Syracuse. *MÉLROME* 101.2: 751–83.

Gifford, J. A.. and G. Rapp Jr. 1985b. In Drake, T., and W. M. Jordan, eds. *Geologists and Ideas: A History of American Geology*. Geological Society of America. 409–421

Ginouvés, R. 1952. Une salle de bains hellénistique à Delphes. *BCH* 76: 541–61.

———. 1972. Note sur les Appareils Romains de Briques à Argos et le Problème de la Datation des Apparreils Romains en Grêce. *Le Theatron à Gradins Droits et l'Odéon d'Argos*. École Française d'Athènes, Études Péloponnésiennes. 6: 217–45.

Ginouvés, R., A.-M. Guimier-Sorbets, J. Jouanna, and L. Villard, eds. 1994. *L'Eau, la santé et la maladie dans le monde grec*. Paris: Actes de Colloque à Paris. 25–27 Nov. 1992. *BCH* Suppl. 4 (entirely devoted to water).

———. 1994. L'Eau dans les Sanctuaires Médicaux (suppl. to the paper of Lambrenoudakis). In Ginouvés, R., A.-M. Guimier-Sorbets, J. Jouanna, and L. Villard, eds. *L'Eau, la santé et la maladie dans le monde grec*. Paris: Actes de Colloque à Paris. 25–27 Nov. 1992. *BCH* Suppl. 4: 237–46.

Glaser, F. 1983. *Antike Brunnenbauten (KPHNAI) in Griechenland*. Vienna: Verlag der Österreichischen Akademie der Wissenschaften.

Gli ipogei per i turisti? Le guide: Ottima idea. *La Sicilia* (newspaper) 22 gennaio, 1994.

Glover, T. R. 1945. *Springs of Hellas and Other Essays*. Cambridge: Cambridge University Press. 1–29.

Goldsberry, M. A. S. 1973. Sicily and Its Cities in Hellenistic and Roman Times. Ph.D. dissertation, University of North Carolina at Chapel Hill.

Gordon, R. L. 1979. The Real and the Imaginary: Production and Religion in the Graeco-Roman World. *Art History*. March: 5–34.

Greco, E. 1996. City and Countryside. In Pugliese-Carratelli, G., ed. *The Western Greeks*. Venice: Bompiani. 233–42.

Gregory, T. 1979. The Late Roman Wall at Corinth. *Hesperia* 48: 264–80 and plates 76–78.

Gregory, T. E., ed. 1993. The Corinthia in the Roman Period. *JRS* Suppl. 8:9–30.

Greene, K. 1986. *The Archaeology of the Roman Empire*. Berkeley: Univeristy of California Press.

Grewe, K. 1988. *Licht am Ende des Tunnels: Planung und Trassierung im antiken Tunnelbau*. Mainz-am-Rhein: Verlag Phillip von Zabern.

Griesheimer, M. 1989. Genèse et developpement de la Catacombe St.-Jean à Syracuse. *MÉLROME* 101.2: 751–83. Extensive bibliography in the notes.

Grund, A. 1906a. Vorläufiger Bericht über physiogeographische Untersuchungen in den Deltagebieten des Grossen and Kleinen Mäanders. *Akadamie der Wissenschaften Sitzungberichte*. 115 Abt. 1 (Feb.): 1757–69.

———. 1906b. Vorläufiger Bericht über physiogeographische Untersuchungen im Deltagebiet des Kleinen Mäander bei Ajasoluk (Ephesus). *Memoires du Museum National D'histoire Naturelle* N.S. SERIE C, *Sciences de la Terre* 3.1: 241–62.

Gualandi, M. L, L. Messei, and S. Settes, eds. 1982 *Aparchai in Honor of Paolo Enrico Arias*. I. Pisa.

Guidoboni, E. 1989. *I terremoti prima del mille in Italia e nell'area Mediterranea*. Bologna.

———. 1990–91. La sismologia storica e lo studio dei terremoti antichi. *Kokalos* 36–37: 269–84. With extensive bibliography.

Guidoboni, E., A. Comastri, and G. Triana. 1994. *Catalogue of Ancient Earthquakes in the Meditarranean Area from Antiquity to the 10th Century* A.D. Rome: Istituto Nazionale di Geofisica. Considered by Ambraseys the best earthquake catalog. Not available to me.

Gülec, K. 1972. The Relationship between the Degree of Weathering and the Physico-mechanical Properties of Marbles. *Bull. Geol. Soc. Turkey* 15.2: 171–79.

Güngör, T., and A. Alkan. 1995. Geology and Ancient Urbanization: Some Data from Priene, Ephesus, and Miletus.. Manuscript in D. P. Crouch collection.

———. 1998. Stratigraphy and Tectonic Evolution of the Menderes Massif in the Söke-Selçuk Region. Ph.D. dissertation, Dokuz Eylül University, Izmir, Turkey.

Hackens, R., N. D. Holloway, and R. R. Holloway, eds. 1983. *Crossroads of the Mediterranean.* Paper Delivered at the International Conference on the Archaeology of Early Italy, Haffenreffer Museum, Brown University, 8–10 May 1981, Louvain. English edition, Brown University, 1994.

Hägg, R. 1987. Geometric Sancutaries in the Argolid. In Piérart, M., ed. *Polydipsion Argos: Argos de la fin des palais mycéniens à la constitution de l'etat classique* (Fribourg, Switzerland). *BCH* Suppl. 22. Paris: De Boccard Édition.

———. 1998. Argos and Its Neighbors in the Protogeometric and Geometric Periods. In Pariente, A., and G. Touchais, eds. *Argos et l'Argolide: Topographie et urbanisme.* École Française d'Athènes. Paris: De Boccard Édition. 131–35.

Hägg, R., N. Marinatos, and G. C. Nordquist, eds. 1988. *Early Greek Cult Practice.* Stockholm: Skrifter Utgivna av Svennsska Inst. i Athen.

Hall, J. M. 1995. How Argive Was the "Argive" Heraion? The Political and Cultic Geography of the Argive Plain, 900–400 B.C. A *JA* 99: 557–613.

Hallam, A. 1989. *Great Geological Controversies.* 2nd ed. New York: Oxford University Press.

Halstead, P., and J. O'Shea, eds. 1989. *Bad Year Economics.* Cambridge: Cambridge University Press.

Hammilakis, Y. 1995. Zooarchaeology at Panakton. Abstracts of the American Institute of Archaeology Meeting of Dec. 1994. *AJA*: 324.

Hammond, M. 1972. *The City in the Ancient World.* Cambridge, Mass.: Harvard University Press.

Harding, S. 1991. *Whose Science? Whose Knowledge?* Ithaca, N.Y.: Cornell University Press.

Harland, W. B., and K. E. Fancett. 1989. *A Geologic Time Scale.* Cambridge: Cambridge University Press.

Hassan, F. A. 1978. Demographic Archaeology. In Schiffer, M. B., ed. *Advances in Archaeological Method and Theory.* Vol. 1. New York: Academic Press: 49–89.

Hawkes, C. 1973. Innocence Retrieval in Archaeology. *Antiquity* 47: 176–78.

Hayward, C. L. 1995. A Systematic Petrographic Study of Ancient Limestone Quarries in the Ancient Corinth Region. Non-circulating paper in collection of Weiner Laboratory, American School of Classical Studies, Athens. Abstracted for the American Institute of Archaeology meeting of Dec. 1995: 324, published in *AJA*, 1996.

Hearty, P. J., G. H. Miller, C. E. Stearns, and B. J. Szabo. 1986. Aminostratigraphy of Quaternary Shorelines in the Mediterranean. Basin. *Geological Society of America Bulletin* 5.97 (July): 850–58.

Heberdey, W. R., and G. Niemann. 1912. Das Theater in Ephesos. *FiE* 2.

Heberdey, W. R., and W. Wilberg. 1923. Der Aquäduct des C. Sextilius Polio. *FiE* 256 ff.

Heichelheim, F. M. 1956. Effects of Classical Antiquity on the Land. In Thomas, W. L., ed. *Man's Role in Changing the Face of the Earth.* Chicago: Wenner-Gren Foundation–University of Chicago Press. 165–82.

Held, W. 1993. Heiligtum und Wohnhaus. Ein Beitrag zur Topographie des klassischen Milet. *Ist. Mitt.* 43: 371–80.

———. 1999. Excavation at Miletus. *AA.*

Hellman, M.-C. 1981. L'Eau des cisternes et la salubrité: Textes et archéologie. In Ginouvés, R., A.-M. Guimier-Sorbets, J. Jouanna, and L. Villard, eds. *L'Eau, la santé et la maladie dans le monde grec.* Paris: Actes de Colloque à Paris. 25–27 Nov. 1992. *BCH* Suppl 4: 273–82.

————. 1990. Treasuries. In Delphes: Races cultes et jeux: 100 ans de fouilles et de decouvertes. *Les Dossiers d'Archaeologie* 15 (July–August): 66–67.

Helly, B., and A. Pollino, eds. 1989. *Tremblement de terre: Histoire et archeologie. 4: Rencontres internationales d'archeologie et d'historie d'Antibes, 2–4 Nov. 1983.* Vallbonne.

Herak, M., and V. T. Stingfield. 1972. Historical Review of Hydrological Concepts. In Herak, M., and V. T. Stingfield, eds. *Karst: Important Karst Regions of the Northern Hemisphere.* Elsevier: Amsterdam. 19–241.

Herz, N. 1988a. Geology of Greece and Turkey: Potential Marble Source Region. In Herz, N., and M. Waelkens, eds. *Classical Marble: Geochemistry, Technology, Trade.* NATO ASI Series E: Applied Sciences, Vol. 153. Dordrecht: Kluwer Academic: 7–10.

————. 1988b. The Oxygen and Carbon Isotopic Data Base for Classical Marble. In Herz, N., and M. Waelkens, eds. *Classical Marble: Geochemistry, Technology, Trade.* NATO ASI Series E: Applied Sciences, Vol. 153. Dordrechtn: Kluwer Academic: 305–14.

Hesse, A. 1988. Une nouvelle hypothèse de trace pour un rempart de la ville. *BCH* 106: 647–51.

Hidiroglu, B. 1988. *Selçuk–Efes–Mergemanon–Pamucak: Yörelerinin Jeolojisi.* Mühendislik Özellikleri ve Anlik Mermes Ocaklar ve Anlik mermes Ocaklaari. Izmir/Bornova: Dokuz Eylül University. Ephesus geological section p. 19.

Higgins, M., and R. Higgins. 1996. *A Geological Companion to Greece and the Aegean.* London: G. Duckworth.

Hill, B. H. 1964. *Corinth I,* pt. VI. *The Springs.* Princeton, N.J.: American School of Classical Studies at Athens.

Hirsch, G. 2000. Proprioception, Reflection, Recognition: The Positive Affect of Paradigmatic Form. Oral paper (citing the GYU-ZHI) presented at the Society for the Anthropology of Consciousness, Tuscon, Arizona, April. Manuscript in D. P. Crouch collection.

Hoepfner, W. 1990. Zum Problem griechischer Holz- und Kassettendechen. In Hoffman A., E.-L. Schwandner, W. Hoepfner, and G. Brands, eds. *Bautechnik der Antike: International Colloquium in Berlin.* Mainz-am-Rhein: Verlag Philipp von Zabern. 90–98.

Hoepfner, W., and E.-L. Schwandner. 1985. *Haus and Stadt im klassischen Griechenland.* Munich: Deutscher Kunstverlag. 2nd ed. 1994.

Hoffmann, A., E.-L. Schwandner, W. Hoepfner, and G. Brands, eds. 1990. *Bautechnik der Antike: International Colloquium in Berlin.* Mainz-am-Rhein: Verlag Philipp von Zabern.

Hogarth, A. C. 1972. Common Sense in Archaeology. *Antiquity* 46: 301–4.

Hohenberg, P. M., and L. H. Lees. 1985. *The Making of Urban Europe 1000–1950.* Cambridge, Mass.: Harvard University Press.

Holliday, V. T. 1992. *Soils in Archaeology: Landscape Evolution and Human Occupation.* Washington, D.C.: Smithsonian Inst. Press.

Homage à Pierre Birot: La mobilité des paysages méditerranéans. Toulouse. 1984. *Revue géographique des pyrénées et du sud-ouest.* Travaux II.

Hopkins, K. 1978a. Rules of Evidence. (A review of F. Millar, *The Emperor in the Roman World.) JRS* 48: 178–86.

————. 1978b. Economic Growth and Towns in Classical Antiquity. In Abrams, P., and E. W. Wrigley, eds. *Towns in Societies: Essays in Economic History and Historical Sociology.* Cambridge: Cambridge University Press. 35–77.

————. 1980. Taxes and Trade in the Roman Empire (200 B.C.–A.D. 400). *JRS* 70: 101–25.

Houel, J. D.L. L. 1782–87. *Voyage pittoresque des iles de Sicile, de Malte et de Lipari.* Paris: Imprimiere de Monsieur.

Howgego, C. 1995. *Ancient History from Coins*. London: Routledge.

Hsü, K. J., and W. B. F. Ryan. 1972. Summary of the Evidence for Extensional and Compressional Tectonics in the Mediterranean. *Deep Sea Drilling*. 1011–19.

Hueber, F. 1997. Zur städtebaulichen Entwicklung des hellenistisch-römischen Ephesos. *Ist. Mitt.* 47: 251–69.

Hughes, J. D. 1983. How the Ancients Viewed Deforestation. In Evans, R. K., ed. *Symposium (1978): Deforestation, Erosion, and Ecology: Contributions 1–2. JFA* 10: 437–44

———. 1994 (revised and expanded). *Pan's Travail: Environmental Problems of the Ancient Greeks and Romans*. Baltimore: Johns Hopkins University Press.

Hulot, J., and F. Fougeres. 1910. *Selinonte: La ville, l'acropole et les temples*. Paris: Schmidt.

Huppert, G. 1970. *The Idea of Perfect History*. Urbana: University of Illinois Press.

Hutton, J. 1788. *Theory of the Earth, Or an Investigation of the Laws Observable in the Composition, Dissolution and Restoration of Land upon the Globe*. Edinburgh: Royal Society of Edinburgh.

Ilhan, E. 1971a. Structural Features of Turkey. In Campbell, A. S., ed. *Geology and History of Turkey*. Petroleum Exploration Society of Libya, 13th Annual Field Conference. 159–70.

———. 1971b. Earthquakes in Turkey. In Campbell, A. S., ed. *Geology and History of Turkey*. Petroleum Exploration Society of Libya, 13th Annual Field Conference. 431–42.

Issar, A, H. Tsoar, and D. Levin. 1989. Climatic Changes in Israel during Historical Times and Their Impact on Hydrological, Pedological and Socio-Economic Systems. In Leinen, M., and M. Sarnthein, eds. *Paleoclimatology and Paleometeorology: Modern and Past Patterns of Global Atmospheric Transport*. Dordrecht: Kluwer Academic. 525–41.

Jacobsthal, P. 1951. The date of the Ephesian foundation-deposit. *JHS* 71: 85–95.

Jacquemin, A. 1990a. Le sanctuaire d'Athéna á Marmaria. In Delphes: Races cultes et jeux: 100 ans de fouilles et de decouvertes. *Les Dossiers d'Archaeologie* 15 (July–August): 51

———. 1990b. Les trépieds delphiques. In Delphes: Races cultes et jeux: 100 ans de fouilles et de decouvertes. *Les Dossiers d'Archaeologie* 15 (July–August): 14–19.

Jacquemin, A., and H. Jacquemin. 1990. La découverte de Delphes. In Delphes: Races cultes et jeux: 100 ans de fouilles et de decouvertes. *Les Dossiers d'Archaeologie* 15 (July–August): 4–8.

Jacquemin, A., M.-C. Hellman, and P. Amardry. 1990. Les Offrandes. In Delphes: Races cultes et jeux: 100 ans de fouilles et de decouvertes. *Les Dossiers d'Archaeologie* 15 (July–August): 58–71.

Jacquemin, A., C. Rolley, R. Queyrel, and A. Bélis. 1990. Les concours. In Delphes: Races cultes et jeux: 100 ans de fouilles et de decouvertes. *Les Dossiers d'Archaeologie* 15 (July–August): 32–37.

Jacquemin, A., G. Rougemont, and A. Bélis. 1990. Le sanctuaire et la cité. In Delphes: Races cultes et jeux: 100 ans de fouilles et de decouvertes. *Les Dossiers d'Archaeologie* 15 (July–August): 20–25.

Jacquemin, A., C. Vaatin, G. Sauron, and V. Déroche. 1990. La Fin de Delphes. In Delphes: Races cultes et jeux: 100 ans de fouilles et de decouvertes. *Les Dossiers d'Archaeologie* 15 (July–August): 72–77; Chronologie, 78–79.

Jacquemin, H. 1990. Géologie et paysage: Un site à la recherche de son equilibre. In Delphes: Races cultes et jeux: 100 ans de fouilles et de decouvertes. *Les Dossiers d'Archaeologie* 15 (July–August): 4–8.

Jacques, F., and B. Bousquet. 1989. Le Cataclysme du 21 Juillet 365: Phenomene Regional ou Catastrophe Cosmique? In Helly, B., and A. Pollino, eds. *Tremblement de terre: Histoire et archeologie.4. Rencontres internationales d'archeologie et d'historie d'Antibes, 2–4 Nov. 1983*. Vallbonne. 183–196.

Jameson, M. H. 1990.Domestic Space in the Greek City-state. In Kent, S., ed. *Domestic Architecture and the Use of Space: An Interdisciplinary Study*. Cambridge: Cambridge University Press. 92–113.

Jansen, G. C. M. 1991. Water Systems and Sanitation in the Houses of Herculaneum. *Mededelingen van het Nederlands Institut te Rome* 50: 145–66.

———, ed. 2000. *Cura aquarum in Sicilia. Annual Papers on Classical Archaeology*, suppl. 6. Leiden, Stichting Babesch.

Japelli, R. 1980. Notizie sui dissesti di Agrigento (The Landslides of Agrigento). *Proceedings of the 14th Congress Nazionale di Geotecnica*. Florence. Vol. 3.

Jarman, M. R., C. Vita-Finzi, and E. S. Higgs. 1970. Site Catchment Analysis in Archaeology. In Ucko, P. J., R. Tringham, and G. W. Dinbleby, eds. *Man, Settlement and Urbanism: Proceedings of the Research Seminar in Archaeology and Related Subjects*. Cambridge, Mass.: Schenkman Publishing. 61–66.

Jashemski, W. E. 1979a. *The Gardens of Pompeii*. New Rochelle, N.Y.: Caratzas Bros.

———. 1979b. Pompeii and Mount Vesuvius. In Sheets, P. D., and D. K. Grayson, eds. *Volcanic Activity and Human Ecology*. New York: Academic Press. 587–622.

Jeftic, L., J. D. Milliman, and G. Sestini, eds. 1992. *Climate Change and the Mediterranean*. London: Edward Arnold.

Jelgersma, S., and G. Sestini, 1992. Implications of a Future Rise in Sea-Level on the Coastal Lowlands of the Mediterranean. In Jeftic, L., J. D. Milliman, and G. Sestini, eds. *Climate Change and the Mediterranean*. London: Edward Arnold. 282–303

Jennings, J. N. 1971. *Karst*. Cambridge, Mass.: MIT Press.

Johns, J. 1980. The Monreale Survey: Indigenes and Invaders in Medieval West Sicily. Papers in Italian Archaeology. *BAR* 2: 215–23.

Johnston, A. W. 1993–94. Emporia, Emporoi, and Sicilians: Some Epigraphical Aspects. *Kokalos* 39–40 I.1: 155–69

Jones, A. H. M. 1940. *The Greek City from Alexander to Justinian*. Oxford: Clarendon Press.

Jost, M. 1985. *Sanctuaires et cultes d'Arcadie*. École Française d'Athènes. Études péloponnésiennes, 9.

Jordan, B., and J. Perlin. 1979. Solar Energy Use and Litigation in Ancient Times. *Solar Law Reporter* 583–94.

———. 1984. On the Protection of Sacred Groves. In Boegehold, A. L., ed. *Studies Presented to Sterling Dow on His 80th Birthday*. Durham, N.C.: Duke University Press.

Jordanova, L. J., and R. S. Porter. 1979/1981. Introduction. In Jordanova, L. J., and R. S. Porter, eds. *Images of the Earth: Essays in the History of the Environmental Sciences*. Monographs 1. British Society for the History of Science. vii–xvii.

Judson, J. n.d. (ca. 1958). Geologic and Geographic Observations at Morgantina. Unpublished manuscript on file in the Morgantina Room, Princeton University.

Judson, S. 1963. Stream Changes during Historic Time in East-Central Sicily. *AJA* 67: 287–89.

Kamen-Kay, M. 1971. A Review of Depositional History and Geological Structure in Turkey. In Campbell, A. S., ed. *Geology and History of Turkey*. Petroleum Exploration Society of Libya, 13th Annual Field Conference. 111–37.

Kandler, M., H. Zabehlicky, and W. Vetters. 1995. Fragile Towns in the North of the Ancient Roman Empire: A Geo-ecological Impact for the Last Quarter of the 2nd Century A.D. and Earthquakes. *Geologia Applicata e Idrogeoloogia*. 30. Atti prino conv. d. Gruppoo Naz. d. Geol. Appl. in connection with Intern. Assn. Engineering Geology (IAEG).

Kapitan, G. 1967–68. Sul Lakkios, porto piccolo di Siracusa nel perioodo greco. *ASS* 13–14: 13–32.

Karatekin, N. 1953. Hydrological Research in the Middle East. *Reviews of Research on Arid Zone Hydrology* (UNESCO) 1: 78–95

Kardulias, P. N. 1994. Archaeology in Modern Greece: Bureaucracy, Politics, and Science. In Kardulias, P. N., ed. *Beyond the Site: Regional Studies in the Aegean Area.* Lanham, Md.: University Press of America. 373–88.

Kardulias, P. N., and T. E. Gregory. 1991. Regional Study in the Korinthia: The Korinthia Exploration Project 1991. *Old World Archaeological Newsletter* 15.2 (winter): 20–26.

Karfakis, Y., and N. Mauriaris. 1988. Remote Sensing and Seismotectonic Inspection of the West Corinthian Gulf: Disasters during Ancient Times and Seismic Hazard. In Marinos, P. G., and G. C. Koukis, eds. *The Engineering Geology of Ancient Works, Monuments and Historical Sites: Proceedings of the Symposium of the International Association of Engineering Geology, Athens.* Rotterdam: A. A. Balkema. 1291–99.

Karlsson, L. 1993. Did the Romans Allow the Sicilian Greeks to Fortify Their Cities? In Bilde, P. G., I. Nielsen, and M. Nielsen, eds. *Aspects of Hellenism in Italy: Towards a Cultural Unity?* Danish Studies in Classical Archaeology, Acta Hyperborea 5. Museum Tusculanum P, University of Copenhagen. 38–45.

Karweise, S. 1985. Koressos: Ein fast vergessener Statteil von Ephesos. *Pro Arte Antiqua: Festschrift für Hedwig Kenner. Sonderschriften.* Vienna: Verlag A. F. Kosta for the Österreichisches Archäologisches Institut. Vol. 2 of 18 vols.: 214–25.

———. 1995. *Gross ist die Artemis von Ephesos: Die Geschichte einer der grossen Städte der Antike.* Vienna: Phoibos Verlag.

———. 1996.The Church of Mary and the Temple of Hadrian Olympius. In Koester, H., ed. *Ephesos: Metropolis of Asia.* Valley Forge, Pa.: Harvard Theological Studies, Trinity Press International. 311–19.

Kastning, E. H. 1977. Faults as Positive and Negative Influennces on Ground Water Flow and Conduit Enlargement. In Dilamarter, R. R.. and S. C. Csallany, eds. *Hydrologic Problems in Karst Regions: Proceedings of the International Symposium, April 1976.* Bowling Green, Ky.: W. Kentucky University Press. 193–201.

Kayan, I. 1988. Late Holocene Sea-Level Changes on the Western Anatolian Coast. *Palæogeography, Palæoclimatology, Palæoecology* 68: 205–18.

———. 1990. Late Holocene Geomorphic Evolution of the Beshige-Troy Area (NW Anatolia) and the Environment of Prehistoric Man. In Bottema, S., G. Entijes-Nieborg, and W. van Zeist, eds. *Man's Role in the Shaping of the Eastern Mediterranean Landscape.* Rotterdam: A. A. Balkema. 69–70.

———. 1991. Holocene Geomorphic Evolution of the Beshik Plain and Changing Environment of Ancient Man. *Studia Troica.*1. Mainz am Rhein: Verlag Philipp von Zabern. 79–92.

———. 1996. Holocene Coastal Development and Archaeology. *Z. Geomorph.* N.S., suppl. 102 (Feb.): 37–59.

Keegan, W. F., and J. N. Diamond. 1987. Colonization of Islands by Humans: A Biogeographical Perspective. *Advances in Archaeological Method and Theory* 10: 49–92.

Keil, J. 1922–24a. Ortygia, die Geburtsstäte der ephesischen Artemis. *JÖAI* 21–22: 113–19.

———. 1922–24b. Zur Topographie und Geschichte von Ephesos. *JÖAI* 21–22: 96–112.

———. 1926. Vorläudifer Bericht über die Ausgrabungen in Ephesos. *JÖAI* 12: 247–300

———. 1931. Vorläudifer Bericht über die Ausgrabungen in Ephesos. *JÖAI* 15:33ff.

———. 1932. Vorläudifer Bericht über die Ausgrabungen in Ephesos. *JÖAI* 16: 5–16.

———. 1959. Erlass des Prokonsuls L. Antonius Albus über die Freihaltung des ephesischen Hafens. *JÖAI* 44: 142–47.

———. 1964. *Führer durch Ephesos.* 5th ed. Vienna: F. Eichler: Österreichisches Archäologisches Institut.

Keller, D. R., and D. W. Rupp, eds. 1983. *Archaeological Survey in the Mediterranean Area*. Oxford: BAR Interrnational Series 155.

Keraudren, B., and D. Sorel. 1987. The Terraces of Corinth (Greece): A Detailed Record of Eustatic Sea-Level Variations during the Last 500,000 Years. *Marine Geology* 77.1/2 (July): 99–107.

Kerenyi, C. 1966. Le divinità ed i templi di Seilinunte. *Kokalos* 12: 3–7.

Kiepert. H. 1872. Map of Valley and Lake "Panormos" at Ephesus. *FiE* 1, 1906.

Kind, A., and M. Henig, eds. 1981. *The Roman West in the Third Century [C.E.]: Contributions from Archaeology and History*. Oxford: BAR International Series 109 (ii).

Kininmouth, C. 1965. *Travelers' Guide: Sicily*. Indianapolis: Bobbs-Merrrill.

Kissel, C, C. Laj, A.M.C. Sengör, and A. Poisson. 1987. Paleomagnetic Evidence for Rotation in Opposite Senses of Adjacent Blocks in Northeastern Aegea and West Anatolia. *Geophysical Research Letters* 14.99 (Sept.): 907–10.

Kleiner, G. 1968. *Die Ruinen von Milet*. Berlin: Walter de Gruyter for the Deutsches Archäologisches Institut.

Klemes, V. 1986. Dilettantism in Hydrology: Transition or Destiny? *Water Resources Research* (August) 22.99: 1775–1885.

Knauss, J. 1987, 1990. KOPAIS II, III. Institut für Wasserbau, Technical University of Munich.

———. 1996. Flussumleiting von Tiryns (17–124); Die Trockenlegung des Sumpfes von Lerna (125–159); Argolische Studien: Alte Strassen; Alte Wasserbauten; Wasserbau und Wassermengenwirt-Schaft. No. 77. Munich: Technical University of Munich.

Knauss, J., B. Heinrich, and H. Kalcyk. 1980. KOPAIS I. Munich: Institut für Wasserbau, Technical University of Munich.

Knibbe, D., and W. Alzinger. 1980. Ephesos vom Beginn der römischen Herrschaft in Kleinasien bis zum Ende der Principatszeit. In ANRW II, 33, 1 (Berlin), 748–830, Abb., Tab.11/55.

Knibbe, D., and G. Langmann, eds. 1993. *Via Sacra Ephesiaca*. 1: Berichte und Materialien, vol. 3. Vienna: Österreichisches Archäologisches Institut.

———. 1995. *Via Sacra Ephesiaca*. 2: Berichte und Materialien, vol. 6. Vienna: Österreichisches Archäologisches Institut

Koenigs, W. 1991. *Westtürkei von Troia bis Knidos*. Munich: Artemis Kunst- und Reisführer. 171–236.

———. 1993. Plannung und Ausbau der Agora von Priene. *Ist. Mitt.* 43: 381–97.

Koenigs, W., F. Runscheid, and S. Westphalen. 1996. Priene 1995. *Arashtirma Sonuçlari Toplantisi* 2: 71–94.

Koester, H., ed. 1996. *Ephesos: Metropolis of Asia*. Valley Forge, Pa.: Harvard Theological Studies, Trinity Press International.

Kolaiti, E., P. Marinos, and M. Kavvadas. 1995. Engineering Geology and Geotechnical Engineering of the Archaeological Site of Delphi, Greece. Manuscript in D. P. Crouch collection.

Kolb, F. 1981. *Agora und Theater, Volks- und Festversammlung*. Berlin: G. Mann Verlag for Deutsches Archäologisches Institut.

Koldwey, R., and O. Puchstein. 1899. *Die Griechischen Templi in Unteritaliienn und Sicilien*. Berlin.

Koloski-Ostrow, A. O. 1995. Latrines, Baths, and Health in Post-Earthquake Pompeii. Abstracts of the 96th Annual Meeting of the American Institute of Archaeology. AJA 99: 317.

———. forthcoming. *Archaeology Sanitation: Water, Sewers, and Latrines* (working title). Chapel Hill: University of North Carolina Press.

Kolota, A. L., and C. R. Ortloff. 1996. Agroecological Perspectives on the Decline of the

Tiwanaku State, In Kolota, A. L., ed. *Tiwanaku and Its Hinterland: Archaeology and Paleoecology of an Andean Civilizsation.* 1: *Agroecology.* Washington, D.C.: Smithsonian Institution Press. 181–202.

Konstantinopoulos, G. 1968. Rhodes, New Finds and Old Problems. *Archaeology* 1: 115–23 and plate 23. Another version of this material is found in W. Hoepfner and E.-L. Schwandner (1985), where the section on Rhodes is Konstantinopoulos' work.

Koroniotis, K. S., A. A. Collios, and A. P. Basdekis. 1988. Stabilisation of the Rock Slopes at the Region of Kastalia Spring at Delphi. In Marinos, P. G., and G. C. Koukis, eds. *The Engineering Geology of Ancient Works, Monuments and Historical Sites: Proceedings of the Symposium of the International Association of Engineering Geology, Athens.* Rotterdam: A. A. Balkema: 207–11.

Koukis, G. 1988. Peloponnesus: History, Geology and Engineering Geology Aspects. In Marinos, P. G. and G. C. Koukis, eds. 1988. *The Engineering Geology of Ancient Works, Monuments and Historical Sites: Proceedings of the Symposium of the International Association of Engineering Geology, Athens.* Rotterdam: A. A. Balkema: 2213–35, esp. Hydrogeological Conditions, 2225–2227.

Kraft, J. C. 1972. *A Reconnaissance of the Geology of the Sandy Coastal Areas of East Greece and the Peloponnese.* Technical Report 9, July. College of Marine Studies, University of Delaware.

————. 1998. Adaption of the Harbors of Ancient Ephesus and the Artemision to Rapid Deltaic Progadation Events, and Engineering Works from 700 B.C. to A.D. 1000. Abstract of a paper presented at IGCPA Project 367/INQUA Conference at Corinth and Samos, Greece, Sept. 10–20, 1998.

Kraft, J. C., and G. R. Rapp Jr. 1975. Late Holocene Paleogeography of the Coastal Plain of the Gulf of Messenia, Greece. *Geological Society of America Bulletin* 86: 1191–1298.

Kraft, J. C., S. E. Aschenbrenner, and I. Kayan. 1980. Late Holocene Coastal Changes and Resultant Destruction or Burial of Archaeological Sites in Greece and Turkey. In Schwartz, M. L., ed. *Proceedings of the Commission of the Coastal Environment, 24th International Geographical Congress:* 13–31.

Kraft, J. C., S. E. Aschenbrenner, and G. Rapp Jr. 1977. Paleogeographic Reconstructions of Coastal Aegean Archaeological Sites. *Science* 195 (11 March): 941–47.

Kraft, J. C., H. Brückner, and I. Kayan. 1999. Paleogeographies of Ancient Coastal Environments, in Environs of the Feigengarten Excavation and the 'Via(e) Sacra(e)' to the Artemision at Ephesus. In Scherrer, P., H. Taeuber, and H. Thür, eds. *Steine und Wege: Festschrift für Dieter Knibbe zum 65. Geburtstag.* Sonderschriften 32. Vienna: Österreichisches Archäologisches Institut. 91–100.

Kraft, J. C., I. Kayan, and S. E. Aschenbrenner. 1985. Geological Studies of Coastal Change Applied to Archaeological Settings. In Rapp, G., and J. A. Gifford, eds. *Archaeological Geology.* New Haven, Conn.: Yale University Press. 57–84.

Kraft, J. C., G. Rapp Jr., I. Kayan, and R. K. Dunn. 1995. Palaeogeographies of the Embayed Coasts of Greece and Turkey. Abstract from the 96th Meeting of the American Institute of Archaeology, 1994. *AJA* 99: 341.

Kristensen, T. M., S. E. Larsen, P. G. Moller, and S. E. Petersen, eds. 1995. *City and Nature: Changing Relations in Time and Space.* Colloquium Odense, 1993.

Kritzas, D. B. 1992. Aspécts de la vie politique et economique d'Argos au Ve siècle av. J.-C. *BCH* Suppl.: 231–40.

Kummer and Wilberg. 1885–89. Plan of Priene after Excavations of 1895–99. German Archaeological Institute, Istanbul.

Kupper, H. 1972. The Language of Sites in the Politics of Space. *American Anthropologist* 74.3 (June): 411–25.

Kuprianov, V. V. 1976. Urban Influences on the Water Balance of an Environment. In

Effects of Urbanization and Industrialization on the Hydrological Regime and on Water Quality: Proceedings of the Amsterdam Symposium, Oct. 1977. International Assn. for Hydraulic Studies. 41–47.

Kuznetsov, V. D. 1999, Early Types of Greek Dwelling House in the North Black Sea. In Tsetskhladze, G. R, ed. *Ancient Greeks West and East.* Leiden: E. J. Brill. 531–64.

La Fleur, R. G., ed. 1984. *Groundwater as Geopmorphic Agent.* Boston: Allen and Unwin.

Lamb, H. H. 1981. An Approach to the Study of the Development of Climate and Its Impact in Human Affairs. In Wigley, T. M. L., M. J. Ingram, and G. Farmer, eds. *Climate and History: Studies in Past Climates and Their Impact on Man.* Cambridge: 291–309;

Lambeck, K. 1995. Late Pleistocene and Holocene Sea-Level Change in Greece and South-Western Turkey: A Separation of Eustatic, Isostatic and Tectonic Contributions. *Geophysical J. International* 122: 1022–44.

Lamprecht, H.-O. 1990. Rationalisiertes Bauen durch *opus caementicium* (Zusammenfassung). In Hoffmann, A., E.-L. Schwandner, W. Hoepfner, and G. Brands, eds. *Bautechnik der Antike: International Colloquium in Berlin.* Mainz-am-Rhein: Verlag Philipp von Zabern. 137–39.

Lance, G. N., and W. T. Williams. 1966. A Generalized Sorting Strategy for Computer Classification. *Nature* 212, 218.

Landon, M. E. 1994. Contributions to the Study of the Water Supply of Ancient Corinth. Ph.D. dissertation, University of California at Berkeley.

———. 1997. A Rediscovered Fountain House in Ancient Corinth. In Abstracts of the Dec. 1996 Meeting of the American Institute of Archaeology. *AJA.*

Landsberg, H. E. 1981. Past Climates from Unexploited Written Sources. In Rotberg, R. I., and T. K. Rabb, eds. *Climate and History: Studies in Interdiscliplinary History.* Princeton, N.J: Princeton University Press. 50–62.

Landucci Gattinoni, F. 1992. L'imagine di una citta' ellenistica: Il caso di Priene. *Contribati dell' Instituto di Storia Antica* (Milan): 83–92.

Langmann, G. 1972. Nymphum des C. *Laekanius Bassus. JÖAI* 50: 72.

Langmann, G., and P. Scherrer. 1993. Agora. *JÖAI*: 12–14.

Langmuir, C. H., R. D. Vocke, G. N. Hanson, and S. R. Hart. 1977. A General Mixing Equation: Applied to the Petrogenesis of Basalts from Iceland and Teykjanes Ridge. *Earth Plan. Sci. Lett.* 37: 380–92.

LaPenna, V., M. Mastrantono, D. Patella, and G. Di Bello. 1992. Magnetic and Geoelectric Prospecting. *Bollettino di Geofisica Teorica ed Applicata* 34 134–35 (June–Sept.): 133–44.

La Rosa, V. 1996. The Impact of the Greek Colonists on the Non-Hellenic Inhabitants of Sicily. In Pugliese-Carratelli, G., ed. *The Western Greeks.* Venice: Bompiani. 523–32.

Laurence, R. 1994. *Roman Pompeii: Space and Society.* London: Routledge.

Le Dinahet, M.-T. 1984. Sanctuaires chthoniens de Sicile de l'epoque archaïque à l'epoque classique. In Roux, G., ed. *Temples et Sanctuaires.* GIS—Maison de l'Orient: 137–52.

Le Dinahet, V. 1979. *Selinunte: Vicende Storiche: Illustrazione dei Monumenti.* Palermo: S. F. Flaccovio Edizioni.

Lee, W. H. K., ed. 1988. *Historical Seismograms and Earthquakes of the World.* New York: Academic Press.

Lefèvre, F. 1995. La chronologie du IIIᵉ siècle à Delphes, d'après les actes amphicioniques. *BCH* 119: 161–206.

LeGrand, H. E. 1973. Hydrological and Ecological Problems of Karst Regions. *Science.* 179: 859–64.

———. 1977. Karst Hydrology in Relationship to Environmental Sensitivity. In Dilamaarter, R. R. and S. C. Csallany, eds. *Hydrologic Problems in Karst Regions: Proceedings of the International Symposium, April 1976.* Bowling Green, Ky.: W. Kentucky University Press. 10–18.

Leighton, R. 1993. *Morgantina Studies. 4: The Protohistoric Settlement on the Cittadella.* Princeton, N.J.: Princeton University Press.

Lentini, F., S. Carbone, S. Catalano, M. Grasso, and C. Monaco. 1991. Presentazione della carta geologica della Sicilia centro-orientale. *Mem. Soc. Geol. It.* 47: 145–56.

Lenzzi, F., and G. Nenzioni. 1991. *Il tempo e la natura, culture e insediamenti preistorici nella zona dei Gessi.* Commune de S. Lazzaro di Savena—Assessorato alla Cultura.

Lepelley, C. 1989. L'Afrique du nord et le seisme du 21 Juillet 365: Remarques methodologiques et critiques. In Helly, B., and A. Pollino, eds. *Tremblement de terre: Histoire et archeologie. 4. Rencontres internationales d'archeologie et d'historie d'Antibes, 2–4 Nov. 1983.* Vallbonne: 199–206.

Lethaby, W. R. 1913. The Sculpture of the Earlier Temple of Artemis at Ephesus. *JHS* 34: 1–16.

———. 1914. The Sculpture of the Later Temple of Artemis at Ephesus. *JHS* 33: 87–96.

———. 1916. Another Note on the Sculpture of the Later Temple of Artemis at Ephesus. *JHS* 36: 25–35.

———. 1917. The Earlier Temple of Artemis at Ephesus: the Sculpture. *JHS* 37: 1–16.

Lewin, J. 1980. Introduction. In Cullingford, R. A., D. A. Davidson, and J. Lewin, eds. *Timescales in Geomorphology.* Chichester: J. Wiley and Sons.1–6.

Limberis, V. 1996. The Council of Ephesos: The Demise of the See of Ephesus and the Rise of the Cult of the Theotokos. In Koester, H., ed. *Ephesos: Metropolis of Asia.* Valley Forge, Pa.: Harvard Theological Studies, Trinity Press International. 321–40.

Linders, T. 1988. Continuity in Change: The Evidence of the Temple Accounts of Delos (Prolegomena to a Study of the Economic and Social Life of Greek Sanctuaries). *Acta Instituti Atheniensis Regni Sueciae.* Series in 4, 38: 267–70.

Lindh, G. 1979. *Socio-Economic Aspects of Urban Hydrology.* Studies and Reports in Hydrology, 27. Paris: UNESCO.

Lloyd, R. V., A. Tranh, S. Pearce, M. Cheeseman, and D. N. Lumdsen. 1988. ESR Spectroscopy and X-ray Powder Diffractometry for Marble Provenance Determination. In Herz, N., and M. Waelkens, eds. *Classical Marble: Geochemistry, Technology, Trade.* NATO ASI Series E: Applied Sciences, Vol. 153. Dordrecht: Kluwer Academic. 369–77.

Lohmann, H. 1997. Survey in der Chora von Milet. *AA* 2: 285–94.

Lolos, Y. A. 1997. The Hadrianic Aqueduct of Corinth, with Appendix on the Roman Aqueducts of Greece. *Hesperia* 66.2: 271–314 and plates 66–76.

Londey, P. 1990. Greek Colonists and Delphi. In Descœudres, J.-P., ed. *Greek Colonists and Native Populations.* Oxford: Clarendon Press. 117–27.

Los Angeles Times. August 30, 1999. Turkey Quake Provides Lessons for California.

———. 5.9 September 8, 1999. Quake Hits Athens, Leaving 36 Dead and Nearly 100 Missing. *Los Angeles Times.* Feb. 7, 2001. A High Calling for Priestesses at Delphi.

Loy, W. S. n.d. (fieldwork 1965–66). *The Land of Nestor: A Physical Geography of the Southwest Peloponnese.* Foreign Field Research Program, No. 34. London: National Academy of Science.

Lüttig, G., and P. Steffens. 1976. Explanatory Notes for the *Paleographic Atlas of Turkey from the Oligocene to the Pleistocene.* I:1,500,000. Hanover.

Lyons, C. L. 1991. Modalità di acculturazione a Morgantina. *Bolletino di Archeologia* 11–12: 1–10.

———. 1996. *Morgantina Studies. 5: The Archaic Cemeteries.* Princeton, N.J.: Princeton University Press.

MacDonald, W. L. 1986. *The Architecture of Imperial Rome.* II. New Haven, Conn.: Yale University Press.

MacKay, P. A. 1967. The Fountain at Hadji Mustapha. *Hesperia* 36: 193–95.

MacKenzie, D. 1972. Active Tectonics of the Mediterranean Region. *Geophysical J. of the Royal Astronomy Society* 30: 109–85.

———. 1980. Sea Level Variations in the Northwest Mediterranean during Roman Times. *Symposium on Evolution and Tectonics of the Western Mediterranean and Surrounding Areas.* Madrid: Internat. Geografico Nacional: 45–97.

MacMullen, R. 1990. *Changes in the Roman Empire: Essays in the Ordinary.* Princeton, N.J.: Princeton Univrsity Press.

Macready, S., and F. H. Thompson. 1987. *Roman Architecture in the Greek World.* Occasional Papers (N.S.) 10. London: Society of Antiquaries of London.

Macro, A. D. 1980. The Cities of Asia Minor under the Roman Imperium. ANRW: 658–97.

Maddoli, G. G. 1982. Sparta e la "Liberazione" degli Empori. In Gualandi, M. L, L. Messei, and S. Settes, eds. *Aparchai in Honor of Paolo Enrico Arias.* I. Pisa. 245–52.

Malkin, I. 1989. Delphoi and the Founding of Social Order in Archaic Greece. *Metis* 4.1: 129–53.

Maniatis, Y., N. Herz, and Y. Basiakos, eds. 1999. *The Study of Marble and Other Stones Used in Antiquity. Archetype.* Toulouse, France: Centre de Recherche en Physique Applique.

Marasco, G., 1986. Alexandro Magno e Priene. *Sileno* 12.1–4: 59–77.

Marchese, R., ed. 1983. *Aspects of Greco-Roman Urbanism: Essays on the Classical City.* Oxford: BAR International Studies.

Marchese, R. T. 1986. *The Lower Mœander Flood Plain: A Regional Settlement Study.* Oxford: BAR International Studies 292 (i and ii).

Marchetti, P. 1993. Recherches sur les mythes et la topographie d'Argos. 1:. Hermès et Apriodite. *BCH:* 211–23.

———. 1994. Recherches sur les mythes et la topographie d'Argos. II. Présentation du site, and III. Le téménos de Zeus. *BCH* 118: 131–41.

———. 1998. Le Nymphée de Argos, In Pariente, A., and G. Touchais, eds. *Argos et l'Argolide: Topographie et urbanisme.* École Française d'Athènes. Paris: De Boccard Édition. 357–72.

Marchetti, P., and K. Kolokotsas. 1993. La reconstruction au 1er siècle. *BCH* 117: 211–23.

———. 1994. La reconstruction au 1er siècle. *BCH* 118: 131–60.

Marchetti, P., and K. Kolokotsas, with C. Abadie-Reynal. 1995. *Le nymphée de l'agora d'Argos.* École Française d'Athènes, Études Péloponnésiennes 21.

Marconi, P. 1929. Studi sull'organizzazione urbana di una città classica. *Agrigento, topografica ed arte.* Rivista d'Inst. Naz. d'Archaeologia e Storia dell'Arte (RIASA) Florence. 2(1930): 7–62.

Marconi Bovio, J. 1957. Inconsistenza di una Selinunte romana. *Kokalos* 3: 70–79.

———. 1991. Problemi di restauro e difficoltà dell'anastylosis del "Tempio E" di Selinunte. *Historia* 40: 85–96 and aerial view p. 108.

Marinos, P. G. 1978. Les conditions hydrogeologiques et la disposition des dechets municipaux dans un haut plateau karstique (Grèce-Poloponnèse centrale). *Implications de l'hydrologie dans les autres sciences de la terre:* 537–51.

———. 1995. Notes and sketches for the hydrogeology of Corinth. Manuscript in D. P. Crouch collection.

Marinos, P. G., and G. C. Koukis, eds. 1988. *The Engineering Geology of Ancient Works, Monuments and Historical Sites: Proceedings of the Symposium of the International Association of Engineering Geology, Athens.* Rotterdam: A. A. Balkema.

Marinos, P. G., and S. Plesas. 1994. The Geology of Water at Argos. Manuscript in D. P. Crouch collection.

Marinos, P. G., G. C. Koukis, G. C. Tsiambaos, and G. C. Stourjaras, eds. 1997. *Engineering Geology and the Environment: Proceedings of the Symposium of the International Association of Engineering Geology, Athens.* Rotterdam: A. A. Balkema.

Mariolakos, I., E. Logos, S. Lozios, and S. Nasopoulou. 1991.Technicogeological Observations in the Ancient Delphi Area (Greece). In M. E. Almeida-Taxeria, ed. *Proceedings of the Course on Natural Hazards and Engineering Geology: Preservation and Control of Landslides and Other Mass Movements. Lisbon, 1990.* Luxembourg: Commission of the European Communities. (Typscript in collection of D. P. Crouch.)

Marsh, G. P. 1874. *The Earth as Modified by Human Action.* New York: Schribner, Armstrong. Reprinted 1970 by Areno and the *New York Times*.

Marshak, A. 1972. *The Roots of Civilization.* New York: McGraw-Hill.

Martienssen, R. D. 1956. *The Idea of Space in Greek Architecture.* Johannesburg: Witwatersrand University Press

Martin, R. 1956. *L'Urbanisme dans la Grece antique.* Paris: A. & J. Picard.

———. 1975. Rapport sur l'Urbanisme de Sélinonte. *Kokalos* 21: 54 ff.

———. 1977. Histoire de Sélinonte d'après les fouilles récentes. *CRAI*: 46–63.

———. 1979. Relations entre Mêtropoles et colonies: Aspects Institutionnales. In *Philias Charin: Misc. in onore di E. Manni.* Vol. 4 of 6 vols. Rome: 1435–45.

———. 1980–81. Recherches sur l'acropole de Sélinonte. *Kokalos* 26–27.II.2: 1009–16

———. 1983. L'Espace civique, religieux et profane dans les cités grecques de l'archaisme à l'epoque hellenistique. *Architecture et Société.* Ecole Française de Rome: 9–42.

Mastoris, K. J, D. G. Monopolis, and S. D. Skayas. 1971. *Hydrogeological Investigation of Korinth-Loutraki Area.* No. 3 of Athens: *Hydrological and Hydrogeological Investigations*, Athens: Institute for Geology and Mineral Exploration. Reprinted 1980. In Greek.

Mauceri, E. 1909. *Siracusa e la vale dell'Anapo.* Bergamo: Instituto Italiano d'Arti Grafiche.

Maugeri, R. 1992. Notes on the Geology of Syracuse, Especially Ortygia, and Map of Tunnel under the Amphitheater. Notes in D. P. Crouch collection.

Maurizio, L. 1995. Anthropology and Spirit Possession: A Reconsideration of the Pythia's Role at Delphi. *JRS* 115: 69–86.

Mazza, M. 1990–91. Cataclismmi e calamita naturali: La documentazione letteraria. *Kokalos* 36–37: 307–30.

Mazzanti, R. 1979. Linee di Costa attuali e fossili fra Castiglioncello e Punta Lillastro. *Atti del congegno "Variazioni della Linea di Riva fra Punta di Castiglioncello e Marina di Castagneto" Cecina.* Provincia di Livorno. 30–33.

McCredie, J. R. 1971. Hippodamus of Miletus. In Pedley, J. G. and J. A. Scott, eds. 1971. *Studies Presented to George M. A. Hanfmann.* Fogg Art Musuem/Harvard University Monographs in Art and Archaeology II. Mainz: P. von Zabern. 95–100.

McKenzie, D. 1972. Active Tectonics of the Mediterranean Region. *Geophysical J. of the Royal Astronomy Society* 30.2: 109–185.

McNally, S. 1985. Art History and Archaeology. In Wilkie, N. C., and W. S. E. Coulson, eds. *Contributions to Aegean Archaeology: Studies in Honor of William A. MacDonald.* Publications in Ancient Studies, 1. Center for Ancient Studies, University of Minnesota. 1–21.

McPhee, J. 1986 *Rising from the Plains.* New York: Farrar, Straus and Giroux.

McPherson, M. B. 1974. *Hydroloogical Effects of Urbanization.* Paris: UNESCO.

Mee, C. 1978. Aegean Trade and Settlement in Anatolia. *Anatolian Studies.* Vol. D 28: 126–37.

Méga, V. 1998. Espaces et fonctions publics dans l'Argos classique. Evolutions et perspectives. In Pariente, A., and G. Touchais, eds. *Argos et l'Argolide: Topographie et urbanisme.* École Française d'Athènes. Paris: De Boccard Édition. 315–23.

Meiggs, R. 1960. *Roman Ostia.* Oxford: Clarendon Press. 2nd ed., 1974

———. 1982. *Trees and Timber in the Ancient Mediterranean World.* Oxford: Clarendon Press. 3rd ed., 1985.

Melhorn, W. N., and R. C. Flemal, eds. 1975. *Theories of Landform Development.* London: George Allen and Unwin.

Meriç, R. 1985. Zur Lage des ephesischen Aussenhafens Panoramos. *Lebendige Altertums Wissenschaft von Hermann Vetters.* Vienna: Verlag Adolf Holzhausens Nfg.: 30–32.

Merritt, B. D. 1939–1950. *The Athenian Tribute Lists.* Cambridge, Mass.: Harvard University Press.

Mertens, D. 1988–89. Le Fortificazioni di Selinunte: Rapporto preliminare (Fino al 1988). *Kokalos* II: 573–94, with plates 74–80, incl. a fold-out plan at 1: 5000.

———. 1989. Castellum oder Ribat? Das Küstenfort in Selinunt. *Ist. Mitt.* 39: 391–98.

———. 1990. Some Principal Features of West Greek Colonial Architecture. In Descœudres, J.-P., ed. *Greek Colonists and Native Populations.* Oxford: Clarendon Press. 373–83.

———. 1993. Note dell'edilizia selinuntina del V sec. a.C. *Studi sulla Sicilia Occidentale in onore di Vicenzo Tusa.* Padua: Bottega d'Erasmo. 1341–38 and plates 35–38.

———. 1996. Greek Architecture in the West. In Pugliese-Carratelli, G., ed. *The Western Greeks.* Venice: Bompiani. 315–523.

Métraux, G. 1978. *Western Greek Land Use and City Planning in the Archaic Period.* New York: Garland.

Mettos, A., D. Galanakis, and C. Georgiou. 1997. Geological-Neotectonic and Morphotectonic Investigation for Urban Planning of Thiva Town [Thebes], Greece. In Marinos, P. G., and G. C. Koukis, eds. *The Engineering Geology of Ancient Works, Monuments and Historical Sites: Proceedings of the Symposium of the International Association of Engineering Geology, Athens.* Rotterdam: A. A. Balkema. 1375–80.

Meulenkamp, G. 1972. Sedimentary History and Paleogeography of the Late Cenozoic of the Island of Rhodes. *Z. Deutsch. Geol. Ges.* (Hanover) 123: 541–53.

Mijatovic, B. 1977. Current Problems in the Rational Exploitation of Karst Water. In Dilamarter, R. R., and S. C. Csallany, eds. *Hydrologic Problems in Karst Regions:, Proceedings of the International Symposium, April 1976.* Bowling Green, Ky.: W. Kentucky University Press .262–78.

Millett, M. 1981. Whose Crisis? The Archaeology of the Third Century: A Warning. In Kind, A., and M. Henig, eds. *The Roman West in the Third Century [C.E.]: Contributions from Archaeology and History.* Oxford: BAR International Series 109 (ii): 525–30.

Milliman, J. D. 1992. Sea-Level Response to Climate Change and Tectonics in the Mediterranean Sea. In Jeftic, L., J. D. Milliman, and G. Sestini, eds. *Climate Change and the Mediterranean.* London: Edward Arnold. 45–57.

Miltner, F. 1955. Vorläufiger Bericht über die Ausgrabungen in Ephesos. *FiE* Annual Reports. 21: 33–37.

———. 1958. *Ephesos: Stadt der Artemis und des Johannes.* Vienna: Verlag Franz Deuticke.

Minissale, A. 1989. Geochemical Characteristics of Greek Thermal Springs. *J. of Volcanology and Geothermal Research* 39: 1–16.

Mirisola, R. ca. 1996. Aspetto geologico. *La Valle dei Templi: Tra iconografia e storia.* Assessorato Regionale Beni Culturali Ambientali di Agrigento. 6–11. WWF. Set. Service

Moens, L., P. Roos, J. de Rudder, P. de Paepe, J. van Hende, and M. Waelkens. 1988a. A Multi-method Approach to the Identification of White Marbles Used in Antique Artifacts. In Herz, N., and M. Waelkens, eds. *Classical Marble: Geochemistry, Technology, Trade.* NATO ASI Series E: Applied Sciences, Vol. 153. Dordrecht: Kluwer Academic Publishers: 243–50. With Fig. 1 of four quarries at and near Ephesos.

————. 1988b. White Marble from Italy and Turkey: An Archaeometric Study Based on Minor- and Trace-Element Analysis and Petrography. *J. of Radiolanalytical and Nuclear Chemistry.* V. 123.1: 333–48.

Mogey, J. 1971. Society, Man and Environment. In Buchanan, R. H., E. Jones, and D. McCourt, eds. *Man and His Habitat.* New York: Barnes and Noble. 82–83.

Morel, J.-P. 1983. Greek Colonizatiuon in Italy and the West: Problems of Evidence and Interpretation. In Hackens, R., N. D. Holloway and R. R. Holloway, eds. *Crossroads of the Mediterranean.* Papers delivered at the International Conference on the Archaeology of Early Italy, Haffenreffer Museum, Brown University, 8–10 May 1981. Louvain. English language ed., Brown University, 1994. 123–61.

Moretti, J.-Ch. 1998. L'Implantation du Théâtre d'Argos. In Pariente, A., and G. Touchais, eds. *Argos et l'Argolide: Topographie et urbanisme.* École Française d'Athènes. Paris: De Boccard Édition. 233–59.

Morfis, A., and H. Zojer, eds. 1986. Karst Hydrology of the Central and Eastern Peloponnesus (Greece). *Steirische Beiträge sur Hydrologie* 37/38. Graz.

Morgan, C. 1990. *Athletes and Oracles: The Transformation of Olympia and Delphi in the Eighth Century B.C.* Cambridge: Cambridge University Press.

Morgan, C., and T. Whitelaw. 1991. Pots and Politics: Ceramic Evidence for the Rise of the Argive State. *AJA* 95: 79–108.

Morgan, C. M. 1939. Excavations at Corinth. 1938. *AJA* 43: 255–64.

Morgan, W. J. 1968. Rises, Trenches, Great Faults, and Crustal Blocks. *J. of Geophysical Research* 73: 1959–82.

Morgantina Excavation Notebooks, on file in the Morgantina Room, Dept. of Art and Archaeology, Princeton University.

Morizot, Y. 1994. Artémis, l'eau et la vie humain. In Ginouvés, R., A.-M. Guimier-Sorbets, J. Jouanna, and L. Villard, eds. *L'Eau, la santé et la maladie dans le monde grec.* Paris: Actes de Colloque à Paris 25–27 Nov. 1992. 201–16.

Motta, S. 1957. Nota descrittiva della geologia della tavoletta di Agrigento con particolare esame della Serie Gesso Solififera in essa esistente. *Bolettino Società Geologia di Italia* No. 45.

Mourtzas, N. D. 1988a. Holocene Vertical Movements and Changes of Sea-Level in the Hellenic Arc. In Marinos, P. G., and G. C. Koukis, eds. *The Engineering Geology of Ancient Works, Monuments and Historical Sites: Proceedings of the Symposium of the International Association of Engineering Geology, Athens.* Rotterdam: A. A. Balkema. 2247–62.

————. 1988b. Neotectonic Evolution of Messara's Gulf: Submerged archaeological constructions and use of the coast during the prehistoric and historic periods. In Marinos, P. G., and G. C. Koukis, eds. *The Engineering Geology of Ancient Works, Monuments and Historical Sites: Proceedings of the Symposium of the International Association of Engineering Geology, Athens.* Rotterdam: A. A. Balkema. 1565–73.

————. 1988c. Neotectonic Development of the Hellenic Arc (Aegean Area) in Tour T3: Field Guide of the Greek Islands; Holocene Vertical Movements and Changes of Sea-Level in the Hellenic Arc. Tour T3: Field Guide of the Greek Islands, International Symposium on Enginering Geology and Protection of Ancient Works, Monuments, and Sites: 2235–45.

Mourtzas, N. D., and P. G. Marinos. 1992. Holocene Sea-Level Changes along the Aegean Coasts (Greece). *27th International Congress of Geology*, Japan. Vol. 29.

———. 1994. Upper Holocene Sea-Level Changes: Paleographic Evolution and Its Impact on Coastal Archaeological Sites and Monuments. *Environmental Geology* 23: 1–13.

Müller, S. 1992. Topographie de la région de Delphes. *BCH* 116: 447–89.

Mumford, L. 1961. *The City in History*. New York: Harcourt Brace and World.

Musso, A., and Ercoli, L. 1988. Monuments and Landslides in the Agrigento Valley In Marinos, P. G., and G. C. Koukis, eds. *The Engineering Geology of Ancient Works, Monuments and Historical Sites: Proceedings of the Symposium of the International Association of Engineering Geology, Athens*. Rotterdam: A. A. Balkema: 113–21.

Nance, J. J. 1988. *On Shaky Ground*. New York: Wm. Morrow.

Naselli, G. 1972. La fortezza e la fornace. *Sicilia Archeologica* 5.17: 21–26.

Nash, D. T., and M. D. Pertaglia. 1987. Burial Formation Processes and the Archaeological Record: Present Problems and Future Requisites. In Nash, D. T., and N. D. Pertaglia. *Natural Formation Processes and the Archaeological Record*. Oxford: BAR International Series 352. 186–203.

Naveh, Z., and J. Dan. 1973. The Human Degradation of Mediterranean Landscapes in Israel. In di Castri, F., and H. A. Mooney, eds. *Mediterranean Type Ecosystems: Origin and Structure*. New York: Springer-Verlag. 373–90.

Neboit, N., J.-J. Dufaure, and M. Julian. 1984. Facteurs naturels et humains dans la morphogenèse méditerranéenne. In *Homage à Pierre Birot: La mobilité des paysages méditerranéans*. Toulouse. Revue Géographique des Pyrénées et du Sud-Ouest. Travaux II. 317–330.

Neeft, C. W. 1981. Observations on the Thapsos Claes. *MÉLROME* 93: 7–88.

Nenci, G. 1979. Le Cave di Selinunte. *ANNPISA* 9.4: 1415–25 and plates 61–71. With extensive early bibliography.

Nicolet, C. 1990. *Space, Geography, and Politics in the Early Roman Empire*. Jerome Lectures, 19. Ann Arbor: University of Michigan Press.

Niemi, T. M., and E. A. W. Finke (Zangger). 1988. The Application of Remote Sensing to Coastline Reconstructions in the Argive Plain, Greece. In Raban, A., ed. *Archaeology of Coastal Changes*. Oxford: BAR International Series 404.2: 119–36.

Nilsson. T. 1983. *Pleistocene Geology: Life in the Quaternary Ice Age*. Dordrecht: D. Reidel.

North, F. J. 1928. *Geological Maps: Their History and Development with Special Reference to Wales*. Cardiff: Natural Museum of Wales.

Norton, P., ed. 1965. *Guide to the Geology and Culture of Greece*. Petroleum Exploration Society of Libya, 7th Annual Field Conference.

Norwich, J. I., ed. 1975. *Great Architecture of the World*. New York: Bonanza Books.

Ogniben, L. 1960. Nota introduttiva dello schema geologico della Sicilia nord-orentale. *Rivista Mineraria Siciliana* 11: 64–65, 183–212.

Oldroyd, E. R. 1990. *The Highlands Controversy*. Chicago: University of Chicago Press.

Oleson, J. P. 1985. Herod and Vitruvius: Preliminary Thoughts on Harbor Engineering at Sebastos. In Raban, A., ed. *Harbour Archaeology*. Oxford: BAR International Series 257: 165–72.

Ortloff, C. R. 1989. A Mathematical Model of the Dynamics of Hydraulic Societies in Ancient Peru. In Rice, D. S., C. Standish, and P. R. Scarr, eds. *Ecology, Settlement and History in the Osmore Drainage*. Oxford: BAR International Series.

———. 1993. Chimu Hydraulics: Technology and Statecraft on the North Coast of Peru, A.D. 1000–1470. In Scarborough, V., and B. Isaac, eds. *Research in Economic Anthropology*, Suppl. 7. Greenwich, Conn.: JSI Press. 327–67.

———. 1996. Engineering Aspects of Groundwater-Controlled Agriculture. In Kolota,

A. L., ed. *Tiwanaku and Its Hinterland: Archaeology and Paleoecology of an Andean Civilizsation.* 1: *Agroecology.* Washington, D.C.: Smithsonian Institution Press. 153–68.

Ortloff, C. R., and D. P. Crouch. 1998. Hydraulic Analysis of a Self-Cleaning Drainage Outlet at the Hellenistic City of Priene. *J. of Archaeological Science:* 1211–20.

———. 2001. The Urban Water Supply and Distribution System of the Ionian City of Ephesus in the Roman Period. *J. of Archaeological Science.*

Ortloff, C. R. and Kolata, A. L. 1993. Climate and Collapse: Agroecological Perspectives on the Decline of the Tiwanaku State. *Journal of Archaeological Science* 20: 195–221.

Ortloff, C. R., M. E. Moseley, and R. A. Feldman. 1982. Hydraulic Engineering Aspects of the Chimu Chicama-Moche Inytervalley Canal. *American Antiquity* 47: 375–95.

Ortloff, C. R., M. E. Mosely, R. A. Feldman, and A. Navarez. 1983. Principles of Agrarian Collapse in the Cordillera Negra, Peru. *Annals of the Carnegie Museum* 52 (16 Sept.): article 13.

Osbourne, P., and C. Tarling, eds. 1996. *The Historical Atlas of the Earth.* New York: Holt.

O'Shea, J., and P. Halstead, 1989. Conclusions: Bad Year Economics. In Halstead, P. and J. O'Shea, eds. *Bad Year Economics.* Cambridge: Cambridge University Press. 123–26.

Oster, R. E. 1987. *A Bibliography of Ancient Ephesus.* Metuchen, N. J.: Scarecrow Press.

Ozis, Ü. 1984. Environment and Underground: Karst Waters of Anatolia. *Proceedings of the 5th Turk-Alman Cevre Muhendisligi Symposium, Izmir, Turkey.* Izmir: V. Turk-Alman Cevre Muhendisligi. 137–44.

———. 1985. Assessment of Karst Water Resources. IWRA 5th World Congress on Water Resources, Brussels. Paper 31a: 95–102.

———. 1987. Ancient Water Works in Anatolia. *Water Resources Development* 3. London: Butterworth. 55–62.

———. 1995. *Çaglar Boyunca Andolu'da su MühenndisligI.* Istanbul Shubesi: TMMOB Inshaat Mühenndisligi Odasi.

———. 1996. Historical Water Schemes in Turkey. *Water Resources Development* 12.3: 347–83.

———. 1997. Historical Water Conveyance Systems to Ephesus. Presened at the First International Symposium of Selçuk. Sept. Also published in the *14th Congress on Civil Engineering in Turkey,* Oct. 1997.

Ozylgit, O. 1991. On the Dating of the City Walls of Ephesos.In Malayu, H., ed. *Erol Atalay Memorial.* Izmir: Arkeoloji Dergisi, Özel Sayi I, Ege University. 137–40.

Pace, B. 1935. *Arte e civilita della Sicilia antica.* Vol. 1 of 4 vols. Milan: Edizioni Dante Alighieri.

———. 1938. *Arte e civilita della Sicilia antica.* Vol. 4 of 4 vols. Milan: Edizioni Dante Alighieri.

Page, L. E. 1969. Diluvialism, and Its Critics in Great Britain in the Early 19th Century. In Schneer, C. C., ed. *Towards a History of Geology.* Cambridge, Mass.: MIT Press. 259–71.

Pallottino, M. 1991. *A History of Earliest Italy.* Trans. M. Ryle and K. Soper. Ann Arbor: University of Michigan Press.

Palmer, A. N. 1984. Geomorphic Interpretation of Karst Features. In LaFleur, R. G., ed. *Groundwater as a Geomorphic Agent.* Boston: Allen and Unwin. 173–209.

Palmer, M. W. 1976. Groundwater Flow Patterns in Limestone Solution Conduits. Ph.D. dissertation, State University at Oneonta, New York.

Panayotopoulou, A. 1998. Thermes d'époque romaine et tardo-romaine à Argos. In Pariente, A., and G. Touchais, eds. *Argos et l'Argolide: Topographie et urbanisme.* École Française d'Athènes. Paris: De Boccard Édition. 373–84.

Pancrazi, G. M. 1751–52. *Antichita Siciliane Spiegate.* Naples: A. Pelecchia. 4 parts.

Papadopoulos, G. A. 1988. Statistics of Historical Earthquakes and Associated Phenomena in the Aegean and Surrounding Regions. In Marinos, P. G., and G. C. Koukis, eds. *The Engineering Geology of Ancient Works, Monuments and Historical Sites: Proceedings of the Symposium of the International Association of Engineering Geology, Athens.* Rotterdam: A. A. Balkema. 1279–83. .

Papageorgakis, J. and E. Kolaiti. 1992. Ancient Limestone Quarries of Profitis Elias near Delfi (Greece). In M. Waelkens, W., Nitterz and L. Moens, eds. *Ancient Stones: Quarrying, Trade & Provenance.* Acta Archaeologica Lovaniensica. Monographics 4. Leuven University Press, 37–39.

Papahatzis, N. 1984. *Ancient Corinth: The Museums of Corinth, Isthmia and Sicyon.* Athens: Ekdotike Athenon.

Papakis, N. J. 1966. *Hydrological Investigation of Ayios Yeoryios Springs (Kiveri-Argos).* Athens: Institute for Geology and Subsurface Research. Geological and Geophysical Research, Vol. 11, No. 3. In Greek with English summary and captions.

Papazachos, B., and Papazachou, K. 1989. *Earthquakes in Greece.* Thessaloniki: Ziti. In Greek.

Pariente, A. 1993. Chronique des Fouilles en 1992. *BCH* 117: 822–23. Review and bibliographical references for 100 years of excavation.

Pariente, A., and G. Touchais, eds. 1998. *Argos et l'Argolide: Topographie et urbanisme.* École Française d'Athènes. Paris: De Boccard Édition.

Pariente, A., M. Piéart, and J.-P. Thalmann, 1998. Les recherches sur l'agora d'Argos. In Pariente, A., and G. Touchais, eds. *Argos et l'Argolide: Topographie et urbanisme.* École Française d'Athènes. Paris: De Boccard Édition. 211–31.

Parzinger, H. 1989. Zur frühesten Besiedlung Milets. *Ist..Mitt.* 39: 415–31.

Paskoff, R. P. 1973. Geomorphological Processes and Characteristic Landforms in the Mediterranean Regions of the World. In di Castri, F., and H. A. Mooney, eds. *Mediterranean Type Ecosystems: Origin and Structure.* New York: Springer-Verlag. 53–61

Peacock, D. P. S. 1980. The Roman Millstone Trade: A Petrological Sketch. *World Archaeology* 12.1: 43–53.

Péchoux, P.-Y. 1977. Nouvelles remarques sur les versants quaternaire du secteur de Delphes. *RGPGD* (2) (Paris) 19.1: 83–92.

Pecírka, J. 1973. Homestead Farms in Classical and Hellenistic Hellas. In Finley, M. I., ed. *Problèmes de la Terre en Grèce ancienne.* Paris. 113–47.

Pedersen, P. 1988. Town Planning in Halicarnassus and Rhodes. In Dietz, S., and I. Papachristodoulou, eds. *Archaeology in the Dodecanese.* Copenhagen: National Museum of Denmark, Dept. of Near Eastern and Classical Antiquities. 98–103.

Pedley, J. G., and J. A. Scott, eds. 1971. *Studies Presented to George M. A. Hanfmann.* Fogg Art Musuem/Harvard University Monographs in Art and Archaeology II. Mainz: P. von Zabern.

Pedley, J. G., and M. Torelli. 1984. Excavations at Paestum 1983. *AJA* 88: 36 ff.

Pelagati, P. 1982 and 1984. Siracusa: Le ultime recherche in Ortygia. *Ann. Sc. Alene,* N.S. 44: 136–38.

Penning-Rousell, E. C., and D. Lowenthal, eds. 1986. *Landscape Meanings and Values.* London: Allen and Unwin.

Pentazos, E. 1988. Studies and Works of the First Ephoria for the Preservation and Protection of the Monuments at Delphi. In Marinos, P. G., and G. C. Koukis, eds. *The Engineering Geology of Ancient Works, Monuments and Historical Sites: Proceedings of the Symposium of the International Association of Engineering Geology, Athens.* Rotterdam: A. A. Balkema: 2195–97.

Perlin, J., and K. Butti. 1980. *A Golden Thread: 2500 Years of Solar Architecture and Technology.* New York: Cheshire/Van Norstrand-Reinhold.

Peschlow-Bindokat, A. 1992. *Die Steinbrüche von Selinunte: Cave di Cusa.* Castelvetrano: Edizioni Mazzotta.

Peschlow-Bindokat, A., and K. Germann 1981. Steinbrüche von Milet und Herakleia am Latmos. *JDAI Istanbul* 96: 215–35.

Philias Charin: Misc. in onore di E. Manni. 6 vols. 1979–85. Rome.

Philippson, A. 1892. *Der Peloponnes: Versuch einer Landeskunde aus geologischer Grundlage.* Berlin: R. Friesländer u. Sohn.

———. 1910. *Reisen und Forschungen im westlichen Kleinasien.* 5 vols. Gotha: Justus Perthes.

———. 1916. Antike Stadtlagen an der Westküste Kleinasiens. *Bonner Jahrbücher* 123: 120–27.

———. 1918. "e": Das Kustengebirge zwischen Kayster und Mäander, Kleinasien. *Handbuch der Regionalen Geologie* 2. Heidelberg: 109–16. Reprinted 1968.

———. 1959. *Grieschen Landschaften.* Vol., 3 pt. 1. Frankfurt-am-Main.

Picard, L. 1957. *Development of Underground Water Resources in Greece.* Report 3 (of 5), prepared for the government of Greece through the U.N. Technical Assistance Program, 1955–58.

Picard, O. 1998. Introduction. In Pariente, A., and G. Touchais, eds. *Argos et l'Argolide: Topographie et urbanisme.* École Française d'Athènes. Paris: De Boccard Édition. 1–3.

Piccioni, G. ca. 1996. Brevi Cenni storico-geografici. In *La valle dei templi: Tra iconografia e storia.* Agrigento: Assessorato Regionale Beni Culturali Ambientali di Agrigento. 14–18. WWF. Set. Siracusa.

Piérart, M. 1988. Eau et sécheresse dans les mythes argiens. *Actes du colloque "L'Eau dans les pays mediterranienne."* Athens: 1–12.

———. 1992. "Argos Assoiffée" et "Argos riches en cavales." In Piérart, M., ed. *Polydipsion Argos: Argos de la fin des palais mycéniens à la constitution de l'etat classique* (Fribourg, Switzerland, 1987). *BCH* Suppl. 22. Paris: De Boccard Édition. 110–156. With app. 1, Sanctuaires et monuments d'Argos, and app. 2, Sites, sanctuaires et monuments du territoire agrien.

———. 1998. L'Itinéraire de Pausanias à Argos. In Pariente, A., and G. Touchais, eds. *Argos et l'Argolide: Topographie et urbanisme.* École Française d'Athènes. Paris: De Boccard Édition. 337–56 (no map).

Pinar, N., and E. Lahn. 1952. *Turkiye Depremleri Izahli Katalogu* (Descriptive Catalog of Turkish Earthquakes). *Bayindirlik Bakanligi, Yapi & Imar Isleri Reisligi Yayinlari* (Ministry of Reconstruction and Settlement). Seri 6, Sayi 236, Ankara. [Cited here as *Earthquake Catalog.*]

Pinna, M. 1977. *Climatologia.* Turin: Edizioni UTET.

Piras, S. A. G. 1996. Cura Aquarum in Campania: Eine Einführung. In De Haan, N., and G. C. M. Jansen, eds. *Cura aquarum in Campania.* Annual Papers on Classical Archaeology, Suppl. 4. Leiden: Stichting Babesch. 87–92.

Pirazzoli, P. A. 1987. Sea-Level Changes in the Mediterranean. *Sea-Level Changes.* Oxford: Basil Blackwell. 152–81.

———. 1988. Sea-Level Changes and Crustal Movements in the Hellenic Arc (Greece): The Contributions of Archaeological and Historical Data. In Raban, A., ed. *Archaeology of Coastal Changes.* BAR International Series 404.2: 163 ff.

Pitéros, C. I. 1998. Contribution à la topographie argienne: le site, les ramparts et quelques problèmes topographiques. In Pariente, A., and G. Touchais, eds. *Argos et l'Argolide: Topographie et urbanisme*. École Française d'Athènes. Paris: De Boccard Édition. 179–210.

Placzek, A., ed. 1982. *Macmillan Encyclopedia of Architects*, vol. 1. New York: Free Press, Macmillan. 413.

Pointon, M. 1979/1981. Geology and Landscape Painting in 19th Century England. In Jordanova, L. J., and R. S. Porter, eds. *Images of the Earth: Essays in the History of the Environmental Sciences*. Monograph 1. London: British Society for the History of Science.

Polacco, M. 1984. Le asimmetrie del teatro antico di Siracusa. NAC 13: 85–94. Based on his dissertation (1981), *Il teatro antico di Siracusa*.

Polignac, F. de. 1984. *La Naissance de la Cité Grecque: Cultes, Espace et Société VIIIe–VIIe Siècles Avant J.-C.* Paris: La Découverte. Also published as Cults, Territory, and the Origins of the Greek City-state. Translator J. Lloyd. Chicago: University of Chicago Press.

Pope, K. O., and T. H. Van Andel. 1984. Late Quarternary Alluviation and Soil Formation in the So. Argolid: Its History, Causes and Archaeological Implications. *J. of Archaeological Science* 11: 281–306.

Porter, R. 1977. *The Making of Geology: Earth Science in Britain 1660–1815*. Cambridge: Cambridge University Press.

Pouilloux, J., and G. Roux. 1963. *Enigmes a Delphes*. Paris: De Boccard Édition.

Praschniker, C., and M. Theuer. 1979. *Das Mausoleum von Belevi*. FiE 6.

Preziosi, D. 1989. *Rethinking Art History: Observations on a Coy Science*. New Haven, Conn.: Yale University Press.

Price, D., and J. De Solla. 1965. Is Technology Historically Independent of Science? A Study in Statistical Historiography. *Technology and Culture* 6.4 (fall): 553–68.

Pritchett, W. K. 1965. *Studies in Ancient Greek Topography*. Berkeley: University of California Press.

Procelli, E. 1989. Aspetti e problemi dell'ellenizazione calcidese nella Sicilia orientale. *MÉLROME* 101: 679–89.

Przeworski, S. 1939. *Die Metallindustrie Anatoliens in der Zeit von 1500–700 vor Chr*. Leiden: E. J. Brill.

Pugliese-Carratelli, G. 1983. L'oggetto storico di Selinunte. In Tusa, V., ed. *La scultura in pietra di Selinunte*. I Cristali series. Palermo: Edizioni Sellerio.17–25.

————, ed. 1985. *Sikanie: Storia e civiltà della Sicilia greca*. Antica Madre series. Milan: Garzanti.

————. 1989. I santuari extramurani. In *Santuario di magna grecia*, vol. 3: 149–58.

————, ed. 1996. *The Western Greeks*. Venice: Bompiani.

Quinlin, F. 1978. Types of Karst with Emphasis on Cover Beds in Their Classification and Development. Ph.D. dissertation, University of Texas at Austin.

Raban, A. 1985. *Harbour Archaeology*. Oxford: BAR International Series 257.

————. 1988. Coastal Processes and Ancient Harbour Engineering. In Raban, A., ed. *Archaeology of Coastal Changes*. BAR International Series 404.2: 185–208.

————. 1990. Man-Instigated Coastal Changes along the Israeli Shore of the Mediterranean in Ancient Times. In Bottema, S., G. Entijes-Nieborg, and W. van Zeist, eds. *Man's Role in the Shaping of the Eastern Mediterranean Landscape*. Rotterdam: A. A. Balkema: chap. 9.

Rabb, T .K. 1981. The Historian and the Climatologist. In Rotberg, R. I., and T. K. Rabb, eds. *Climate and History: Studies in Interdiscliplinary History*. Princeton, N.J.: Princeton University Press. 251–57.

Radt, W., 1995. Landscape and Greek Urban Planning: Exemplified by Pergamon and Priene. In Kristensen, T. M., S. E. Larsen, P. G. Moller, and S. E. Petersen, eds. *City and Nature: Changing Relations in Time and Space: Colloquium Odense, 1993.* 201–209.

Raepsaet, G., and M. Tolley. 1993. Le diolkos de l'Isthme à Corinthe: Son trace, son fonctionnement. *BCH* 117: 233–62.

Raffiotta, S. 1991. *C'era una volta Morgantina.* Catania: Banca Agricola Etnea.

Rallo, A. 1984. Nuovi aspetti dell'urbanistica Selinuntina. *ANNSCATENE* 62 (N.S. 46): 81–91 and 92–96. With V. Di Grazia, App., Note sull'applicazione di metodologie di rilevamento di restituzione nel comprensorio aecheologico di Selinunte.

Rallo Franco, A. 1978. Le importazioni freco-orientli a Selinunte a segoto deo recemto scabvo. *Les ceramiques de la Grèce de l'est et leur diffusion de occident.* Naples: Centro J. Berard. 99–103.

Ramsay, W. M. 1890. *The Historical Geography of Asia Minor,* vol. 4. London: Royal Geographical Society.

Raphael, C. N. 1978. The Erosional History of the Plain of Elis in the Peloponnesos. In Brice, W. C., ed. *The Environmental History of the Near and Middle East Since the Last Ice Age.* London: Academic Press: 56–63.

Rapp, G., and J. A. Gifford. 1982. *Troy: The Archaeological Geology.* Cincinnati, Ohio: University of Cincinnati and Princeton University Press.

———, eds. 1985. *Archaeological Geology.* New Haven, Conn.: Yale University Press.

Rapp, G. Jr, and J. C. Kraft. 1994. Holocene Coastal Change in Greece and Aegean Turkey. In Kardulias, P. N., ed. *Beyond the Site: Regional Studies in the Aegean Area.* Lanham, Md.: University Press of America. 69–90.

Regione Siciliana, Assessorato del Territorio e dell'Ambiente. 1987. *Piano regionale di risanamento delle acque.*

Renfrew, C. 1970. Patterns of Population Growth in the Prehistoric Aegean. In Ucko, P. J., R. Tringham, and G. W. Dinbleby, eds. *Man, Settlement and Urbanism: Proceedings of the Research Seminar in Archaeology and Related Subjects.* Cambridge, Mass.: Schenkman Publishing. 383–99.

Reynal, G. 1986. L'archéologie dans la ville: Le cas d'Argos. *La Recherche* 180.17 (Sept.): 1016–26.

Rice, D. S., C. Stanish, and P. R. Scarr, eds. 1989. *Ecology, Settlement, and History in the Osmore Drainage Area.* Oxford: BAR International Series 457: 5545.

Richardson, L. Jr. 1988. *Pompeii: An Architectural History.* Baltimore: Johns Hopkins University Press. App. 1, The Stones.

Richardson, L., Jr. 1988. *Pompeii, an Architectural History.* Baltimore and London: The Johns Hopkins University Press. Appendix I. The Stones.

Richardson, R. B. 1900a. Pirene. *AJA* 204–309.

———. 1900b. The Fountain of Glauke at Corinth. *AJA* 458–75.

———. 1902. Sacred Spring. *AJA* 306–20.

———. 1910. Pirene [Peirene]. *AJA* 204–309.

Riedmüller, G., and J. Zojer. 1986. Karst Hydrogeology of the Central and Eastern Pelponnessus (Greece). *Steirische Beiträge zur Hydrogeologie* 37/38. Graz.

Ringel, J. 1988. Literary Sources and Numismatic Evidence of Maritime Activity in Casearea during the Roman Period. *Mediterranean Historical Review* 3.1 (June): 63–73.

Rinne, K. 1996. Aquae Urbis Romae: An Historical Overview of Water in Public Life of Rome. In De Haan, N., and G. C. M. Jansen, eds. *Cura aquarum in Campania.* Annual Papers on Classical Archaeology, Suppl. 4. Leiden: Stichting Babesch.145–52.

Robert, L. 1978. Documents d'Asie Mineure. *BCH* 102: 395–408.

Roberts, N.1990. Human-Induced Landscape Change in South and Southwest Turkey during the Later Holocene. In Bottema, S., G. Entijes-Nieborg, and W. van Zeist, eds. *Man's Role in the Shaping of the Eastern Mediterranean Landscape*. Rotterdam: A. A. Balkema. 53–66.

Roberts, S. C. 1988. Active Normal Faulting in Central Greece and Western Turkey. Ph.D. dissertation. Cambridge University.

Robertson, D. S. 1964. *A Handbook of Greek and Roman Architecture*. Cambridge: Cambridge University Press.

Robinson, D. M. 1935, The Third Campaign at Olynthus, including 'The Fountain House.' *AJA* 39: 210–47, esp. 219–20.

Robinson, D. M., and J. W. Graham. 1938. *Excavations at Olynthus*. Part 8: *The Hellenic House*. Baltimore: Johns Hopkins University Press.

Robinson, H. S. 1962. Excavations at Corinth: The Baths of Aphrodite, 1960. *Hesperia* 31.1: 120–30. With 2 figures and 37 plates.

————. 1965. *The Urban Development of Ancient Corinth*. Athens: American School of Classical Studies.

————. 1969. A Sanctuary and Cemetery in Western Corinth. *Hesperia* 38.1: 2–35.

————. 1976. Excavations at Corinth: Temple Hill, 1968–72. *Hesperia* 45.3: 203–239, plates 45–58.

Roda, C. 1968. Geologia della tavoletta Pietraperzia (Provinca di Caltanissetta e Enna, F. 268, III-NE). *Atti della Accademia Gioenia de Scienze Naturali*. Catania. Ser. 6, 19: 145–254.

Rodda, J. C., and N. C. Matalas, eds. 1987. *Water for the Future: Hydrology in Perspective*. Publication No. 164. Wallingford, Oxfordshire, U.K.: IAHS.

Roebuck, C. 1945. A Note on Messenian Economy and Population. *Classical Philology*. 40: 150–65.

————. 1951. *Corinth XIV: The Asklepieion and Lerna*. Princeton, N.J.: American School of Classical Studies.

————. 1972. Some Aspects of Urbanization in Corinth. *Hesperia* 41: 96–127.

————. 1979. Stasis in Sicily in the 7th c b.c. In *Philias Charin: Misc. in onore di E. Manni*. 6 vols. Rome.

Rogers, G. M. 1991. *The Sacred Identity of Ephesos*. London: Routledge.

————. 1992. The Constructions of Women at Ephesos. *ZPE* (Bonn) 90: 215–23.

Rolley, C. 1990. Les débuts du sanctuaire. In Delphes: Races cultes et jeux: 100 ans de fouilles et de decouvertes. *Les Dossiers d'Archaeologie* 15 (July–August): 8–13.

Romano, D. G. 1993. Post-146 Land Use in Corinth and Planning of the Roman Colony of 44 b.c. In Gregory, T. E., ed. *The Corinthia in the Roman Period. JRS* Suppl. 8: 9–30.

Romano, D. G., and B. C. Schoenbrun. 1993. A Computerized Architectural and Topographical Survey of Ancient Corinth. *JFA* 200.2 (summer): 177–90.

Romano, I. B. 1994. A Hellenistic Deposit from Corinth. *Hesperia* 63: 62–64.

Rorig, R. 1967. *The Medieval Town*. Berkeley: University of California Press.

Rossi-Manaresi, R., and C. Ghezzo. 1978. The Biocalcarenite of the Agrigento Greek Temples: Causes of the Alteration and Effectiveness of Conservation Treatments. *International Symposium on Deterioration and Protection of Stone Monuments*. Paris: UNESCO-Rilem. 1–31.

Rostovtzeff, M. I. 1941. *The Social and Economic History of the Hellenistic World*. 3 vols. Oxford: Clarendon Press.

Rotberg, R. I., and T. K. Rabb, eds. 1981. *Climate and History: Studies in Interdiscliplinary History*. Princeton, N.J.: Princeton University Press.

Roux. G. 1976. *Delphes, son oracle et ses dieux*. Paris: Belles Lettres.

Rudwick, M. J. S. 1969. Lyell on Etna, and the Antiquity of the Earth. In Schneer, C. C., ed. *Towards a History of Geology*. Cambridge, Mass.: MIT Press. 288 ff.

————. 1976. The Emergence of a Visual Language for Geological Science 1760–1840. *History of Science* 14: 149–95.

————. 1979/1981. Transposed Concepts from the Human Sciences in the Early Work of Charles Lyell. In Jordanova, L. J., and R. S. Porter, eds. *Images of the Earth: Essays in the History of the Environmental Sciences*. Monograph 1. London: British Society for the History of Science. 67–83.

————. 1985. *The Great Devonian Controversy: The Shaping of Scientific Knowledge among Gentlemanly Specialists*. Chicago: University of Chicago Press.

Ruggieri, G., and A. Greco. 1967. Distribuzione dei macrofossili nel Calabriano inferiore di Agrigento. *Atti della Accademia Gioenia de Scienze Naturali in Catania*. Series 6, Vol. 18: 319–27.

Ruggieri, G., and M. Unti. 1978. Dati prelininari sulla neotettonica del foglio 265 (Mazara del Vallo) *Contributi preliminari alla realizzazione della carta neotettonica d'Italia*. Projetto Finalizzttato Geodinamice. Publ. No. 251. Centro Nazionale delle Richerche: 69–77 .

————. 1979. Dati preliminari sulla neotettonica del foglio 257 (Castelvetrano). *Contributi preliminari alla realizzazione della carta neotettonica d'Italia*. Projetto Finalizzttato Geodinamice. Publ. No. 251. Centro Nazionale delle Richerche: 96–10.

Ruggieri, G., A. Unti, M. Unti, and M. A. Moroni. 1975. La calcarenite di Marsala (Pleistocene Inferiore) é i Terreni Contermini. *Bollettino della Società Geologia Italiana* (Rome) 94.6: 1623–57 and 2 plates

Ruggieri, R. 1996. *Gli ipogei di Agrigento: Valorizzazione e fruizione*. Atti del Convegno Studi, Agrigento, May 16. With new map of the *ipogei* of Girgenti (historical center of Agriento).

Rumschied, F. 1994. *Untersuchungen zur Kleinasiatischen Bauornamentik des Hellenismuus*. 2 vols. Mainz: P. von Zabern.

Runnels, C. N. 1981. Stone Artifacts from Surface Sites. In Keller, D. R,. and D. W. Rupp, comps. *Colloquium on Archaeological Surveying in the Mediterranean Area*. Athens: American School of Classical Studies and Canadian Archaeological Institute at Athens.

————. 1995. Environmental Degradation in Ancient Greece. *Scientific American* Mar.: 72–75.

Russell, R. J. 1954. Alluvial Morphology of Anatolian Rivers: Mæander Delta. *Annals of the American Association of Geographers* 44 (Dec.): 374–76.

Rutter, J. 1979. The Last Mycenaeans at Corinth. *Hesperia* 48: 389–92.

Ryan, W. B. F., D. J. Stanley, V. B. Herseeu, D. A. Fahlquist, and T. D. Allan. 1970. The Tectonics and the Geology of the Mediterranean Sea. In Maxwell, A. E., gen. ed. *The Sea: Ideas and Observations on Progress in the Study of the Seas*. New York, Wiley-Interscience. 387–492.

Rykwert, J. 1976. *The Idea of a Town*. London: Faber and Faber.

Sakellariou, M. 1996. The Metropolises of the Western Greeks. In Pugliese-Carratelli, G., ed. *The Western Greeks*. Venice: Bompiani. 177–188.

Sakellariou, M., and N. Faraklas. 1972. Corinthia-Cleonaea, with a map of the Lechaeum port. In Doxiadis, C. A. *The Method for the Study of the Ancient Greek Cities*. 5 vols. Athens: Ekistics Center. 151.

Salinas, A. 1884–85. Relazione . . . sugli acquidodtti di Selinunte e sulle lucerne trovate nella vasca di Bigini presso Castelvetrano. Trovate nella vasca di Biggini pressso Castelvetrano. *NOTSC* Luglio. *Atti della r. Academia dei Linci*, Serie 4, *Memoire.* 463–65 and Tab. 2.

Salmon, J. B. 1984. *Wealthy Corinth: A History of the City to 338 B.C.* Oxford: Clarendon Press.

Sanders, G. D. R. 1984. Reassessing Ancient Populations. *BSA* 251–262.

Sanders, G. D. R., and I. K. Whitbread. 1990. Central Places and Major Roads in the Peloponnese. *BSA* 85: 333–61.

Sanders, W. T., J. R. Parsons, and R. S. Stanley 1989. *The Basin of Mexico: Ecological Processes in the Evolutions of a Civilization.* New York: Academic Press.

Santayana, G. 1905. *The Life of Reason*, vol. 1. New York: Scribner's. Ch. 12.

Sartori, E. 1996. The Constitutions of the Western Greeks. In Pugliese-Carratelli, G., ed. *The Western Greeks.* Venice: Bompiani. 215–22.

Scarborough, V. 1994. *Water Management at Midea and Adjacent Areas, Greece.* Report to Prof. G. Walberg and the Classics Dept., University of Cincinnati.

Scarborough, V., and B. Isaac, eds. 1993. *Research in Economic Anthropology*, Suppl. 7. Greenwich, Conn.: JAI Press.

Scherrer, P. 1994. Agora, Serapeion. *JÖAI:* 11–14.

———, ed. 1995. *Ephesos: Der Neue Führer.* Vienna: Österreichisches Archäologisches Institut, and Seljuk: Efes Museum,.

———. 1996. The City of Ephesos from the Roman Period to Late Antiquity. In Koester, H., ed. *Ephesos: Metropolis of Asia.* Valley Forge, Pa.: Harvard Theological Studies, Trinity Press International. 1–25.

———. 1999. Bemerkungen sur Siedlungschichte von Ephesos vor Lysimachos. In Friesinger, H., and F. Krinzinger, eds. *100 Jahre österreichische Forschungen in Ephesos: Akten des Symposiums Wien 1995.* Archäologische Forschungen 260: 379–87, and tables 61 ff.

Scherrer, P., H. Taeuber, and H. Thür, eds. 1999. *Steine und Wege: Festschrift für Dieter Knibbe zum 65. Geburtstag.* Sonderschriften 32. Vienna: Österreichisches Archäologisches Institut.

Schilirò, F., G. Bruno, A. Cannata, and E. Renna. 1996. La sinclinale di Morgantina quale sito archeologico ellenistico. *Proceedings of the II Symposium on Geotopes, 20–22 May, Rome.*

Schindler, A. 1906. Carte der Umgebung von Ephesos *and* Bemerkungen zur Karte. *FiE* 1: 235–36. Map created in 1896.

Schmiedt, G. 1957. Applicazione della fotografia aerea in richerche estensive di topografia antica in Sicilia. *Kokalos* 3: 18–30 and plates 5–6.

Schmiedt, G., and P. Griffo. 1958. *Agrigento antica dalle fotografie aeree e dai recenti scavi.* Florence: L'Universo.

Schneer, C. C., ed. 1969. *Towards a History of Geology.* Cambridge, Mass.: MIT Press.

Schneider, C. 1997. Grabung an der Stademauer 1995. *AA* 2: 134–63.

Schneider, P. 1990. Zur Herstellung eines archaischen Tondaches. In Hoffmann, A., E.-L. Schwandner, W. Hoepfner, and G. Brands, eds. *Bautechnik der Antike: International Colloquium in Berlin.* Mainz-am-Rhein: Verlag Philipp von Zabern. 197–207.

———. 1997. Die Arkadiane in Ephesos: Konzept einer Hallenstrasse. *Stadt und Umland: Bauforschungskolloquium Berlin 1997.* Mainz: 120–22. (Diskussionen zur archäologischen Bauforschung, 7.)

Schove, D. J. 1955. The Sunspot Cycle, 649 B.C. to A.D. 2000. *JGEOPHYS* 60: 127–46.

Schröder, B. 1990. Archäologiebegleitende Geologie zur Grabung Milet 1989. *Ist. Mitt.* 40: 62–68.

Schröder, B., and Ü. Yalçin. 1991. Geologische Begleituntersuchungen. *Ist. Mitt.* 41: 149–62.

———. 1992a. Untersuchungen im Umfeld von Milet. *Ist. Mitt.* 42: 109–16.

———. 1992b. Geology of the Non-metamorphiic Formations around Milet (Western Turkey). *Geology of the Aegean: 6th Congress of the Geological Society of Greece, Athens, May 25–27*: 108. abstract (only).

Schröder, B., H. Brückner, H. Stümpel, and Ü. Yalçin. 1994. Geowissenschaftliche Umfelderkundung. *JDAI Istanbul*: 238–75.

Schubring, J. 1865a. Die Topographie der Stadt Selinus. *NGG*: 401–41.

———. 1865b. Die Bewässerung von Syrakuse. *Philologus* 22: 577–638, with map. In von Leutsch, E., *Zeitschrift fur das Klassische Alterthum*.

———. 1870. Historische Topographie von Akragas in Sicilien. *Philologus*: 38–44 plus map; also in Italian as *Topografia storica di Agrigento*, Turin, 1887–88, reprinted as No. 86 of Bibl. Istorica dell' Antica e Nuova Italia, A. Forni Edizioni, n.d., ca. 1990.

———. 1873. Die Neuen Entdeckungen von Selinunt. *Arch. Zeitung* 39: 97–103; see also plan of the spring of Gaggera on p. 70 of that volume.

Sciortino, A. ca. 1993. Climatological and Hydrogeological Considerations [of Morgantina]. Manuscript in D. P. Crouch collection.

Scranton, R. L. 1951. *Corinth I.3: Monuments in the Lower Agora and North of the Archaic Temple*. Princeton, N.J.: American School of Classical Studies.

Scullard, H. H. 1967. *The Etruscan Cities and Rome*. London: Thames and Hudson.

Secord, J. A. 1986. *Controversy in Victorian Geology: The Cambrian Silurian Dispute*. Princeton, N.J.: Princeton University Press.

Seiterle, G. 1964–65. Ephesos: Lysimachische Stadtmauer. *JÖAI* 47: 8–11.

Semple, E.C. 1912. Geographic Factors in the Ancient Mediterranean Grain Trade. *AAAG* 11: 47–74.

———. 1931. *The Geography of the Mediterranean Region: Its Relationship to Ancient History*. New York: Henry Holt. Reprint of 1912 ed.

Seve, M. 1998. L'Urbanisme Argien au XVIIIe et au début du XIXe siécle. In Pariente, A., and G. Touchais, eds. *Argos et l'Argolide: Topographie et urbanisme*. École Française d'Athènes. Paris: De Boccard Édition. 417–37.

Sguaitamatti, M. 1993. Les metamorphoses du Temple M ou l'art de Noyer des données de Fouille dans une fontaine. *Studie della Sicilia Occidentale in onore di Vincenzo Tusa*. Padua: Bottega d'Erassmo (Feb.): 151–57, 2 plates

Shaw, B. D. 1984. Water and Society in the Ancient Maghrib: Technology, Property and Development. *Antiquities Africaines*. 20: 121–73.

Sheets, P. D., and D. K. Grayson, 1979. Volcanic Disasters and the Archaeological Record. In Sheets, P. D., and D. K. Grayson, eds. *Volcanic Activity and Human Ecology*. New York: Academic Press. 623–32.

Sherwin-White, S. M. 1985. Ancient Archives: The Edict of Alexander to Priene: A Reappraisal. *JHS* 69–89.

Simkin, T., L. Siebert, L. McClelland, D. Bridge, C. Newhall, and J. H. Latter. 1981. *Volcanoes of The World*. Straudsburg, Pa.: Hutchinson Ross Publishing for the Smithsonian Institution. 69–89.

Siracusano, A. 1983. *Il santuario rupestre di Agrigento in localita S. Biagio. Sicilica*. Serie Archeologica, 2. Rome: G. Bretschneider Edizioni.

Siviglia, I. 1979. Il periplo della Sicilia secondo Strab(o). VI 2,1 in *Philias Charin: Misc. in onore di E. Manni*. 6 vols. Rome: 1999–2013.

Sjöqvist, E. 1973. *Sicily and the Greeks*. Ann Arbor: University of Michigan Press.

Skayias, S. D., K. Trippler, and H. Vierhuff. 1983. Evaluation of Geochemical Data of Spring Water from Peloponnes, Greece. Part 1. Proceedings of the 8th Salt Water Intrusion Meeting. *Geologia Applicata e Idrogeologia* 18. 49–63.

Smith, N. 1975. *Man and Water: A History of Hydrotechnology*. New York: Scribner's.

Smith, P. F. 1977. *The Syntax of Cities*. London: Hutchinson.

————. 1979. *Architecture and the Human Dimension*. London: G. Godwin.

Smith, T. K. 1994. Water Management in the Late Bronze Age Argolid, Greece. Ph.D. dissertation, University of Cincinnati.

Snell, L. J. 1963. Effect of Sediment on Ancient Cities of the Aegean Coast, Turkey. *IASH Bull.* 8.4 (Dec.): 71–73.

Snodgrass, A. M. 1987. *An Archaeology of Greece: The Present State and Future Scope of a Discipline*. Berkeley: University of California Press. Chap. 2: Archaeology and History, 71–194.

Sofianos, A. I., C. V. Costantinidis, and J. Christodoulias. 1988. Rockfall Analysis at the Ancient Region of Argos. In Marinos, P. G., and G. C. Koukis, eds. *The Engineering Geology of Ancient Works, Monuments and Historical Sites: Proceedings of the Symposium of the International Association of Engineering Geology, Athens*. Rotterdam: A. A. Balkema: 213–16.

Soren, D. 1988. The Day the World Ended at Kourion: Reconstructing an Ancient Earthquake. *National Geographic* 174.1 (July): 30–53.

Spawforth, A. J., and S. Walker. 1986. The World of the Panhellenion. II: Three Dorian Cities. *JRS* 76: 88–105.

Squier, E. 1877. *Peru: Incidents of Travel and Exploration in the Land of the Incas*. New York: Harper and Brothers. Reprinted 1973 by AMS Press for the Peabody Museum.

Squier, E. G., and E. H. Davis. 1848. *Ancient Monuments of the Mississippi Valley*. Contributions No. 1. Washington, D.C.: Smithsonian Institution.

Squyres, C. H. ed. 1975. *Geology of Italy*. Earth Sciences Society of the Libyan Arab Republic. Includes Vita-Finzi, C. Late Quaternary Alluvial Deposits in Italy, 329–80.

Steel, E. W., and T. J. McGhee. 1979. *Water Supply and Sewerage*. New York: McGraw-Hill.

Steiner, A. 1987. Private and Civic Cult at Corinth. Abstract from the 89th American Institute of Archaeology Meeting, 1986. *AJA*.

Stent, G. S. 1972. Prematurity and Uniqueness in Scientific Discovery. *Scientific American*. 272.11 (December): 84–93.

Stier, H.-E., E. Kirsten, H. Quyirin, W. Trillmich, and G. Czybulka. 1967. *Westermanns Atlas zur Weltgeschichte*. Braunschweig: G. Westermann Verlag.

Stillwell, R. 1948. *Corinth XV. 1: The Potter's Quarter*. Princeton, N.J.: American School of Classical Studies.

————. 1952. *Corinth XV. 2. The Potter's Quarter: The Terracottas*. Princeton, N.J.: American School of Classical Studies.

Stillwell, R., W. L. MacDonald, and M. McAllister. 1976. *Princeton Encyclopedia of Classical Sites*. Princeton, N.J.: Princeton University Press.

Stiros, S. C. 1988. Earthquake Effects on Ancient Constructions. New Aspects of Archaeological Science in Greece. Proceedings of a Meeting Held at the British School at Athens, January 1987 (Occasional Paper of the Fitch Laboratory). Athens: 1–6.

————. 1998. Archaeology: A Tool to Study Active Tectonics. *EOS* (Dec. 23): 1633 and 1639.

Stone, S. C., III. 1981. Roman Pottery from Morgantina in Sicily. Ph.D. dissertation, Princeton University.

————. 1993. Sextus Pompiy, Octavian, and Sicily. *AJA* 87: 11–22.

Storia della Sicilia. 1979. Società editrice storia di Napoli del mezzogiorno continentale e della Sicilia. R. Romeo, director. Esp. Part 2: La città greche di Sicilia, Vol. 1: 229–780

Storoni Mazzolani, L. 1970. *The Idea of the City in Roman Thought from Walled City to Spiritual Commonwealth*. Trans. S. O'Donnell. Bloomington: Indiana University Press.

Stringfield, V. T. and J. R. Rapp. 1977. Progress of Knowledge of Hydrology of Carbonate Rock Terranes: A Review. In Dilamarter, R. R., and S. C. Csallany, eds. *Hydrologic Problems in Karst Regions: Proceedings of the International Symposium, April 1976*. Bowling Green, Ky.: W. Kentucky University Press. 1–9.

Stümpel, H., C. Bruhn, F. Demirel, M. Gräber, M. Panitzki, and W. Rabbel. 1997. Stand der geophysikalische Messungen im Umfeld von Milet. AA 2: 124–34.

Stümpel, H., F. Demirel, S. Lorra, and S. Wende. 1994. Lion Harbor. *JDAI Istanbul*: 238–53.

Svarlien, D.A. 1990/91. Epicharmus and Pindar at Hieron's Court. *Kokalos* 36–37: 103–10.

Symeonoglou, S. 1985. *The Topography of Thebes from the Bronze Age to Modern Times*. Princeton, N.J.: Princeton University Press.

Tahal (Water Planning) Ltd. 1962. Preliminary Report on Possible Remedial Measures and Additional Irrigation Development in the Argos Plain. Tahal.

Tainter, J. A. 1988. *The Collapse of Complex Societies*. Cambridge: Cambridge University Press.

Tanriöver, A. 1974. Unpublished paper on the City walls of Priene. In Ünal Özis collection, Doküz Eylül University, Izmir, Turkey.

Tarrant, H. 1990. The Question of "Alien Influences." In Descœudres, J.-P., ed. *Greek Colonists and Native Populations*. Oxford: Clarendon Press. 625–28.

Taylor, W. W. 1948. A *Study of Archaeology*. American Anthropological Association. Reprinted 1964 and 1967/71, Carbondale, Ill.

Tennyson. L. C., and C. D. Settergren. 1977. Subsurface Water Behavior and Sewage Effluent Irrigation in the Missouri Ozarks. In Dilamarter, R. R., and S. C. Csallany, eds. *Hydrologic Problems in Karst Regions: Proceedings of the International Symposium, April 1976*. Bowling Green, Ky.: W. Kentucky University Press. 411–18.

Theodorescu, D. 1975. Remarques préliminaires sur la topographie urbaine de Sélinonte. *Kokalos* 21: 108–20.

Thirgood, J. V. 1981. *Man and the Mediterranean Forest: A History of Resource Depletion*. London: Academic Press.

Tholfsen, T. R. 1967. *Historical Thinking*. New York: Harper and Row.

Thomas, W., ed. 1956. *Man's Role in Changing the Face of the Earth*. Chicago: University of Chicago Press.

Thornes, J. B., and D. Brunsden. 1977. *Geomorphology and Time*. New York: Wiley.

Thrower, N. J. W., and D. E. Bradbury. 1973. The Mediterranean Type Region. In di Castri, F., and H. A. Mooney, eds. *Mediterranean Type Ecosystems: Origin and Structure*. New York: Springer-Verlag. 40 ff.

Thür, H. 1989. Das Hadrianstor in Ephesos. *FiE* 11.1: 25–28.

———. 1990. Ausgehölte römische Bauglieder. In Hoffmann, A., E.-L. Schwandner, W. Hoepfner, and G. Brands, eds. *Bautechnik der Antike: International Colloquium in Berlin*. Mainz-am-Rhein: Verlag Philipp von Zabern. 238–45.

———. 1995. Der ephesische Ktistes Androklos und (s)ein Heroon am Embolos. Sonderdruck. *JÖAI* 64: 63–103.

———. 1996. The Processional Way in Ephesos as a Place of Cult and Burial. In Koester, H., ed. *Ephesos: Metropolis of Asia*. Valley Forge, Pa.: Harvard Theological Studies, Trinity Press International. 157–87.

———, ed. 1997. *Und verschönertre die Stadt*. Vienna: Österreichisches Archäologisches Institut.

———. 1999a. Die spätantike Bauphase der Kuretenstrasse. *Akten des Symposiums "Efeso Paleochristiana e Bizantina."* Rome: 140 ff.

———. 1999b. Der Embolos: Tradition und Innovation anhand seines Erscheinungsbildes.

In Friesinger, H., and F. Krinzinger, eds. *100 Jahre österreichische Forschungen in Ephesos: Akten des Symposiums Wien 1995.* Archäologische Forschungen 260: 421–28.

Tolle-Kastenbein, R. 1990. *Antike Wasserkultur.* Munich: C. H. Beck.

————. 1991. Doppelstollen: Ein Phänomen griechischer Wasserleitung. *Kultur und Technik* 2: 38–43.

Tolman. C. F. 1937. *Groundwater.* New York: McGraw Hill.

Tomlinson, R. A. 1972. *Argos and the Argolid: From the End of the Bronze Age to the Roman Occupation.* London: Routledge and Kegan Paul.

Touchais, G. 1998a. Argos à l'époque mesohelladique. In Pariente, A., and G. Touchais, eds. *Argos et l'Argolide: Topographie et urbanisme.* École Française d'Athènes. Paris: De Boccard Édition. 71–84.

————. 1998b. Conclusions de la Table Ronde. In Pariente, A., and G. Touchais, eds. *Argos et l'Argolide: Topographie et urbanisme.* École Française d'Athènes. Paris: De Boccard Édition. 479–80.

Touchais G., and N. Divari. 1998. Argos du Néolithique à l'époque géometrique. In Pariente, A.. and G. Touchais, eds. *Argos et l'Argolide: Topographie et urbanisme.* École Française d'Athènes. Paris: De Boccard Édition. 9–21.

Toynbee, A.J. 1946–47. A *Study of History.* New York: Oxford University Press. [Somervell's abridgement of Vols. 1–4.]

Trenhaile, A. S. 1987.*The Geopmorphology of Rock Coasts.* Oxford: Clarendon Press.

Tsakirgis, B. 1984. The Domestic Architecture of Morgantina in the Hellenistic and Roman Periods. Ph.D. dissertation, Princeton University.

————. 1989. The Decorated Pavements of Morgantina. 1: The Mosaics. *AJA* 93: 395–416.

————. 1990 The Decorated Pavements of Morgantina. 2: The Opus Signinum. *AJA* 94.3: 425–44.

Tsetskhladze, G. R. 1994. Greek Penetration of the Black Sea. In Tsetskhladze, G. R., and F. De Angelis, eds. *The Archaeology of Greek Colonisation.* Monograph 40. Oxford: Committee for Archaeology. 111–135.

————. 1999. Between West and East: Anatolian Roots of Local Cultures of the Pontus. In Tsetskhladze, G. R, ed. *Ancient Greeks West and East.* Leiden: E. J. Brill. 469–96.

Tuan, Y.-F. 1968. *The Hydrologic Cycle and the Wisdom of God: A Theme in Geotheology.* Toronto: University of Toronto Dept. of Geology Research Publications.

Türk, N., S. Çakici, D. M. Uz, S. Akça and K. Geyik, 1988. The Geology, Quarrying-Technology and Use of Beylerköy Marbles in West Turkey. In Herz, N., and M. Waelkens, eds. *Classical Marble: Geochemistry, Technology, Trade.* NATO ASI Series E: Applied Sciences, Vol. 153. Dordrecht: Kluwer Academic: 83–89.

Türkoglu, S. 1978. Aysoluk (Selçuk Ephesus): Recent Archaeological Research in Turkey. *Anatolian Studies* 28: 13–14.

Tusa, V. 1962. L'irradiazione della civiltà greca nella Sicilia Occidentale. *Kokalos* 8: 156–58.

————. 1967. Le divinità ed i templi di Sélinunte (1). *Kokalos* 13: 186–93.

————. 1971. Selinunte punica, *Rivista del Institito Nazional d'Archrologia et storia dell'arte.* N.S. 18: 47–68.

————. 1972. Selinunte: Muro id cinta dell'Acropoli. Restauro dell'angolo di Nord-est. *Sicilia Archeologica* 5: 43–48.

————. 1979–82. Edifici sacri in centri non Greci della Sicilia occidentale. *Philias Charin:* 2129–2137 and plates 2–7.

————. 1980–81. Le divinità ed i templi di Sélinunte (2). *Kokalos:* 1009–16.

————. 1982. Selinunte: La cinta muraria dell' Acropoli. *La fortification dans l'histoire du monde grec*. CNRS Colloque Int. 614. (Dec.): 111–119. With dates of walls; plates 43–55.

————, ed. 1983. *La scultura in Pietra di Selinunte*. I Cristali series. Palermo: Edizioni Sellerio.

Tuttahs, G. 1995. Wasserbauliche Problemfelder am Grabungsplatz Milet: Zustand und Aufgaben. *JDAI Istanbul*: 265–271.

————. 1997. Vorbericht zur Wasserversorgung Milets im Einzugsgebiet des Nymphäum Aquäduktes in Romischer Zeit (1 and 2 Jh n. Chr.). *Archäologischer Anzeiger*. 2: 163–79.

————. 1998. *Milet und das Wasser: Ein Beispiel Für die Wasserwirtschaft einer antiken Stadt*. Forum Siedlungswasserwirtschaft und Abfallwirtschaft, Universität GH Essen, Vol. 12.

Ucko, P. J., R. Tringham, and G. W. Dinbleby, eds. 1970. *Man, Settlement and Urbanism: Proceedings of the Research Seminar in Archaeology and Related Subjects*. Cambridge, Mass.: Schenkman Publishing.

Ulzega, A. 1993–94. Condizioni geografiche dei mari e delle coste della Sicilia. *Kokalos* 39–40.2: 1–8, map.

Ünlüoglu, B. B. M. 1998. Plan of Ephesus Theater North Corridor. In the D. P. Crouch collection.

Vagnetti, L. 1996. The First Contacts between the Minoan-Mycenaean World and the Western Mediterranean Worlds. In Pugliese-Carratelli, G., ed. *The Western Greeks*. Venice: Bompiani. 117–20

Vallet, G. 1962. La colonisation Chalcidienne et l'hellenisation de la Sicilie Orientale. *Kokalos* 8: 30–51.

————. 1967. La cité et son territoire dans les colonies grecques d'Occident. *Actes du 7ᵉ Congrès d'etudes sur la grand Grèce*: 68–140.

————. 1983. Urbanisation et organisation do la chora coloniale grecque en Grande Grèce et en Sicile. Forme di contatto e processi di trasformazionne nelle società (Cortona 24–30 May 1981). Pisa: 937–56.

Van Andel, T. H. 1994. Geo-archaeology and Archaeological Science: A Personal View. In Kardulias, P. N., ed. *Beyond the Site: Regional Studies in the Aegean Area*. Lanham, Md.: University Press of America. 45–68.

Van Andel, T. H., and E. Zangger. 1990a. Landscape Stability and Destability in the Prehistory of Greece. In Bottema, S., G. Entijes-Nieborg, and W. van Zeist, eds. *Man's Role in the Shaping of the Eastern Mediterranean Landscape*. Rotterdam: A. A. Balkema. 139–57.

————. 1990b. Land Use and Soil Erosion in Preshistoric and Historical Greece. *JFA* 17: 379–96.

Van Andel, T. H., C. N. Runnels, and K. O. Pope. 1986. 5000 Years of Land Use and Abuse in the South Argolid, Greece. *Hesperia* 56: 103–28.

Van Andel, T. H., E. Zangger, and C. Perissoratis. 1990. Quaternary Trangressive/Regressive Cycles in the Gulf of Argos, Greece. *Quaternary Research* 34: 317–29.

Van Der Leeuw, S. E. 1987. Revolutions Revisited. In Manzanilla, L., ed. *Studies in the Neolithic and Urban Revolutions: V. G. Childe Colloquium, Mexico 1986*. Oxford: BAR International Series 349: 217–43.

Varti-Matarangas, M., and D. Matarangas, 1994. Lithofacies Determination and Geological Origin of Building Stones at the Asklepieion at Epidavros: Their Correlation to the Ancient Quarries. *PACT* 45: 15–24.

Varti-Matarangas, M., N. Beloyannis, V. Nitsaki-Hafner, J. Katsikis, and E. Pantelias. 1995. Lithofacies Determination, Origin and Decay of the Building Stones in the Monuments of the Ancient Olympia area. In Maniatis, Y., N. Herz, and Y. Basiakos, eds. *The Study of Marble and Other Stones Used in Antiquity*. Archetype. 295–302.

Veblen, T. 1899. *The Theory of the Leisure Class*. New York: Macmillan.

Vermule, C. 1982. Transmission of Roman Historical Relief throughout the Empire, with Special Reference to Southern Italy and Sicily. In Gualandi, M. L, L. Messei, and S. Settes, eds. *Aparchai in Honor of Paolo Enrico Arias*. I. Pisa.: 635–40.

Vernant, J. P. 1983. *Myth and Thought among the Greeks*. London: Routledge and Kegan Paul.

Vetters, W. 1985. Die Kästenversschiebngen Kleinasiens: Eine Konsequenz tektonischer Ursachen. *Lebendige Altertums wissenschaft Festschrift zur Vollendung des 70 Lebens Jahres von Hermann Vetters*. Vienna: Verlag Adolf Holzhausens Nfg.: 33–37 and plates 6–8.

———. 1988. Ancient Quarries around Ephesus and Examples of Ancient Stone Technologies. In Marinos, P. G., and G. C. Koukis, eds. *The Engineering Geology of Ancient Works, Monuments and Historical Sites: Proceedings of the Symposium of the International Association of Engineering Geology, Athens*. Rotterdam: A. A. Balkema. 2067–78.

———. 1994:. Der Taupo and das Klima um 200 A.D. in Europa. *Markomannenkrige: Ursachen und Wirkungen*. Brno: Archäologisches Institut der Akademie der Wissenschaften der Tschechischen Republik. 457–61.

———. 1996. Cultural Geology: Examples from Antiquity and Modern Times. Russia and West Europe: Interaction of Industrial Cultures 1700–1950. Presented at the International Conference, Nizhny Tagil, Russia, Aug. 15–18.

———. 1997. Hydrologie von Ephesos. 1997. Manuscript in D. P. Crouch collection.

———. 1998. Zur Geologie der antiken Steinbrüche von Ephesos: Erläurterungen zu der geologischen Übersichtskarte der Umgebung von Ephesos. Manuscript in the D. P. Crouch collection.

———. 2000. Ein Beitrag zur Hydrologie von Epehsos. In Jansen, G. C. M., ed. *Cura aquarum in Sicilia*. Annual Papers on Classical Archaeology, suppl. 6. Leiden, Stichting Babesch. 85–89.

Vigo, L. 1827: reprinted 1883. *Lettera a Nicolò Palmerii "Sogli ipogei e catacombe di girgenti."* Palermo: Opera Omnia Palmeri.

Villard, F. 1994. Les sièges de Syracuse et leur pestilences. In Ginouvés, R., A.-M. Guimier-Sorbets, J. Jouanna, and L. Villard, eds. *L'eau, la santé et la maladie dans le monde grec*. Paris: Actes de Colloque à Paris 25–27 Nov. 1992. BCH Suppl. 4: 337–44.

Vine, F. J., and H. H. Hess. 1970. Sea Floor Spreading. *SEA* 4.2–3: 587–622.

Viret Bernal, F. 1992. Argos, du palais à l'agora. *Dialogues d'Historie Ancienne* 18.1: 61–88.

Vita-Finzi, C. 1966. The New Elysian Fields. In Archaeological Notes, *AJA* 70: 175–58.

———. 1969. *The Mediterranean Valleys: Geological Changes in Historical Times*. Cambridge: Cambridge University Press.

———. 1975. Late Quaternary Alluvial Deposits in Italy. In Squyres, C. H., ed. *Geology of Italy*. Earth Sciences Society of the Libyan Arab Republic. 329–80.

Voigtländer, W. 1985. Zur Topographie Milets: Ein neues Modell zur antiken Stadt. *AA* 77–91.

Vollgraff, W. 1956. Le sanctuaire d'Apollo pythéen à Argos. Etudes Peloponnesienne I.

Von Gerkan, A. 1924. *Griechische Städteanlagen*. Berlin and Leipzig.

———. 1959. *Von antiker Architektur und Topographie*. Stuttgart.

Von Graewe, V. 1995. Milet 1992–93. AA D.A.I. Vol. 2: 195–333.

———. 1997/98. Neue Ausgrabungen und Forschungen im archaischen Milet. *Nürnberger Blätter zur Archäologie*. 4: 73 ff.

————. 1999. Millet 1996–97, and Funde aus Milet V. *AA* D.A.I: 373 ff and 415 ff.

Voutsake, S. 1995. Social and Political Processes in the Mycenaeaan Argolid: The Evidence from the Mortuary Practices. In Laffineur, R., and W.-D. Niemeier, eds. *Politeia: Society and State in the Aegean Bronze Age: Proceedings of the 5th International Aegean Conference, 10–13 April, 1994.* Université de Liège and University of Texas at Austin. 55–63. From his Cambridge University dissertation of 1993.

Voza, G. 1976–77. Siracusa. *Kokalos* 22–23.2: 551–61.

————. 1979. Problemetica archeologica. *Storia della Sicilia.* Società Edditrice Storia di Napoli e della Sicilia: 5–42.

Vullard, F. 1994. Les sieges de Syracuse et leurs pestilences. *BCH* Suppl.: 337–44.

Waelkens, M., N. Herz, and L. Moens. 1992. *Ancient Stones: Quarrying, Trade, and Provenance: Interdisciplinary Studies on Stones and Stone Technology in Europe and Near East from Prehistory to Early Christian Times.* Leuven, Belguim.

Wagstaff, J. M. 1975. A Note on Settlement Numbers in Ancient Greece. *JHS* 95: 163–68.

————. 1981. Buried Assumptions: Some Problems in the Interpretation of the "Younger Fill" Raised by Recent Data from Greece. *J. of Archaeological Science* 8: 2477–64.

Walker, S. 1981. The Burden of Roman Grandeur: Aspects of Public Building in the Cities of Asia and Achaea. In Kind, A., and M. Henig, eds. *The Roman World in the Third Century.* Oxford: BAR International Series 109(i): 189–97.

Wallace, P. W. 1969. Strabo on Acrocorinth. *Hesperia* 38: 115–23.

Walters, J. C. 1996. Egyptian Religions in Ephesos. In Koester, H., ed. *Ephesos: Metropolis of Asia.* Valley Forge, Pa.: Harvard Theological Studies, Trinity Press International. 281–309.

Wasowicz, A. 1983. Le programme urbain de la *Polis* grecque. *Archéologie et Société*: 87–91.

Webb, T. III. 1981 The Reconstruction of Climatic Sequences from Botanical Data. In Rotberg, R. I.,and T. K. Rabb, eds. *Climate and History: Studies in Interdiscliplinary History.* Princeton, N.J.: Princeton University Press. 169–92.

Webster, T. B. L. 1973. *Athenian Culture and Society.* Berkeley: University of California Press.

Weinberg, S. S. 1948. Cross-section of Corinthian Antiquities. *Hesperia* 17: 197–241.

Welch, K. 1996. The Roman Arena in the Greek East: Athens and Corinth. Abstracts of the Annual Meeting, 1995. *AJA* 99: 329.

Wells, B., C. Runnels, and E. Zangger. 1993. In the Shadow of Mycenae. *Archaeology* (Jan.–Feb.): 54–63.

Wendell, C. V. 1969. Land-Tilting or Silting? Which Ruined Ancient Harbors? *Archaeology*: 322–24.

Wenner, D. B., S. Havert, and A. Clark, 1988. Variations in Stable Isotopic Compositions of Marble: An Assessment of Causes. In Herz, N., and M. Waelkens, eds. *Classical Marble: Geochemistry, Technology, Trade.* NATO ASI Series E: Applied Sciences, Vol. 153. Dordrecht: Kluwer Academic: 325–38.

Wertime, T. A. 1983. The Furnace versus the Goat: The Pyrotechnologic Industries and Mediterranean Deforestation in Antiquity. In Evans, R. K., ed. *Symposium (1978): Deforestation, Erosion, and Ecology*: Contributions 1–2. *JFA* 10., 445–57.

Wertime, T., and S. Wertime, eds. 1982. *Early Pyrotechnology: The Evolution of the Fire Using Industries.* Washington, D.C.: Smithsonian Institute.

Westaway, R. 1994. Evidence for Dynamic Coupling of Surface Processes with Isostatic Compensation in the Lower Crust during Active Extension of W. Turkey. *JGEOPHYS* 99.B10 (Oct. 10): 20, 203–20, 223.

Whitby, A. 1984. The Union of Corinth and Argos. *Historia* 33: 307–8.

White, D. 1963. A Survey of Millstones from Morgantina. *AJA* 67: 199–206 and plates 47–48.

———. 1964. Demeter's Sicilian Cult as a Political Instrument. *Greek, Roman, and Byzantine Studies* 5:4: 273–77.

———. 1967. The Post-Classical Cult of Malophoros at Selinus. *AJA* 71: 335–52, plates 101–06.

White, L. M. 1996. Urban Development and Social Change in Imperial Ephesos. In Koester, H., ed. *Ephesos: Metropolis of Asia*. Valley Forge, Pa.: Harvard Theological Studies, Trinity Press International. 31–50.

White, W. B. 1977. Conceptual Models for Carbonate Aquifers: Revisited. In Dilamarter, R. R., and S. C. Csallany, eds. *Hydrologic Problems in Karst Regions: Proceedings of the International Symposium, April 1976*. Bowling Green, Ky.: W. Kentucky University Press. 176–87

Whitehouse, R. C. and J. B. Wilkins. 1985. Magna Graecia before the Greeks: Towards a Reconciliation of the Evidence. *Papers in Italian Archaeology IV*. Part IV: *Classical and Medieval Archaeology*. Oxford: BAR International Series 246: 89–109.

Wiegand, T. 1899. *Milet. Ergebnisse der Ausgrabungen und Untersuchungenseit dem Jahre.* Istanbul: DAI.

Wiegand, T., and K. Krause. 1929. *Milet.* II.2, with map by Wilske, 1899–1900, at 1:50,000. Istanbul: DAI.

Wiegand, T., and H. Schrader. 1904. *Priene. IV: Die Wasseranlagen.* Berlin: Reimer Verlag.

Wigley, T. M. L., M. J. Ingram, and G. Farmer, eds. 1981. *Climate and History: Studies in Past Climates and Their Impact on Man.* Cambridge: Cambridge University Press.

Wilberg, W. 1923. Ephesos. I. Die Agora; II. Torbauten am Hafen; III. Wasserleitunng; V. Das Brunnenhaus am Theater. *FiE:* No. 19 & 25, 265–73.

Wille, M. 1995. Pollenanalysen aus dem Löwenhafenn von Milet: Vorlâufige Ergebnisse. In Milet 1992–93. *AA* 22: 330–33.

Williams, C. K. II. 1972a. Corinth 1970: Forum Area. *Hesperia* 41: 1–51.

———. 1972b. Corinth 1971: Forum Area. *Hesperia* 41: 143–84, plates 19–29.

———. 1974. The Route of Pausanias. *Hesperia* 43: 25–29.

———. 1976. Corinth 1975: Forum Southwest. *Hesperia* 45: 99–162.

———. 1977. Corinth 1976: Forum Southwest [including the Centuar Bath]. *Hesperia* 46: 40–81, plates 19–32.

———. 1980. Corinth 1979: South Stoa. *Hesperia* 49: 107–34

———. 1987. The Refounding of Corinth: Some Roman Religious Attitudes. In Macready, S., and F. H. Thompson. *Roman Architecture in the Greek World.* Occasional Papers (N.S.) 10. London: Society of Antiquaries of London. 26–37.

Williams, C. K. II., and J. E. Fisher. 1971. Corinth 1970: Forum Area. *Hesperia* 40: 1–51.

———. 1973. Corinth 1972: Forum. *Hesperia* 42: 1–44, plates 1–12.

———. 1975. The Route of Pausanias. In Corinth 1974: Forum Southwest. *Hesperia* 44: 25–29.

Williams, C. K. II., and P. Russell. 1981. Corinth: Excavations of 1980. *Hesperia* 50: 21–29, plates 1–9.

Williams, C. K. II., and O. H. Zervos 1982. Corinth 1981: East of the Theater. *Hesperia* 51.2: 116–24, plates 37–46.

———. 1984. Around the Fountain House of Glauke, and The Route of Pausanias, in Corinth 1983: The Route to Sikyon. *Hesperia* 53: 97–101, 101–4.

———. 1991. Corinth, 1990: Southeast Corner of Temenos E. *Hesperia* 60: Centaur Bath, 1–6. Also reported in *Hesperia* 45 and 46

Williams-Thorpe, O. 1988. Provenancing and Archaeology of Roman Millstones from the Mediterranean Area. *JARCHSCI* 15: 253–305.

Wilson, A. T. 1981a. Role of the Sun. In Rotberg, R. I., and T. K. Rabb, eds. *Climate and History: Studies in Interdiscliplinary History*. Princeton, N.J.: Princeton University Press. 162–66.

———. 1981b. Isostope Evidence for Past Climate and Environmental Change. In Rotberg, R. I., and T. K. Rabb, eds. *Climate and History: Studies in Interdiscliplinary History*. Princeton, N.J.: Princeton University Press. 215–32.

Wilson, R. A. 1979. On the Date of the Roman Amphitheater at Syracuse. In *Philias Charin*. Vol. 6. 2219–30, plates1–3.

Wilson, R. J. A. 1983. Archaeology in Sicily (1981–82). *JHS* 84–105.

———. 1988. Archaeology in Sicily (1982–87). *JHS* 105–150.

———. 1990. *Sicily under the Romans*. Warminster, Wiltshire, U.K.: Aris and Phillips Ltd.

———. 1996. Archaeology in Sicily (1988–1995). *JHS* 59–123.

Winckelmann, J. 1764. *Geschichte der Kunst in Alterthum* (History of Art in Antiquity).

Wiplinger, G., and G. Wlach. 1996. *Ephesus: 100 Years of Austrian Research*. Vienna: Böhlau Verlag.

Wiseman, J. 1967. Excavations at Corinth: The Gymnasium Area. *Hesperia* 5.36: 4002–28.

———. 1970. The Fountain of the Lamps. *Archaeology* 23.2 (Apr.): 130–37.

———. 1972. The Gymnasium Area at Corinth 1969–1970. *Hesperia* 41: 1–42, esp. The Fountain of the Lamps, 9–16, plates 1–11.

———. 1978. *The Land of the Ancient Corinthians*. Studies in Mediterranean Archaeology, 50. Goteborg: P. Astrom.

———. 1979. Corinth and Rome I: 228 B.C.–A.D. 267. ANRW Berlin: Walter de Gruyter. 438–547.

Wittvogel, 1957. *Oriental Despotism*. New Haven, Conn.: Yale University Press.

Wood, H. O., and F. Neumann. 1931. Modified Mercali Intensity Scale of 1931. *Bulletin of the Seismological Society of America* 21.4: 277–83.

Wood, J. T. 1877. *Discoveries at Ephesus, Including the Sites and Remains of the Great Temple of Diana*. London: Longmans, Green.

Wright, G. E. 1975. The "New" Archaeology. *Biblical Archaeologist* 38.3–4: 104–15.

Wurch-Kozeli, M. 1988. Methods of Transporting Bocks in Antiquity. In Herz, N., and M. Waelkens, eds. *Classical Marble: Geochemistry, Technology, Trade*. NATO ASI Series E: Applied Sciences, Vol. 153. Dordrecht: Kluwer Academic: 55–64.

Wycherley, R. E. 1962. *How the Greeks Built Cities*. London: Macmillan. Revised from 1949 ed.

———. 1978. *The Stones of Athens*. Princeton, N.J.: Princeton University Press.

Xeidakis, G. S., and P. G.Marinos. 1997. Geoarchaeology Problems in Greece: An Overview with Some Case Histories. In Marinos, P. G., G. C. Koukis, G. C. Tsiambaos, and G. C. Stourjaras, eds. *Engineering Geology and the Environment: Proceedings of the Symposium of the International Association of Engineering Geology, Athens*. Rotterdam: A. A. Balkema. 3275–86.

Yalçin, Ü. 1993. Archäometallurgie in Milet: Technologiestand der Eisenverarbeitung in archaischer Zeit. *Ist. Mitt.* 43: 361–70.

Yeats, R. S., K. Sieh, and C. R. Allen. 1997. *The Geology of Earthquakes*. New York: Oxford University Press.

Yegül, F. K. 1986. Review of J.C. Biers, *Corinth. XVII: The Great Bath on the Lechaion Road*. (Princeton, N.J.: American School of Classical Studies. 1985) and of W. Martini, *Das Gymnasion von Samos. AJA* 90: 496–99.

Zabehlicky, H. 1991. Ein spätantiker Bau in Südlichen Hafenbereich von Ephesos (Vorbericht). In Malay, H., ed. *Erol Atlay Memorial*. Izmir: Arkeoloji Dergisi, Özel Sayi I, Izmir: Ege University. 198–205.

———. 1994. Kriegs- oder Klimafolgen in archäologischen Befunden? *Markomannenkrige —*

Ursachen und Wirkungen. Brno: Archäologisches Institut der Akademie der Wissenschaften der Tschechischen Republik. 463–69.

———. 1996. Amphoren–Anker–Aphroditen: Die Graungen in Hafen von Ephesos, 1987–89. *Archäologie Österreichs* 7.1: 64–70.

Zampilli, M., with C. Carocci. 1979. Le sviluppo processuale dell'edilizia di base. *Proceedings of the 23rd European Seismological Commission*, Prague, 7–12 Sept 1992: 38–46.

Zangger, E. 1991. Prehistoric Coastal Environments in Greece: The Vanished Landscapes of Dimini Bay and Lake Lerna. *JFA* 18: 1 ff.

———. 1992. Neolithic to Present Soil Erosion in Greece, In Bell, Martin and J. Boardman, eds. *Past and Present Soil Erosion: Archaeological and Geographical Perspectives*. Oxbow Monograph 22. 133–47.

———. 1993. *The Geoarchaeology of the Argolid*. Athens: German Archaeological Institute, and Berlin: Gebr. Mann Verlag.

———. 1994. Landscape Changes around Tiryns during the Bronze Age. *AJA* 98.2: 189–212.

Examples of Film, Video, and Internet Resources

Crane, G., creator and editor. 1997 on. The Perseus Project. http://medusa.perseus.tufts.edu. A library of 27 Greek authors (in Greek and English) with lexica of classical Greek with links to texts and secondary sources, such as databases on architeture, sculpture, etc. Perseus for the Roman world is under development as of 2000.

Kombouras, A., and D. Loukoloulos. 1997. *Miletus: A City in Four Dimensions*. Documentary film. Cultural Center, 254 Pieros St., 17778 Athens, Greece.

Nur, A. *The Walls Came Tumbling Down*. Visual tour of a fault running from crusader fortifications in northern Israel through abandoned cities near the Sea of Galilee to Jerusalem, Jericho, Masada, Sodom, and Gomorrah. The film combines aerial footage, computer simulations, and ancient art to show the role of earthquakes in this area.

Interviews, Letters, Field Excursions

A. Bammer and U. Muss, 1998

E. Bongiovanni, 1996

H. Brückner, 1996, 1998

A. Ibraham, 1998

S. Karwiese, 1997, 1998

J. C. Kraft, 1998, 2000

C. Micceché, 1994

D. Mulliez, 1999–2000

W.-D. Neimeier, 2000

A. Pariente, 1990–98

E. Rice, 1986

D. G. Romano, 1994

R. Ruggieri 1995

P. Scherrer, 1998

H. Thompson, 1970, 1985–89

H. Thür, 1997, 1998

G. and E. Trasatti, 1985, 1987, 1998

W. Vetters, 1998, 2000

K. and B. Weber, 1985

G. Wiplinger, 1998

Index

archaic (*continued*)
   215–217 (Fig. 5.13), 218, 220, 221 (Fig. 5.15),
   226, 228, 232 (Fig. 5.19), 233, 238, 267
architecture 76, 84, 88, 89, 123–125, 135, 143,
   151, 163, 189, 206, 208, 239, 240, 248, 250,
   258
architect 87
Arethusa fountain 92 (Figs. 3.30, 3.31), 103
Argos 15, 26, 37, 78, 110–129 (Figs. 4.1–4.7)
   130, 147, 180, 214, 248, 249–251, 254, 255,
   257; Argive Plain 110, 111 (Fig. 4.1), 121, 139;
   Argolid 15, 110, 111, 121, 123, 128
armor, 130, 159, 185
Artemis temple (Artemiseion) 15, 216–218
   (Fig. 5.13), 221 (Fig. 5.15), 224, 226, 228, 235,
   236, 238, 240; Artemis (goddess) 164, 218;
   Artemis harbor 221 (Fig. 5.15)
artesian well 67, 162, 235, 236
ashlar 82, 88, 120, 135, 163, 170, 171
Askelpios 46, 125, 144, 147, 166 (Fig. 4.24), 172
Aspis 112, 115, 116 (Fig. 4.1), 117 (Fig. 4.5), 120,
   123, 127, 128, 129, 248
Athena 45, 123, 161, 171, 188–192 (Fig. 5.5),
   200, 201, 206, 208–210 (Fig. 5.11), 212, 214
Athena Harbor (Miletus) 181 (Fig. 5.2:A), 186
Athens 94, 126, 128, 145, 174, 186, 208, 235,
   257
Attalos I of Pergamon 166, 167 (Fig. 4.24), 175
Augustus, 131, 209; Augustan 235
avalanche 163, 166, 171, 172, 178
axis (Fig. 3.32), 144, 178, 203 (Fig. 5.7), 206,
   249
Ayasolak 216–217 (Fig. 5.13), 218, 220 (Fig.
   5.15), 226, 228, 232, 240

Bafa Sea 179 (Fig. 5.2: Bafa Gölü), 187, 189,
   192
Balat 177, 183, 189, 192–194 (Fig. 5.6)
Barone (quarry) (Cave di Barone) 82, 85
   (Figs. 3.22: E, 3.23), 139
barriers 101, 171, 233, 234, 248
basalt 95, 101, 179 (Fig. 5.1), 251
Basdegirmen Pinarlari (spring) 222 (Fig. 5.16),
   236. See also Kenchiros spring
bases 146 (Fig. 4.15), 175; basis 230
basilica 142, 179, 227
basin 141, 142 (Fig. 4.14), 168–169 (Fig. 4.25),
   171, 187, 195, 196–224, 234, 239
bath (-ing) 95, 97, 116, 117 (Fig. 4.5), 125, 127–
   29, 137, 143, 147, 148 (Fig. 4.16), 149, 152,
   172, 190–191 (Fig. 5.5), 193, 194, 197, 209,
   220, 235, 236, 237 (Fig. 5.20), 240, 248, 249,
   250; bathroom 170; bath-temple 128; hot
   baths 169

bay 177, 186, 190–191 (Fig. 5.15), 197, 199–201,
   211, 215, 220, 224, 241, 247, 259
beach 75, 91–92 (Fig. 3.27: Qm), 141, 186, 197,
   234; beach rock 134, 138, 146
beauty 14, 45–46, 79, 152, 155 (Fig. 4.18), 251,
   254, 263
bedding planes 100, 173
Belevi (spring and quarry) 22 (Fig. 5.16), 226,
   232; Belevi creek 224
Belice Valley 70 (Fig. 3.22), 89
berth 96, 136, 186, 238
Bigini (spring) 82, 83 (Fig. 3.23), 85
biocarlcalcarenite 37, 42, 86
Black Sea (area) 69, 88, 185, 186 (Fig. 5.4)
blocks 64, 75, 135, 136, 139, 142, 144, 161–64,
   178, 180, 188, 202, 206, 210 (Fig. 5.10), 214,
   227, 230; block failure 199 (Fig. 3.29)
boats 141, 186, 227, 236. 238
boulders 155 (Fig. 4.18), 156 (Fig. 4.19), 235
bouleuterion, 67, 167 (Fig. 4.24), 188, 201, 207
   (Fig. 5.10), 208, 212
boundaries 129, 144, 200, 201, 229, 231
breccia 97, 106, 134, 137, 152, 155 (Fig. 4.18),
   161, 169, 201, 245, 225–227, 251
brick 58, 95, 118 (Fig. 4.5b), 125, 127, 153, 169,
   211, 257. See also adobe
bridge 85, 96, 230, 247, 261
Bronze Age 35, 47, 67, 120, 128, 129, 142, 185,
   221 (Fig. 5.15), 267; gilded bronze 164;
   bronze items 138, 186, 240, 262
buildings (-ers) 128, 135, 159, 151, 159, 164, 166,
   167 (Fig. 4.24), 171, 189, 190, 207–211, 214,
   246, 248, 250, 251, 254, 256, 257; building
   period 205 (Fig. 5.9); building type 220;
   rebuilding 146, 163, 164, 170, 176, 195, 202,
   218, 220, 229, 240, 248, 249, 258
Bülbül Mountain 216, 221 (Fig. 5.15: Preon),
   224, 246–228, 230, 235, 236, 238, 240
burial, buried 131, 229, 235, 238
buttress 62, 64, 68
Byzantine (-ium) 38, 88, 127, 133, 141 (Fig.
   4.13), 160, 183, 193, 194 (Fig. 5.6), 197, 209,
   212, 216–217 (Fig. 5.13), 222 (Fig. 5.6), 223,
   224, 228, 239 (Fig. 5.21), 240, 248, 259

C14 (dating) 231, 256, 267
calcarenite 28, 40, 41, 46 (Fig. 2.11), 55, 70,
   72, 73 (Fig. 3.19), 74, 81, 84, 86–88, 91 (Fig.
   3.27: Mc, Qc, Qm), 96, 102, 103, 108, 175,
   251, 256
calcareous series 101, 103; calcareous muds
   142
calcium carbonate 25, 228. See calcarenite;
   calcite; dolomite; limestone